New Techniques in Amino Acid, Peptide, and Protein Analysis

New Techniques in Amino Acid, Peptide, And Protein Analysis

edited by

A. Niederwieser
and
G. Pataki

ann arbor science publishers inc.

Second Printing 1973

© 1971 by Ann Arbor Science Publishers, Inc.
Drawer No. 1425, Ann Arbor, Michigan 48106

Library of Congress Catalog Card No. 75–109131
SBN 250–39992–X
Printed in the United States of America

The Editors

A. **Niederwieser,** Department of Chemistry, Universitäts-Kinderklinik, Zürich, Switzerland

G. **Pataki,** Pharmaceutical Department, Sandoz AG., Basle, Switzerland

Contributors

J. V. **Benson, Jr.,** Hamilton Company, Whittier, California

J. R. **Coulter,** Medical Research Division, Institute of Medical and Veterinary Science, Adelaide, South Australia

B. C. **Das,** Institut de Chimie des Substances Naturelles, C.N.R.S., Gif-Sur-Yvette, France

C. S. **Hann,** Medical Research Division, Institute of Medical and Veterinary Science, Adelaide, South Australia

S. **Hjertén,** Institute of Biochemistry, University of Uppsala, Uppsala, Sweden

H. **Hydén,** Institute of Neurobiology, Faculty of Medicine, University of Göteborg, Göteborg, Sweden

B. G. **Johansson,** Department of Clinical Chemistry, University of Lund, General Hospital, Malmö, Sweden

B. **Kolb,** Bodenseewerk Perkin Elmer & Co. GmbH., Überlingen, W. Germany

P. W. **Lange,** Institute of Neurobiology, Faculty of Medicine, University of Göteborg, Göteborg, Sweden

E. **Lederer,** Institut de Chimie des Substances Naturelles, C.N.R.S., Gif-Sur-Yvette, France

J. A. **Patterson,** Sondell Scientific Instruments, Inc., Palo Alto, California

Kazuo Shibata, Laboratory of Plant Physiology, The Institute of Physical and Chemical Research (Rikagaku Kenkyusho), Yamato-Machi Saitama Pref., Japan

H. **Sund,** Department of Biology, University of Konstanz, Konstanz, W. Germany

C. W. **Wrigley,** Commonwealth Scientific and Industrial Research Organization, Wheat Research Unit, North Ryde, N.S.W., Australia

Preface

The exponential increase of scientific activity in the present decades forces the individual researcher to obtain a great part of his information from responsible reviews instead of reading the whole multitude of highly specialized papers. This holds also in the field of amino acid, peptide, and protein analysis. Every few years powerful new techniques are developed which are able to replace the older ones. For instance, automatic ion-exchange chromatography of amino acids according to Stein and Moore, which completely replaced microbiological assays, paper chromatography, and the other quantitative methods used before 1958, has now begun to find competition in gas chromatography. Polypeptide-sequence determination, a laborious task using conventional methods, can now be performed automatically by the Edman sequenator, and the sequence of smaller peptides can be determined most satisfactorily by means of mass spectrometry.

There is insufficient space in a single volume to treat comprehensively all the innovations of recent years. This book presents, instead, selected developments which are believed to be important for anyone working in this large field. All of the chapters have been written by experts who are able to summarize and pass judgment on current work in their specialties. The articles deal with amino acid and peptide analysis by ion-exchange chromatography (J. V. Benson and J. A. Patterson), gas chromatography of amino acids (J. R. Coulter and C. S. Hann), gas chromatography of peptides (B. Kolb), mass spectrometry of peptides (B. C. Das and E. Lederer), molecular-sieve chromatography (S. Hjertén), thin-layer gel filtration (B. G. Johansson), microelectrophoretic determination of proteins in the nanogram range (H. Hydén and P. W. Lange), electrofocusing of proteins (C. W. Wrigley), chemical accessibility and environment of amino acid residues in native proteins (K. Shibata), and methods for investigation of the quaternary structure of proteins (H. Sund).

We are confident that the book will be found useful and that the many efforts involved will be justified.

Zurich and Basel   A. Niederwieser
August, 1971   G. Pataki

Contents

Chapter 2   75

J. R. Coulter and C. S. Hann

Gas Chromatography of Amino Acids

Chapter 5   227

S. Hjertén

Molecular-Sieve Chromatography of Proteins

Chapter 6   249

B. G. Johansson

Thin-Layer Gel Filtration and Related Methods

H. Hydén and P. W. Lange

Microelectrophoretic Determination of Protein and Protein Synthesis in the 10^{-7} to 10^{-9} Gram Range

Chapter 8   291

C. W. Wrigley

Electrofocusing of Proteins

Chapter 9   341

Kazuo Shibata

Chemical Accessibility and Environment of Amino Acid Residues
in Native Proteins

H. Sund

Methods for Investigation of the Quaternary Structure of Proteins

Chapter 1

Chromatographic Advances in Amino Acid and Peptide Analysis using Spherical Resins and Their Applications in Biochemistry and Medicine

by J. V. Benson, Jr., and J. A. Patterson

The massive work in ion-exchange resin chromatography, which still appears to be in the logarithmic stage of growth, rests firmly on the now classical works of Moore and Stein. An ingenious series of articles[1-13] starting in 1949 precipitated such an abundance of innovations in chromatographic technology that any attempt to cover all of them would far exceed the space limitations of this article. Many excellent reviews[14-25] and monographs[26-30] have been written covering history, theories, methods, and applications.

This chapter, therefore, will be selectively restricted to a discussion of those systems and techniques which, at present, appear to be not only highly satisfactory but also widely applicable. Included in this discussion are the variables which may be manipulated to achieve the results desired for specific analyses. "Playing" with several of these variables simultaneously may produce new and unique results for specific needs.

Finally, specific methodology is detailed to enable the reader to start acquiring data without spending a great deal of time experimenting with methods. While a great number of methods are present in the literature, the only ones included here are those with which we have had extensive experience in our own laboratory and found to be most satisfactory.

1.1 Applications

Ion-exchange column chromatography has been applied to a very diverse and large number of projects to provide basic biochemical

1

information and to aid in the diagnosis and management of biomedical disorders. The extent of the range of uses can be surmised by a perusal of several of the listed reviews.[14-34]

1.1.1 Biochemical Applications

Ion-exchange column chromatography has been applied to three very important areas: (1) qualitative and quantitative analyses of amino acids in peptides and proteins have permitted a useful characterization of molecules. They can be used as a means of detecting some of the specific differences among proteins; (2) the extra-protein amino acid composition of biological fluids not only supplies this basic information but allows one to monitor the changes as influenced by many factors such as environment, physiological state and genetic expression; and (3) the primary structure of proteins has assumed an overwhelming importance in biochemistry today. Many investigators are engaged in determining the amino acid sequence of a great variety of proteins. This will provide the actual chemical structure and permit studies relative to function.

There are many procedures available for accurate quantitative analyses of amino acids and peptides (see Section 1.3). Usually, the proteins or peptides are hydrolyzed to their constituent amino acids by treatment with acid or base, while biological fluids are usually deproteinized.

When studies are made to determine the amino acid sequence, partial hydrolysis using enzymes (such as pepsin, trypsin, chymotrypsin) or chemicals which are quite specific in their cleavage is employed. However, analyses of these hydrolysates provide incomplete information regarding sequence. Additional information can be obtained by C-terminal and N-terminal analyses. By producing different-sized peptides using a variety of enzymatic and chemical techniques, much of the sequence may be deduced from overlapping amino acid residues.

In one of Edman's sequencing procedures[35] the N-terminal amino acid is coupled with phenyl isothiocyanate and cleaved from the peptide with an anhydrous acid as the thiazolinone. The terminal amino acid is identified either by the subtractive method, in which the remaining peptide is sampled, hydrolyzed and analyzed chromatographically, or by converting the thiazolinone to a PTH (phenylthiohydantoin)-derivative which can be identified chromatographically. After cleavage the remaining peptide has a new N-terminal amino acid which can again be coupled and cleaved.

A number of interesting uses of such techniques have been described by Smyth,[36] Benisek,[37] and others.[38-41]

1.1.2 Biomedical Applications*

The identification and rapid screening techniques of metabolic malfunctions which can be therapeutically controlled are crucial for progress in the treatment of certain mental disorders. Many of the screening techniques used to detect abnormalities related to amino acid metabolism appear to lack adequate specificity and sensitivity for the quantitation of the amino acids involved. Although the conventional and accelerated methods of ion-exchange chromatography do provide the necessary sensitivity, precision, and resolution, these standard methods are too time-consuming to be used as a screening tool or for chemical monitoring during therapy.

The work done in the detection and treatment of phenylketonuria (PKU) has stimulated biochemical research pertaining to other inborn errors of amino acid metabolism associated with both mental retardation and other incapacitating disorders. Berlow[42] reported on the biochemical findings of a child whose urine at first had given a "false positive" reaction to a ferric chloride test, indicating an elevated phenylpyruvic acid level usually associated with PKU. This child was subsequently found to be suffering from histidinemia. Other cases of disturbed histidine metabolism have been reported by Ghadimi[43] and La Du.[44] Although there have been only a few reported cases of cystathioninuria, further reports are to be expected[45] due to both the improved chromatographic techniques which permit specific amino acid analysis and the large-scale surveys being conducted on serum and urine. In homocystinuria, reported by Perry,[46] elevated levels of methionine and homocystine are found in the blood and homocystine is present in the urine.[47] The detection of the amino acids important in branched-chain ketoaciduria (maple syrup urine disease) has been extensively documented.[48–50] Phenylketonuria detection in the newborn infant as a routine hospital screening procedure has been described by Guthrie.[51] Phenylketonurics inherit the lack of enzymes which convert phenylalanine to tyrosine by hydroxylation. Sacks[52] reports on studies of phenylalanine metabolism comparing "normal" control subjects, mental patients, and phenylketonurics. Additional investigative reports cover aspects of amino acid metabolism in PKU and other amino-acidopathies,[53] chemical and metabolic studies on phenylalanine in oligophrenics,[54] and management of malignant disease by phenylalanine restriction.[55]

Other amino acid metabolic disturbances of current interest are

---

*For reviews see also References 41a and 41b.

cystinuria, described by Stein,[3] Dent,[56] Rosenberg,[57] and Crawhall;[58] Wilson's disease by Bell *et al.*,[59] hyperprolinemia,[60] and hyperglycinemia and hyperglycinuria.[61]

In other areas of clinical interest, Salzer and Balis[62] report on those amino acids associated with DNA of tumors. It is indicated that tumor-DNA contains less of the basic amino acids than found in normal tissue and that a substance is present which is chromatographically similar to the aromatic amino acids that appear in a number of small tumors.

Studies have been made to determine amino acid content in the serum of patients affected with myocardial infarction and other vascular diseases.[63] It was found that there were significant alterations of some amino acids in the serum of patients suffering from myocardial defects.

The free amino acids of human breast cancer have been studied by Ryan;[64] the amino acid, homocitrulline, and two unidentified peptides were found in the tumors.

The effects of Vitamin B6 upon the amino acids in Kwashiorkor patients have been described.[65] Quantitation of 27 amino acids and other ninhydrin-positive compounds was determined in the urine of patients with psoriasis.[66] Amino acid excretion during pregnancy was studied by Armstrong and Yates[67] who found that threonine increased threefold, that the amount of taurine excreted daily increased until the eighth week, and that the urea and ethanolamine levels appeared to remain constant. Further studies on amino-acidurias are reported by Brown.[68] Elevated levels of cystine, ornithine, arginine, and lysine resulted when cycloleucine was administered to cancer patients. The endogenous renal clearance rates of free amino acids in prepubertal children has been studied using an accelerated procedure for the chromatography of the basic amino acids.[69] The excretion of phosphoethanolamine and its relationship to adult and infantile hypophosphatasia has been reported[70] as has the multivariate analysis of amino acid excretions.[71]

Accompanying the increase in knowledge of amino acid metabolism as related to clinical disease has been a need not only for faster chromatographic analyses of amino acids but also for quantitative data related to the amino acid content of human cerebrospinal fluid,[72] the free amino acids of human spinal fluid,[73] the free amino acids of blood and urine,[6, 74, 75] and the free amino acids in the plasma of the newborn infant.[76]

A study on β-alaninuria in patients with tuberculosis[77] and the importance of amino acid uptake by the intestine has been reported by Agar *et al.*[78]

Accelerated chromatographic procedures of amino acids, associated not only with the many inborn errors of metabolism mentioned in the publications referred to above but also with supplying rapid analyses of the many amino acids associated with physiological fluids, have been developed by several researchers. A short ion-exchange column was used for the estimation of methionine and cystine by Lewis.[79] A 40 × 1.5-cm resin column of Zeocarb 225, >200 mesh resin, was used for the separation at room temperature and after oxidation of the sample. At a buffer flow rate of 20 ml per hour, cysteic acid is eluted at 36–40 ml and methionine sulfone at an elution volume of 110–120 ml. From these quantities cystine and methionine are calculated.

A rapid, short-column chromatographic procedure for amino acids is described by Shih.[80] His group achieved a rapid separation on short columns of some amino acids required for diagnosis and monitoring of clinical disorders related to amino acid metabolism. A 20 × 0.9-cm resin column was used at a buffer flow of 60 ml per hour. All the analyses were completed in less than 2.5 hours.

The accelerated chromatography of amino acids associated with phenylketonuria, leucinosis, and other inborn errors of metabolism have been described by Benson.[81] These procedures have been applied to a single column of Type PA-35 spherical cation-exchange resin which can also be used for the chromatographic separation of those basic amino acids commonly found in physiological fluids.

In the over 200 reported metabolic disorders are many areas of immediate interest for amino acid chromatography as related to amino acid function and human disorders. These will be outlined in Tables Ia and Ib.

1.2 Chromatographic Systems
1.2.1 *One-Column Systems*

In the process of separating the amino acid-containing samples on a resin column, the acidic amino and the hydroxy amino acids have the weakest affinities for the cation-exchange resin. The ionic bonding of the neutral amino acids is next, and the aromatic amino acids have a stronger affinity for the resin than the other neutral amino acids. The strongest affinities for the resin are possessed by the basic amino acids.

The procedure of Piez and Morris uses a single-column gradient elution technique to obtain a complete analysis of the acidic, neutral, and basic amino acids in one run.[82] The methodology employs a gradient elution system which eliminates the regeneration step necessary in the two-column system, since all of the amino acids are eluted from the column. The use of solenoid valves is not required as it would be with automated discrete buffer-change methods, and the elimination of other equipment (pumps) is possible.

Table Ia

Amino Acids Affected in Some Metabolic Diseases*

Disorder	Urine	mg/day	Plasma	
Hartnup disease	Ala↑	600	Val↑	
	Val↑	300	Leu↑	
	Leu↑	150	Ile↑	
	Ile↑	150		
	Ser↑	600		
	Thr↑	500		
	Phe↑	500		
	Tyr↑	500		
	Trp↑	150		
	Asn↑			
	Glu↑	800		
	His↑	650		
	Cit↑			
Histidinemia	Ala↑		His↑	6–8 × normal
	Thr↑			
	His↑	4–10 × normal		
Hyperglycinemia	Ala↑		Gly↑	
	Gly↑			
Osthouse disease	Ala↓		Val↑	
	Leu↑		Leu↑	
	Phe↑		Ile↑	
	Tyr↑		Phe↑	
	Met↑		Tyr↑	
			Met↑	
Galactosemia	Ala↑	Thr↑		
	Gly↑	Phe↑		
	Val↑	Tyr↑		
	Ser↑	(Cys)$_2$↑		
	Glu↑	Gln↑		
	His↑	Lys↑		
	MeHis↑			
Branched-chain ketoaciduria (maple syrup urine disease)	Ala↓	Ile↑	Val↑	
	Val↑	Ser↓	Leu↑	
	Leu↑	Thr↓	Ile↑	10 × normal
	Gln↓	allo-Ile↑	(Cys)$_2$↓	
	Tau↓		allo-Ile↑	
Hyperprolinemia	Gly↑	Hyp↑	Pro↑	
	Pro↑			

Table Ia (*continued*)

Disorder	Urine		Plasma	
		mg/day		
Joseph's syndrom	Gly↑	Pro↑		
	Hyp↑			
Wilson's disease	Gly↑	5–10 × normal		
	Val↑	5–10 × normal		
	Ser↑	5–10 × normal		
	Thr↑	20 × normal		
	Phe↑	2–4 × normal		
	Tyr↑	5–10 × normal		
	(Cys)$_2$↑	20 × normal		
	Pro↑			
	Asn↑	5–10 × normal		
	Gln↑	5–10 × normal		
	His↑	2–4 × normal		
	Lys↑	5–10 × normal		
	Tau↓			
	Orn↑	2–4 × normal		
	Cit↑			
	MeHis↓			
Phenylketonuria	Phe↑		Gly↓	0.2 mg/100 ml
			Phe↑	15 × normal
			Tyr↑	
			Cys↓	0.4 mg/100 ml
			Glu↓	0.4 mg/100 ml
			Arg↓	0.7 mg/100 ml
Tyrosinosis			Tyr↑	
Cystinuria	(Cys)$_2$↑Lys↑		Lys↓	
	Orn↑Arg↑			
Homocystinuria	Met↑		Met↑	
Hydroxyprolinemia	Pro↑		Pro↑	3.9 mg/100 ml
Citrullinuria	Cit↑		Cit↑	
Cystathionuria	Cyst↑	1000 mg/day	Cyst↑	0.45 mg/100 ml

↑increased
↓decreased

°Kurtz, D. J. *Automatic Analyzer for Screening Inborn Errors of Metabolism* Bethesda, Maryland: Office of Program Analysis, National Institute of Neurological Disease and Blindness, 1964).

Table 1b
Normal Values of Free Amino Acids in Human Urine and Plasma

Amino Acid	Mol. Wt.	Urine (mg/day) Mean[1]	Adult Mean[2]	Adult Range[3]	Plasma (mg/100 ml) Newborn Mean[4]	Newborn Range[4]	Adult Mean[5]	Adult Range[5]
Taurine	125	156	123	35–300	1.76	0.93– 2.70	0.79	0.34– 2.10
Urea	60				16.4			
Hydroxyproline	131				(0.42)			
Aspartic acid	133	<10	8	3–29	0.11	Tr.– 0.22	0.10	0– 0.32
Threonine	119	28	17	2–50	2.59	1.36– 3.99	1.54	0.94– 2.30
Serine	105	43	42	25–75	1.72	0.99– 2.55	1.21	0.77– 1.76
Asparagine	132	54			(0.60)		(0.58)	0.54– 0.65
Glutamine	146		73	40–103	11.16	7.86–14.01	8.30	6.07–10.15[6]
Proline	115	<10			2.13	1.23– 3.19	2.12	1.17– 3.87
Glutamic Acid	147	<10		7–40	0.76	0.30– 1.57	0.85	0.21– 2.82
Citrulline	175				0.28	0.15– 0.50	0.53[6]	0.21– 0.97[6]
Glycine	75	132	104	53–200	2.58	1.68– 3.86	1.78	0.90– 4.16
Alanine	89	46	22	5–71	2.94	2.10– 3.65	2.99	1.87– 5.89
α-Aminoadipic acid	161	10	8	5–13				
α-Amino-n-butyric acid	103				0.15	0.06– 0.30	0.21	0.08– 0.36
Valine	117	<10	10	4–17	1.60	0.94– 2.88	2.50	1.65– 3.71
Cystine	240	10	14	3–33	1.47	0.85– 2.02	1.05	0.20– 2.02

Amino acid							
Methionine	149	<10	7	5–11	0.44	0.13– 0.61	0.34	0.09– 0.59
Isoleucine	131	18	15	5–30	0.52	0.35– 0.69	0.83	0.48– 1.28
Leucine	131	14	11	5–25	0.95	0.61– 1.43	1.45	0.98– 2.30
Tyrosine	181	35	19	7–50	1.26	0.76– 1.80	0.94	0.39– 1.58
Phenylalanine	165	18	13	8–31	1.30	0.69– 1.82	0.88	0.61– 1.45
β-Alanine	89		6	3–10	(0.13)		(0.08)[7]	
β-Aminoisobutyric acid	103		22	6–37				
Ornithine	132		1	0–4	1.21	0.65– 2.00	0.79	0.39– 1.40
Ethanolamine	61				0.32	0.16– 0.56	0.016[6]	Tr.– 0.076[6]
Lysine	146	19	7	0–48	2.93	1.67– 3.93	2.24	1.21– 3.48
1-Methylhistidine	169	180	73	9–210				
Histidine	155	216	138	20–320	1.19	0.76– 1.77	1.15	0.49– 1.66
3-Methylhistidine	169		65	33–87				
Tryptophan	204				0.65	Tr.– 1.37	0.98[6]	0.51– 1.49
Arginine	174	<10	6	0–14	0.94	0.38– 1.53	1.30	0.37– 2.40

[1] Average of eight normal male adults between 25 and 40 years of age. Diets not controlled. From Reference 4.
[2] Six normal male adults. According to P. Soupart, Clin. Chim. Acta 4, 265 (1959).
[3] R. C. Westall, in Amino Acid Pools. Ed. by J. T. Holden (Amsterdam: Elsevier Publishing Co., 1962), p 200.
[4] Data on 9 male and 16 female. According to Dickinson et al.[76]
[5] Data from nine laboratories on 39 male and 37 female. Compiled by Dickinson et al.[76]
[6] Three male and five female. According to Dickinson et al.[76]
[7] Single plasma.

The use of a single sample is also an advantage of the single-column operation. The principal shortcomings of this procedure include preparation of the gradient solutions and subsequent cleaning of the gradient device, the elimination of contaminating ammonia, and the progressive change in the base line value during analysis. A later accelerated, single-column procedure by Miller[83] describes a method for eliminating this base line change. Cysteic acid (8.3 μmole per liter), which is not chromatographed in this procedure, is added to the starting buffer, which increases the blank value at the start of the analysis and which is also equal to those blank values at the completion of the run. A single-column gradient elution system is also described by Thomson and Miles[84] using 10% methanol reagent in the buffer to maintain the resolution of threonine and serine. A single 125×0.636-cm column, high-resolving, fully automatic procedure of Hamilton[85] uses discontinuous buffer changes. Improved sensitivity and resolution have been achieved by studying the details of resin characteristics, column design, cuvette design, and recorder characteristics.

Dus *et al.*[86] describe the application of a simple, inexpensive, rotary valve mechanism to automate a stepwise four-buffer elution program and the sequential column selection. A buffer flow rate of 70 ml per hour was used to chromatograph the amino acids on a 60×0.9-cm resin column.

1.2.2 *Two-Column Systems*

The procedure of Spackman, Stein and Moore[12] uses two discrete buffers which are changed midway in the analysis for the elution of the acidic and neutral amino acids with a 150×0.9-cm column. A third buffer with a higher ionic strength and pH is used for the chromatographic analysis of the basic amino acids with a second column. A 15×0.9-cm column is used for the resolution of the basic amino acids normally found in a protein or peptide hydrolysate. This procedure requires the use of one sample for each column; however, there is great versatility in the system. If only a few basic amino acids are of interest, particularly in a physiological fluid, then these amino acids can be chromatographed without having to wait for the elution of the acidic and neutral amino acids which will be chromatographically eluted first by the single-column systems.

A second elution scheme is available, which does provide greater resolution for biological samples and which is capable of separating a greater number of amino acids. The same 150×0.9-cm column is used; however, a 50×0.9-cm resin column is used to separate the basic amino acids (see Section 1.3).

1.2.3   Resins

The ion-exchange resin ordinarily used for the chromatographic separation of amino acids, peptides and simple related compounds in physiological fluids is a copolymer of styrene and divinylbenzene in bead form. The resin is generally characterized by the percentage of divinylbenzene or degree of cross-linking, producing a three-dimensional aromatic network (raw polymer). This product requires the addition of functional groups to produce either a cation- or anion-exchange resin. Sulfonation by the addition of an excess of sulfuric acid or chlorosulfonic acid plus a catalyst introduces 8–10 SO_3H groups for every ten free aromatic rings, to produce a strongly acidic cation exchanger. Chloromethylation (chloromethyl ether) of the raw polymer beads plus a catalyst followed by treatment with a tertiary amine (trimethylamine) produces the strongly basic quaternary nitrogen group. It is extremely important that when these functional groups are added to the polymer, side reactions be carefully controlled. It is possible to introduce sulfone cross-linking into the strong acidic resin to produce a product that appears to be more highly cross-linked. The same "high cross-linked" effect can also be observed with the anion exchangers if the chlorine from the chloromethyl group on one ring and the hydrogen from an adjacent ring are not eliminated.[87] It is therefore important that the polymerization process and the addition of functional groups be closely controlled for chromatographic reproducibility. As indicated above, the functional group of the cation-exchange resins are $-SO_3Na$ (when using sodium citrate buffers) and $-N(CH_3)_3^+$ OH^- for the anion-exchange resins.

Commercial resins were developed for water-conditioning applications; Dowex, Amberlite, Permutit, Aminex (Dowex), and Duolite were made in the 20–50-mesh size range.* Early studies on the chromatographic techniques of amino acid and peptide chromatography indicated that a smaller particle size was necessary to reduce analysis times by attaining a faster approach to equilibrium of the amino acid with the surrounding solution. (Even using minus-400 mesh material presented problems of reproducibility from batch to batch.) Grinding these spherical polymers to give smaller pulverized particles did improve resolution and allowed faster analysis times. However, the yield of usable material was extremely small and required extensive cleaning

---

*Manufacturers: Dowex, Dow Chemical Company, Midland, Michigan; Amberlite, Röhm and Haas Company, Philadelphia, Pennsylvania; Aminex, Bio-Rad Laboratories, Richmond, California (distributor for Dowex Resins); Duolite, Chemical Process Company, Redwood City, California.

of foreign material from the resin. Spherical resins, specifically manufactured from the monomers, have allowed further reduction in analysis times with improved resolution.[88]

Amino acids are organic compounds having at least one carboxyl (acid) group and an amino (basic) group positioned alpha to the carboxyl group. When an amino acid is chromatographed on the cation-exchange resin column which has been equilibrated with the sodium citrate buffer ($-SO_3^- Na^+$), the amino acid molecule (in acid solution) is attracted to the $-SO_3^-$ group of the resin through ionic forces by means of its positively charged amino groups. Thus, at various pH's, the amino acid is:

pH 2.0	pH 4.0	pH 7.0	pH 10.0
COOH	COO$^-$	COO$^-$	COO$^-$
$\mid$	$\mid$	$\mid$	$\mid$
R$-$NH$_3^+$ $\rightleftarrows$ H$^+$ +	R$-$NH$_3^+$ $\leftrightarrows$ H$^+$ +	R$-$NH$_3^+$ $\leftrightarrows$ H$^+$ +	R$-$NH$_2$
$\mid$	$\mid$	$\mid$	$\mid$
COOH	COOH	COOH	COO$^-$
Net charge: 1	0	-1	-2

in various isoelectric forms.[89-90] The amounts of amino acid bound to the resin relative to that remaining in solution is expressed as their distribution coefficients (K). It is these distribution coefficients which help separate the various amino acids chromatographically, and which are dependent on the R side-groupings of the amino acids. The amino acid with the lowest distribution coefficient is moved down the column by the buffer until it comes in contact with resin that is less saturated with that amino acid, and there it is adsorbed to set up a new distribution coefficient. The amino acid with the next lowest distribution coefficient left up above on the resin is now released by a new portion of buffer, free of any amino acid, and is carried further down the column to repeat the equilibrium cycle. The fact that there are different distribution coefficients means that each amino acid will travel at different rates down the column.

The variable factors which affect the separation of amino acids are outlined in Section 1.4. In some cases the anionic form of the resin offers certain advantages which aid in the separation of amino acids and peptides due to their greater binding potential as their ionized carboxylic group. The use of anion-exchange resin for the chromatography of amino acids is reported by Partridge,[91] and under these conditions those amino acids and peptides with the least tendency to carry a negative charge are the least tightly bound to the resin in alkaline solutions. The use of the borate complexes of neutral sugars also have found an application with anion-exchange resins (Ohms,[92] Green[93]); separation of mixtures of bases, nucleosides, and nucleotides on a single anion-exchange column have also been achieved.[94]

A number of considerations must be resolved before the extensive use of anion-exchange techniques: (a) stability of the functional group on the resin, (b) resin cross-linking or polymer matrix stability, and (c) controlled reproducibility of batches. The loss of active site stability affects the chromatographic results in two areas: first, a decrease in the number of theoretical plates on the resin column[95] results in a change in the elution-pattern sequence of the chromatographic band; and second, the decomposition products of the resin are, in many cases, ninhydrin positive, which will result in base line variation on the chromatogram (chemical noise). Reproducibility of resins is required so that each chromatographic technique that is developed and published can be repeated to avoid efforts in this field becoming individual scientific curiosities.

1.2.4 Resin Columns

The conventional column support used for the ion-exchange resin beds are precision-bore, heavy-wall, borosilicate glass tubing. Internal precision-bore glass is required for the following reasons: (a) a constant linear flow of eluant to prevent band mixing of the amino acids, (b) a known specification to which internal seals and components can be made, *i.e.*, sealed end-fittings and resin retaining screens. A heavy wall glass is required to prevent undue breakage at the normal operating pressures, 200–600 psi. However, too thick a column wall can interfere with the desired temperature gradient during temperature-change operations. Glass composition does offer somewhat of a problem in two general areas of operation: (a) an increase in operational pressure as a result of glass-resin binding force and (b) impaired resolution with small samples[96–97] due to wall channeling or binding of sample at the glass wall. The state of the art with respect to the composition of glass is at present dependent on empirical testing. However, as the pressure due to increases in the flow rate of buffer and decreased analysis time approaches the limit of operation, the deleterious effects noted at lower operational pressures diminish. Consistent with the above, the techniques presented are for 0.9-cm columns rather than those of smaller inside-diameter columns in order to minimize the glass-wall effect.

Precision-bore glass is manufactured by shrinking raw glass tubing over precision mandrels using heat and vacuum. The quality of the precision-bore tubing is dependent upon the quality of the raw glass. The resin columns used are a borosilicate glass which is relatively resistant to chemical attack and which, because of its property of low thermal expansion, is resistant to thermal shock. To maximize these effects, the raw material for glass requires a high silica content. Due to the many uncontrolled variables introduced into the manufacture of glass columns (slight variations in composition from batch to batch and among vendors), there are surface variations which have an effect on the glass-resin interface.

1.2.5 *Elution Fluids*

To understand the function of the buffers used in chromatography of amino acids, it is important to understand that as the amino acids in solution at the top of the column pass down through the resin bed, an equilibrium is established at each theoretical plate or zone in the column. The separation of the amino acids in the sample is dependent on many controllable variables: resin-particle size, chemical nature of the resin, degree of resin cross-linking, diameter and length of the resin bed, column temperature, the charge and size group of the amino acid, and the ionic strength, pH, and flow rate of the eluting buffer.

It is the composition and pH of the buffers and developers that largely determine the elution rate of the amino acids and peptides. Stein and Moore[1] reported on the use of hydrochloric acid of increasing normalities to elute the amino acids from columns of Dowex-50 (a sulfonic acid cation-exchange resin). They found in later studies that sodium citrate and sodium acetate buffers had the advantages of less decomposition and loss of amino acids of sensitive compounds, and easier analysis of the neutral effluent from the column with ninhydrin. They have also shown that the properties of the eluent can be changed by the addition of organic solvents. A 1% solution of benzyl alcohol added to the buffers sharpened the peaks of the aromatic amino acids. Propyl alcohol accelerated preferentially those amino acids with the larger nonpolar side chains.[2] Buffers used in liquid-column chromatography require consideration of a number of aspects which are not common to the other areas of buffer applications: (a) purity of reagents; (b) molar salt concentration, including those salts for buffering capacity; (c) gas-free solutions; (d) mold- and mold spore-free solutions; (e) dust- or colloid-free solutions; and (f) H-ion determination with minimum interference from other cation activity.

Since the buffer solutions will be passed in relatively large volume over long periods of time through an ion-exchange resin bed, certain metal cations will adsorb on the resins which will not be eluted by normal regeneration procedures. This will cause a gradual loss in resin capacity. However, the original capacity can usually be restored by prescribed acid cleanup procedures. In anion-exchange resin chromatography some of the metal complexes in trace amounts, found in the buffers, result in irreversible reaction. Their loss in capacity cannot be regenerated.

Normally the exact molar salt concentrations are not extremely critical in buffer applications. However, in column chromatography the cation concentration (other than H^+) is very active in elution of the amino acids or peptides from the column as adsorbed cations. For this reason the accurate control of the buffer cation is required. The type and concentration of the counter-ion (anions in this case) is also important, because a spectrum of weak complex ions is formed with amino acids and peptides at their respective operating pH ranges.

Mold, mold spores, and dust, all of which are present in buffers, can be treated as particulate matter contamination of the elution fluids. These contaminants, like dissolved gas, will be filtered out in the resin bed and could result in build-up of column backpressure to a point of column rupture. It has been the practice to use mold inhibitors, caprylic acid, pentachlorophenol, or phenol in the buffers. However, it was found that the use of caprylic acid produced an unknown peak which coincided with the emergence of the ornithine peak. Phenol also produced a peak which elevated the base line in the glycine–alanine area, using the hydrolysate procedure. A UV germicidal lamp was also placed in the buffer reservoir to eliminate mold growth. There was a further increase in area of the unknown peak mentioned above, which suggests that it might be a degradation product of caprylic acid. It is possible to minimize mold formation and to remove other contaminants by passing a freshly prepared solution through a Micule filter and refrigerating the buffers. The actual quantities of reagents required for making those specific buffers needed for the chromatographic techniques are referred to in Section 1.6. Recommended methodology is presented in Tables II–VI (see below).

Studies investigating the optimum elution fluid for the chromatography of peptides present problems of a different nature than those normally experienced with the amino acid buffers. A variety of buffers have been studied for the chromatography of peptides.[99-100] Initially, sodium citrate or sodium acetate buffers of the type used for chromatographing amino acids as described by Moore and Stein[2] were used for peptides. When the peptide fractions are chromatographed with this method the buffer salts are contaminated by the peptides. Peptides have also been separated by using ammonium acetate or formate buffers, which can be removed by sublimation.[98] Volatile buffers have been used for the separation of peptides on a cation-exchange resin using pyridine–acetic acid[101] and on anion exchangers using pyridine–collidine–acetic acid eluents.[102] It has been reported that the reaction of pyridine–acetic acid eluents and the ninhydrin reagent is incomplete, apparently due to the low pH of the reaction mixture.[100] This problem was corrected by increasing the ratio of ninhydrin (which contains a strong sodium acetate buffer) to the pyridine–acetic acid eluent[103] to bring the pH of the mixture closer to 5 for the optimal color development.[104]

The purity of the eluents, particularly the pyridine reagent, is important if a stable base line is to be expected (due to ninhydrin-positive compounds, probably primary amines).

In addition to the brief description of the buffer and eluent systems used for the elution of amino acids and peptides from resin columns, specific attention will be given to the chromatographic variables in Section 1.4. The magnitude of these variables gives an indication of degree to which the chromatographic operating conditions can affect the elution patterns of amino acids and peptides.

1.2.6   Detectors

Many photometric methods have been reported for the determination of amino acids and peptides eluted from a chromatographic column. Several colorimetric methods have been investigated. The use of a stabilized indametrione hydrate reagent for the determination of amino acids in food was reported by Jacobs,[105, 106] with a 0.08% solution of stannous chloride used to stabilize the colored complex. β-Naphthoquinone sulfonic acid and ninhydrin were also reported.[140, 107] An interesting report describes the reaction of amino acids and peptides using an improved cobalt–phenol reagent read at 750 nm.[108] Yemm and Cocking have examined the reaction of amino acids with ninhydrin and found it possible to obtain a stoichiometric reaction[109] for most of the amino acids with a quantitative yield of diketohydrindylidenediketo-hydrindamine (DYNA), the probable end product of the reaction.

The reaction between α-amino acids and ninhydrin has been extensively studied and it has been established that colored compounds are formed with amino acids, peptides, proteins, and other compounds having free amino groups.

It has also been demonstrated[110] that ninhydrin can be used to detect cationic complexes of nickel, cobalt and chromium. The ethylenediamine complexes of nickel and cobalt respond more intensely to the ninhydrin than the ammonia-containing complexes of nickel and cobalt. In paper chromatography these stains were reported to be stable for several months.

Ruhemann[111] showed that when an aqueous system of triketohydrindene hydrate (ninhydrin) was treated with hydrogen sulfide, hydrindantin was produced. Hydrindantin also dissolves in sodium carbonate to form a deep red solution that will precipitate on the addition of dilute hydrochloric acid.

Troll[112] demonstrated that the organic solvents dioxane, alcohol, methyl Cellosolve, pyridine, and phenol accelerate the development of color to varying degrees. Ten of the naturally occuring amino acids gave theoretical yields at room temperature. At 100°C all except tryptophan and lysine reacted quantitatively. Phenol (80%) in absolute alcohol and a KCN–pyridine reagent are used as the most effective solvent for the ninhydrin–hydrindantin reaction mixture. Other anomalous reactions of ninhydrin have been presented by Schelling.[113]

A "ninhydrin-positive" peak had been eluted early in a chromatogram and it was suggested that levulinic acid was present as the artifact. There is other supporting evidence that ninhydrin can be used to detect non-amino compounds eluted from the column of the amino acid analyzer.[114]

Zacharius[115] revealed an unidentified ninhydrin peak which, when isolated and characterized, was found to be a mixture of fructose and glucose. Further investigation indicated that there were a number of carbohydrates and non-nitrogenous compounds that will give ninhydrin-positive peaks on the amino acid chromatogram.

Other modifications of the ninhydrin reagent have been reported by Moore and Stein.[11] One of the modifications was elimination of the stannous chloride which was earlier added to produce hydrindantin. Hydrindantin was added directly to prevent the possibility of precipitating the tin salts. In addition, a detailed discussion on the chemical composition of the ninhydrin and α-amino acid reactants which form a colored product called Ruhemann purple and on the various factors affecting the stability of the ninhydrin reagent is suitably presented by Blackburn.[34]

The use of hydrindantin in the reagents is a possible cause of some problems arising from its low solubility and instability in air.[116] A stable reagent using sodium cyanide instead of hydrindantin, and acceptable to automatic methods, is described by Rosen, Knight, and Thomas.[117-119]

Kirschenbaum[120] used a mixture of methyl Cellosolve–dimethylsulfoxide (70:30) as a solvent in the ninhydrin reagent. To improve the composition of the reagent, which would eliminate methyl Cellosolve, Moore[121] used dimethylsulfoxide and a lithium acetate buffer to obtain a completely soluble mixture of ninhydrin and hydrindantin. The color yields with this reagent are reported to be 1–2% higher than those obtained using only methyl Cellosolve–sodium acetate reagents.

When used with the recommended precautions, the ninhydrin reagent appears to be the most sensitive and reproducible agent for the quantitation of amino acids and peptides. In normal operation, the buffered ninhydrin reagent is added to an aliquot of the column effluent and the mixture is then heated to 100°C in a reaction bath. To allow the reaction to proceed reproducibly to completion, the minimum amount of hydrindantin must be present, due to its instability in air and limited solubility in the methyl Cellosolve. Also, to maximize operating conditions, heat and light must be excluded from the reagent. It is therefore very important that the exact amount of stannous chloride be added to the ninhydrin reagent (the stannous chloride is used as an agent for reducing an equimolar amount of ninhydrin to hydrindantin). The use of a nitrogen atmosphere in the reagent bottles helps prevent oxidation.

Use of the reagents reported above for detection of the amino acids and peptides requires a colorimeter or photometer to quantitatively detect the absorbance changes associated with the reactions. Since ninhydrin is the reagent most widely used for detection of amino acids and proteins, it will form the basis for further discussion of the colorimeter detector.

Accurate continuous-flow colorimeters are not uncommon for quantitative colorimetric analysis. However, for applications to chromatography they should meet the following requirements: (a) long-term thermal stability of components, (b) elimination of liquid leaks under pressure, (c) minimum mixing of peak band frontals, (d) rapid gas bubble clearance, (e) light-path clearance of colloidal precipitates, and (f) ease of realignment.

Long-term thermal stability is required since the chromatograms can require as long as 0.25 to 6.0 hours. (Some procedures take 18.0 hours or longer.) During these runs, standardization cannot be accomplished to maintain the required accuracy of $\pm 2.5\%$ if a failure of the above exists. Leak-free conditions are required so that accurate flow time is the means by which the chromatographic peaks are fingerprinted for identification and quantitation. Mixing of the flowing stream in the colorimeter cuvette will destroy the leading and trailing chromatographic peak edges. Where two or more peaks are resolved in the minimum volume possible, mixing will destroy the degree of resolution.

The thermal reactions at 100°C of ninhydrin–hydrindantin and amino acids generate carbon dioxide (CO_2).[104] If gas bubbles migrate in the cuvette, the light-path can become blocked. Therefore, gas bubbles from the chemical reaction or other sources must be accommodated by rapid clearance. Similar to the bubble problem, colloidal precipitation, especially from the tin salts and possibly from the methyl Cellosolve (see Ninhydrin discussion), must be cleared from the cuvette light-path. The density of these colloids is high enough in some cases to overcome the buffer flow-lifting velocity, unlike the low density gas bubbles. Therefore, the colloidal accommodation must be an independent consideration in system design.

Cuvette fogging or blockage requires disassembly of the cuvette-colorimeter to clean. To facilitate this simple repair function, the assembly should be self-aligning.

Detection systems have been described which depend on methods other than ninhydrin or other color-developing reagents for quantitation of amino acids and peptides. The polarographic estimation of amino acids as their copper complex is based on the formation of two molecules of amino acid with one cupric ion to give a deep blue-colored complex. This complex then passes through a polaragraphic cell where the amount of copper is determined.[122] The amino acids can also be estimated by using fluorodinitrobenzene and photometrically determining the DNP-amino acids at 420 nm.[123]

Hupe and Bayer[124] have developed a liquid chromatography detector which is based upon the change in heat of solution (ΔH) as a band is absorbed on an exchanger. By calorimetry control, these ΔH bands can be standardized to allow nonrestrictive quantitative analyses of a chromatographic pattern. This type of detector is not limited to the analysis of amino acid or peptide profiles but can be adjusted to any species sequence which can be chromatographed using liquid column elution.

A detector which is an improvement on the ΔH detector is one which senses the change in osmotic or solution activity as a function of a synthetic matrix pressure (ΔP)[125] when an eluted band passes through a packed cell. Like the ΔH detector, the system must be free of variation in flow pressures and must be controlled to a constant temperature. The sensitivity of the ΔP detector is a factor of 3 to 5 greater than that of a ΔH detector system.

While both ΔP and ΔH detectors are highly sensitive to eluting chromatographic bands, they have a common possible problem in that because they are nonspecific detectors they may detect bands which cause interference with the identification of the desired components.

Improvements in the classic ninhydrin-positive reaction detectors will be in the areas of nonpulsed flow, reduction of gas bubbles, and improved photo-sensing components.

Anderson *et al.*[126] describe a regulated pressurized chromatographic system with feedback controls sensing the pressure after the detector cuvettes. From the description of advancement in system chromatography, he anticipates that there will be: (a) a reduction in size and weight of amino acid analyzers, (b) improved chromatographic resolution by reduction of mixing within the coil, and (c) a lower signal-to-noise ratio to allow a greater electronic amplification of the signal. We can anticipate that the liquid chromatographic systems will be reduced from 2.7 to 0.014 m³ without sacrificing analysis accuracy or time.

1.2.7   Column Temperature Control

The exchange kinetics in the ion-exchange chromatography of an amino acid or a peptide are highly temperature-sensitive. To obtain a reproducible sequence pattern of peaks against a time base for amino acid or peptide identification and to resolve certain critical areas, reproducibily controlled column temperatures are required. This controlled condition is usually accomplished by temperature-controlled water, circulating in a jacket surrounding the resin columns. A sensor tests the water temperature, and heat is added as required to a water reservoir. The heat capacity of the reservoir, temperature sensing, and pumping rate should be sufficient to maintain the jacket at $\pm$ 0.5°C in the 30–70°C range. One of the subtle variables of a temperature program is the slope of the time-temperature gradient in changing from one temperature to a second, as required in many chromatographic procedures. Where methodology for a given instrument calls for a temperature change, and the change condition requires a 20-minute gradient, any other time gradient may induce undesirable results. Therefore, it is not surprising that some methodologies cannot be duplicated by similar instruments when the gradient conditions are not equivalent. It is preferable that controls for instrumentation of known temperature gradients be incorporated into the basic design of the analyzer.

1.2.8   Automation

Automatic amino acid analyzers are usually of the type developed by Spackman, Moore and Stein in 1958. These analyzers automated and programmed certain functions of the instrument required for completing an analysis. Earlier methodologies required more than a normal workday to complete an analysis of the acidic, neutral, and basic amino acids. During unattended periods of the analysis, an automatic system was required to make buffer and column temperature changes. In addition, it is essential that the ninhydrin reagent be flushed from the reaction coil at the end of the analysis. At shut-down of the instrument the ninhydrin pump and the recorder are automatically turned off and at a later, preselected interval of time, the buffer pump is switched off. During this time of buffer pump operation, the reaction coil has been free of the ninhydrin reagent.

In line with present accelerated methods for amino acid analysis and the large number of samples required by many researchers, the time needed for sample application and computation of chromatographic results requires that some of these steps be automated or at least simplified. Flexibility of the chromatographic systems allows degrees of automation. Some methodologies operate at single temperature and use only one buffer; in such a case an analysis can be completed in 18 minutes (*e.g.*, analysis of tyrosine and phenylalanine; see Section 1.6.7). For these analyses much of the automation can be eliminated; however, it might be convenient to have automatic sample injection to facilitate the operation.

Injectors should have the following required characteristics: (a) quantitative transfer of sample to the column; (b) no introduction of air or gas into the chromatographic resin bed; (c) no introduction of the sample in a manner that will destroy the chromatographic peak (channeling); and (d) a fail-safe method to ensure against sample loss. At present there are two types of sample injectors which can meet these requirements: (1) liquid loops and (2) preloaded resin samples or columns. While both types are capable of handling hydrolysate and physiological samples, liquid loops find their optimum application in handling protein or peptide hydrolysate samples since the samples are liquid which contain no colloidal impurities that can be deleterious to the resin bed (*i.e.*, lipids, cells, etc.). On the other hand, preloaded resin samples or columns have their optimum application in handling the physiological free amino acids, *i.e.*, from plasma. The columns of resin extract the amino acids to be chromatographed and function as a filter bed for the removal of the impurities found in this type of sample. The eluting buffer, which then transfers the sample to the chromatographic resin column, also washes occluded amino acids from possible protein colloids.

Sample injection also reduces column free space (that liquid volume above the resin bed). A simple, adjustable screw-type fitting is currently used to eliminate the void space by extending the top fitting down to the resin bed. An automatic free-space eliminator which accomplishes this function has been developed, thus preventing possible sample dilution during unattended periods; it adjusts to this free space variation, however, by floating with any slight compression or expansion of the resin bed.

Other areas of chromatographic procedure that would lend themselves to automation include the development of techniques for calculating the area under a chromatographic peak and for translating the output from the recorder to a form that can relate the concentration of a particular peak to its concentration in the sample.

As each colored band of column effluent–ninhydrin reaction mixture flows through the cuvette in the colorimeter, under constant flow rate conditions, an absorption record is made by the decreased voltage signal (from the photovoltaic cells) to the recorder. Usually the output voltage of the colorimeter to the recorder is 0–5 mV. At maximum color development (highest optical density, OD) the voltage is 0, and with no color present other than the "blank" color of the reagents the voltage is 5 mV, which is used as the "base line." To record the chromatographic results accurately, the recorder should: (a) accurately respond only to the signal from the cuvette-photovoltaic cell system; (b) permit long-term operation with minimal deviation due to thermal and electrical noise; and (c) employ a constant-speed chart drive to ensure an accurate time base for chromatographic peaks. If the chromatographic pattern is recorded on linear chart paper, then an OD overlay is required to quantitate the amino acid or peptide peaks. If the recorder has OD or logarithmic paper, the absorbance readings can be read directly. Other auxiliary features which are associated with the chromatographic read-out and considered a component part of the recorder system are range expansion, linear-log conversion, peak integration, and print-out and computer interfacing. Range expansion is a method by which the full-scale read-out of the recorder can be easily changed from 0–5.0 mV to 4.0–5.0 mV or 4.5–5.0 mV. In this manner, low concentrations of amino acids can readily be displayed and quantitated. Linear-log conversion is the electronic conversion of the colorimeter signal (OD) to a linear read-out form as the electrical-log function of the signal. This electronic operation eliminates the use of OD paper or an OD overlay for manual computation of the amino acid or peptide peak concentrations and allows a direct, simple method of comparative (to a standard calibration pattern) quantitation. In addition, linear-log conversion is usually required if the chromatograph output is to be automatically integrated. At present there are two accepted modes of integration—mechanical and electronic. Both modes give accuracy within the limits of that obtainable from the rest of the system. The integrated results by either mode can be recorded by available print-out systems, which also give direct numerical response to digital computers.

1.3   Chromatographic Techniques

There are two basic procedures available for the chromatographic analysis of amino acids. One method is to analyze the amino acids by a hydrolysate procedure and the second, by a physiological procedure. The descriptions of these procedures follow.

1.3.1   *Procedures*

1.3.1.1   Protein or peptide hydrolysates

The amino acids present in a sample to be analyzed determine the chromatographic procedure to be used, and for the desired results it is important to classify the sample properly. Therefore, those amino acids normally chromatographed and quantitated by the simple hydrolysate procedure are:

tryptophan	proline
lysine	glycine
histidine	alanine
ammonia	cystine
arginine	valine
cysteic acid	methionine
aspartic acid	isoleucine
threonine	leucine
serine	tyrosine
glutamic acid	phenylalanine

1.3.1.2 Physiological fluids

This analytical chromatographic procedure was originally developed for the analysis of urine, blood plasma, tissue extracts, and other physiological materials. However, this procedure can also be used for other materials that contain similar amino acids. The physiological fluid procedure is really an extended mode of the simple hydrolysate procedure of analysis, permitting the separation and quantitation of more than 50 amino acids* or related compounds compared to the usual 20 amino acids of the hydrolysate procedure. Therefore, in determining the correct procedure to use in analyzing the sample, the identity or composition of the sample and not the source of the material should be the consideration. The test procedure chosen should be the one that completely quantitates all of the amino acids suspected of being present in the sample. Those amino acids generally chromatographed by the physiological procedures include:

Acidic and Neutral Amino Acids

phosphoserine	sarcosine
phosphoethanolamine	proline
taurine	glutamic acid
urea	citrulline
hydroxyproline	glycine
aspartic acid	alanine
threonine	α-aminoadipic acid
serine	α-amino-n-butyric acid
asparagine	valine

---

*Hamilton reports 35–150 or 160 components in urine using different chromatographic techniques.[127]

cystine   tyrosine
cystathionine   phenylalanine
methionine β-alanine
isoleucine β-aminoisobutyric acid
leucine

Basic Amino Acids

DL-allohydroxylysine   histidine
γ-aminobutyric acid   3-methylhistidine
ornithine   anserine
ethanolamine   creatinine
ammonia   carnosine
lysine   arginine
1-methylhistidine

1.3.2. Cation-Exchange Resins

1.3.2.1 Pulverized or ground resins

The chromatographic procedures of Moore and Stein[2] were concerned with the analysis of a protein hydrolysate in 1951. In 1958 the chromatographic procedures of Spackman, Moore, and Stein[13] permitted an amino acid analysis of approximately 4 mg of protein (hydrolysate) in 24 hours using Amberlite IR-120.* An accelerated system using blended resins has been reported for the analysis of a protein hydrolysate in 6-⅓ and 4 hours using buffer flow rates of 40 and 60 ml per hour respectively. These resins were ground and fractionated by a procedure similar to that reported by Hamilton.[128]** When the two-column procedure of Spackman[129] was used, the physiological procedure required 16 hours for a total analysis using 50-ml-per-hour buffer flow rate. The use of pulverized resins for the chromatography of amino acids has the disadvantage of generating "fines" as they are used. As reported earlier,[103] during pulverization of the resin cleavage probably takes place along planes of lower cross-linkage in a nonhomogeneous matrix. Also, those pieces which are fractured during the pulverization process can break apart by attrition during use and produce higher operating pressures. Because of the lack of reproducibility from batch to batch, it was necessary to blend fractions to achieve optimum results; thus, specific blends of resin are not available.

---

*Röhm and Haas Company, Philadelphia, Pennsylvania.
**Similar blended resins may be available from Beckman-Spinco, Palo Alto, California.

1.3.2.2. Spherical resins

To minimize the disadvantages mentioned above, it was found advantageous to manufacture resins to definite specifications. A Type AA-15 spherical resin* was produced which would allow a total analysis of amino acids by the hydrolysates procedure in 4 hours, using a 56-cm resin column for the acidic and neutral amino acids and a 5-cm resin column for the basic amino acids employing Type PA-35 resin.[130] A Type PA-28 resin was later developed for analysis of those amino acids normally found in samples such as blood plasma or urine by a physiological procedure requiring a little over 5 hours for the acidic and neutral and almost 6 hours for the basic amino acids using the Type PA-35 resin mentioned above.[88] These resins gave a dramatic increase in peak heights indicating that the amino acids were being eluted from the spherical resin in narrower bands, improving peak resolution. Still later, a Type UR-30 resin became available which allowed the hydrolysate and physiological methods to be used for analysis of the acidic and neutral amino acids. Not only did this resin have the advantage of performing a dual function, but it did so at lower operating pressures and offered improved resolution over what had been previously reported. These resin columns required 56×0.9 cm of resin for optimal separations of the acidic and neutral amino acids.

Much of this acceleration in analysis time is due to improvements in design and reliability of the amino acid analyzers. These advantages have also aided in the development and design of spherical ion-exchange resins used in specific chromatographic systems. Improved resins have permitted the use of shorter resin columns while maintaining sufficient theoretical plates in the column to allow separation of the many different amino acids normally chromatographed by the more complex physiological procedures. That the analysis of acidic and neutral amino acids of the more complex physiological fluids approaches the time required for simpler hydrolysate procedures has become a reality.

A newly available resin, Type UR-40, has been designed which will allow analysis of the acidic and neutral amino acids by the physiological procedure in 170 minutes on a 26-cm resin column. The basic amino acids can also be chromatographed on the same column by the physiological procedure in 210 minutes. A single-column hydrolysate procedure is also outlined in Section 1.6. The use of shorter resin columns

---

*Resins referred to in this paragraph are available from Beckman-Spinco, Palo Alto, California.

and improved peak resolution has been made possible by eliminating specific mixing areas present in most equipment (such as reaction coils, mixing manifolds, valves, and cuvettes).

1.3.2.3 Pellicular resins

Since the resin bed is the heart of any liquid chromatographic system, it is to be expected that this area will demand future improvements. Improvements in specifically designed resin phases have been recently reported.[131] The "pellicular" resins to which we refer are controlled active matrices which have been stabilized on inert spherical surfaces. In this manner a resin with a high kinetic rate of exchange is possible without proportionate loss in volume capacity. With these characteristics, reduced sample sizes would be required and a reduction in chromatographic time of analysis would result.

Further future improvements in the resin matrix will involve specifically the design of active resin sites. These sites will be highly selective in their exchange capacity. As an example, this type of resin could have as its active site a specific antigen[132,133] which would then remove and chromatograph specific antibodies from plasma samples.

1.3.3. Anion-Exchange Resins

1.3.3.1 Spherical resins

Methods are described for the stepwise[99] and gradient elution[134] of peptides from an anion-exchange resin. A Dowex 1-X2, strongly basic resin of low cross-linkage was used and should complement the cation-exchange resin procedures for separating peptides. This low-cross-linked resin has the disadvantage of being very compressible, resulting in a sharp increase in column operating pressure. These conditions often require that the column be repacked. Anion exchangers also have a limited chromatographic operating "life" resulting in a change of elution pattern of the chromatographic peaks due to instability of the tertiary and quaternary amine functional groups. Some of these disadvantages have been eliminated by a newly available anion-exchange resin, Type 1-S,[92] which has been used for the chromatography of neutral sugars.

1.3.4. Ligand Chromatography

This method of liquid chromatography is based on the metal-complex form of the amino acid as the mobile species rather than on the usual free amino acid in one of its amphoteric forms. Metal complexes of cadmium, cobalt, copper, nickel, and zinc have been reported.[26,135] These metals function as weak bases in the reaction with amino acids to form the metal complex

$$Zn^{++} + [AA]^- \rightleftarrows Zn\,[AA]^+$$

This metal-ion complex is then chromatographed on a cation-exchange resin which has a weak basic metal ion adsorbed on it.

$$\underset{\text{(zinc complex of amino acid)}}{Zn[AA]^+} + \underset{\text{(sulfonate resin; R = resin)}}{RSO_3^-} \rightleftarrows RSO_3Zn[AA]$$

During the chromatographing process, the amino acids are in dynamic equilibrium with the metal-ion concentration in the resin. Using a buffered eluant to reverse the amphoteric charge on the amino acid, we have

$$RSO_3Zn[AA] \xrightarrow[H^+,\ pH\ 3.2]{Zn^{++}} RSO_3Zn^+ + [AA]^+$$

When the charge is unbalanced, it is restored to equilibrium by the increase in hydrogen ion (H^+) concentration as a result of the lower pH.

Essentially the ligand system converts the cation-exchange resin ($RSO_3^-Na^+$) to an anion-exchange resin for improved chromatography of the amino acids in their carboxylic form, because the free-acid form of the amino acid (pKa) has a greater distribution coefficient than the free-base form (pKb) of the amino acids.

In addition, the use of a metal ligand in the resin phase stabilizes the hydrational state of the resin. In conventional ion-exchange systems, the cationic form of the resin (Na^+, K^+, Li^+, etc., which is changing) determines the degree of resin hydration. However, in the Zn^{++} ligand system, the Zn^{++} is not exchanged and there is no hydrational flux. Also, the hydrational changes of the amino acids have been decreased. The amount of water exchanged by the amino acids as an anion has been minimized.

Although ligand chromatography offers the improvement of a greater pKa spectrum, to allow better separations of the amino acids there are some areas of concern. Of major importance is the stability of the chromatographic resin, both chemically and physically. The purity of the buffer system is also to be considered. Maintaining equilibrium of metal ion between the resin phase and the eluting buffers is essential if reproducible amino acid patterns are to be obtained (identification and fingerprinting of peaks). In addition, the requirements of buffer flow rate and operating temperatures are very important. Improvements in resin stability, through their manufacture, and control of metal ion in the system will remove many touchy problems to enable the full possibilities of this chromatographic technique to be realized.

1.3.5   In-Stream Hydrolysis of Peptides

The reaction of peptides with ninhydrin reagent usually results in low color development because the peptide reacts only at the free amino group. If the peptides undergo alkaline hydrolysis, then all of the amino acids will be liberated, reacting with ninhydrin to produce a higher color yield. An instream hydrolysis procedure, described by Catravas,[136] will give an approximation of the number of amino acids present in a peptide to indicate the size.

1.4   Variable Factors

1.4.1   Sample Preparation

Preparation of the sample for quantitative chromatographic analysis requires exacting techniques to obtain completeness of amino acid recovery.

To analyze for the total amino acids in a biological material, the sample must first be hydrolyzed. This procedure will break all the peptide bonds, releasing the amino acids. Portions of the hydrolysate solution can then be analyzed on the resin column for those amino acids of interest.

If only the free amino acids and related compounds are of interest, it is necessary that the sample be deproteinized. Generally, very small amounts of plasma proteins do not interfere with the analysis, but they can be hydrolyzed on the column and interfere with subsequent analysis. Large amounts of protein in a sample are to be avoided, because the protein will adhere to the ion-exchange resin and block the buffer flow.

Four methods—sulfosalicylic acid, picric acid, ultrafiltration, and centrifugation—have been used for the deproteinization of biological samples.[6, 76, 137–140] The sulfosalicylic acid method of deproteinization will allow the centrifugate to be applied to the resin column. If picric acid is used, the excess can be removed by extraction using an anion-resin column or by a small-batch process in which amine resins are added to the centrifugate, followed by filtration or decantation after the resin has settled.

The use of a Visking dialysis sac for overnight ultrafiltration at 40°C under positive nitrogen pressure proved to be less satisfactory than the sulfosalicylic acid method of DeWolfe.[138] Hamilton[85] indicated that 3% w/v sulfosalicylic acid when added to plasma can then be centrifuged (13,500 rpm) to pack the precipitate. The clear supernatant containing the amino acids can then be placed directly on the column.

Gerritsen[139] presents a method for the determination of the free amino acids in serum and tissue extracts of which only a part of the soluble protein is removed by ultracentrifugation before the sample is placed on the ion-exchange columns. It is claimed that this method is more convenient and faster than deproteinizing with either picric or sulfosalicylic acid.

Increased sensitivity of the amino acid analyzers (by electrical amplification) and improved resolution of the chromatographic peaks have permitted small sample loads to be used for chromatographic analysis. Although it is possible to quantitate 10–20 nanomoles of an amino acid (0.1 ml serum), sample preparations are outlined (in Section 1.6.1) for quantities somewhat larger to allow for repeat analyses if necessary.

1.4.2 Column Packing

The excellence of resolution of amino acids in the sample depends primarily on the ion-exchange resin column. To prepare columns which exhibit a high degree of resolution, it is essential that care be exercised in packing the column, preventing variations which will result in erratic column performance. There are two principal causes of poor column performance: trapped air in the resin column and contaminants on the packed resin column. One of the more common problems preventing successful column performance is the accumulation of air on the surface of the resin beds or formation of an air pocket. Occasionally, a gas bubble will work its way down through the resin bed and lodge on top of the resin support screen. This will effectively limit buffer flow and smear any amino acid peak emerging from the column. During the regeneration cycle, trapped air is sometimes dislodged as the sodium hydroxide front moves down the column. If any appreciable amount of gas is trapped in the resin column (causing channeling of the buffer), not only might resolution be destroyed but the trapped gas may cause a significant increase in column operating pressure. The second cause, the presence of dirt, mold or proteinaceous material, will prevent proper "layering" or application of the sample to the column, resulting in smeared or asymmetrical peaks as well as increases in operating pressures.

Either of the above conditions which impede the normal flow of buffer through the resin will result in a steadily increasing pressure upon the resin bed. Usually under these conditions the resin bed compresses; if the pressure continues to increase, the resin bed becomes "locked," producing excessive operating pressures. As the liquid channels between the resin particles are reduced and contact between buffer and resin is reduced, there will be a decrease in the resolving capability of the resin column. It will then become necessary to repack the column. The procedure we recommend for packing a resin column is outlined in Section 1.6.

1.4.3 *Effect of Buffer pH*

1.4.3.1 Hydrolysate procedures

Four-hour methodology, UR-30 resin, PA-35 resin. An increase of just a few hundredths of a pH unit of the first buffer of the acidic and neutral amino acids, pH 3.25 (0.20N), improves resolution between the threonine–serine peaks. An increase in the second buffer pH from 4.25 to 4.41 (0.20N) decreases the resolution between isoleucine–leucine peaks. An increase in pH of the pH 5.25 (0.35N) buffer elutes histidine faster than the other basic amino acids.

Two-hour methodology, UR-30 resin for acidic and neutral amino acids. Increasing the pH of the first buffer, 3.488 (0.20N), causes a loss of resolution between the twin peaks threonine–serine. There is also a loss between serine–glutamic acid and between glycine–alanine. Cystine is preferably eluted sooner to give improved separation from glycine.

Single-column methodology, UR-30 resin. With column temperature, sodium and citrate ion concentration, and buffer flow rate held constant, the pH of the first buffer is varied. Increasing the pH from 3.28 to 3.31 decreases the separation between the threonine and serine peaks. Cystine is eluted sooner but separation between glycine and alanine is sacrificed. Increasing the pH of the second buffer from 4.30 to 4.50 with other conditions held constant reduces separation of both methionine from isoleucine and isoleucine from leucine. Increasing the pH of the third buffer allows histidine to be eluted sooner.

Single-column methodology, UR-40 resin. With the column temperature, sodium and citrate ion concentration, and buffer flow rate held constant, the pH of the first buffer is varied. Dropping the pH from 3.545 to 3.515 decreases the separation between threonine and serine peaks (valley-to-peak height ratio 0.07; for definition see Section 1.5.1). Cystine is eluted from the column later; however, there is an improvement in the separation of serine–glutamic acid of approximately 0.19. Increasing the pH of the second buffer from 4.25 to 4.30 with other conditions held constant reduces the separation of the leucines. Increasing the pH of the third buffer causes histidine to be eluted sooner.

1.4.3.2 Physiological procedures

Acidic and neutral amino acids, UR-30 resin, sodium citrate buffers. An increase in buffer pH from 3.20 (0.20N) to pH 3.26 (0.20N) improves the separation between the aspartic acid–threonine peaks approximately 0.10 (valley-to-peak height ratio). Cystine is eluted sooner and glutamic acid moves forward into proline. The separation between glycine and alanine is sacrified. Increasing the buffer pH has the general effect of eluting the amino acids from the column sooner.

Acidic and neutral amino acids, UR-30 resin, lithium citrate buffers. As the pH of the first buffer (other variables remaining constant) is increased from pH 2.80 to 3.13, asparagine and glutamic acid are eluted as one peak. There is also a loss of resolution between the proline peak and the α-amino adipic acid peak. A further increase in the buffer pH to 3.27 results in a sacrifice of resolution between the taurine and phosphoethanolamine peaks. Asparagine and glutamic acid are still not separate but the separation between citrulline and α-amino-n-butyric acid is improved. An increase in pH generally elutes most amino acids from the column sooner.

Acidic and neutral amino acids, UR-40 resin. With column temperature, sodium and citrate ion, and buffer flow rate held constant, the increase of pH of the first buffer, pH 3.175 ($0.18N$ Na^+, $0.133M$ citrate) sodium citrate, has the effect of improving the separation between aspartic acid and threonine 0.045 (valley-to-peak height ratio) per 0.010 pH units. Increasing the pH by 0.011 units moves the cystine peak forward 1 minute and has the effect of eluting the glutamic acid peak earlier, but it also compresses the triplet peaks, proline–glutamic acid–citrulline. The separation between glycine and alanine is sacrificed with an increase in buffer pH.

Basic amino acids, UR-40 resin. Decreasing the buffer pH from $4.148(0.40N)$ to $4.080(0.40N)$ while maintaining column temperature, buffer concentration, and buffer flow rate constant has the following effects: γ-amino-n-butyric acid is delayed about 3 minutes, the separation between ethanolamine–ammonia is improved to approximately 0.09 (valley-to-peak height ratio), and lysine-1-methylhistidine decreases to 0.50. Histidine is greatly affected by pH and will be eluted later when the pH is decreased.

Basic amino acids, PA-35 resin. Decreasing the pH of the first buffer from 4.25 to 4.20 ($0.38N$) improves the separation between ethanolamine–ammonia by a valley-to-peak height ratio of 0.09. There is also a loss of resolution between lysine and 1-methylhistidine of > 0.50. A change in the pH of the second buffer affects the separation of tryptophan and carnosine. An increase in pH will cause not only a loss of resolution between these two peaks but also an overlapping of these peaks with those of anserine and creatinine.

1.4.4 Effect of Na+ Concentration

1.4.4.1 Hydrolysate procedures

Four-hour methodology, UR-30 resin. A drop in Na^+ concentration below $0.20N$ tends to broaden the peaks and later will elute the amino acid peaks from the column. There is a slight improvement in resolution of the peaks eluted last on the chromatogram but not at a sacrifice in analysis time.

Two-hour methodology, UR-30 resin. An increase in the sodium ion concentration tends to elute the amino acid peaks from the column too soon, resulting in a loss of resolution between the threonine–serine and serine–glutamic acid peaks.

Single-column methodology, UR-30 resin. With citrate concentration, temperature, and buffer flow rate held constant, the sodium ion concentration is varied in the third buffer. When the normality is decreased from 1.4 to 1.0N, tryptophan is eluted sooner and arginine is delayed. Also with a decrease in normality, the base line shift usually experienced with the third buffer is reduced. Generally, reducing the concentration delays the elution of most peaks.

Single-column methodology, UR-40 resin. With column temperature, citrate concentration, and buffer flow rate held constant, reduction of the sodium ion concentration delays the elution of most peaks. Decreasing the normality of the *third* buffer from 1.8 to 1.4 not only delays the elution time of the basic amino acid peaks but it greatly reduces the base line shift experienced with this buffer.

1.4.4.2   Physiological procedures

Acidic and neutral amino acids, UR-30 resin, sodium citrate buffers. Increasing the sodium ion concentration elutes most of the amino acids sooner. An increase by 0.01N decreases the elution time of the triplet peaks, proline–glutamic acid–citrulline.

Acidic and neutral amino acids, UR-30 resin, lithium citrate buffers. Generally, as the ionic concentration of the buffer is increased, the analysis time is decreased. Increasing the Li^+ concentration from 0.20 to 0.35N advances glutamic acid until it is eluted with asparagine. Proline and glycine are not resolved and cystine is eluted with the valine peak. Methionine and cystathionine are not separated and β-alanine is eluted with phenylalanine.

Acidic and neutral amino acids, UR-40 resins. With column temperature, buffer flow rate, and citrate concentration held constant, the sodium ion concentration is varied in both the first and second buffers. Usually, reducing the concentration delays the elution of most amino acid peaks; however, the most notable effects are noted here when the Na^+ concentration is reduced to allow spreading of the proline–glutamic acid–citrulline peaks and improved separation of isoleucine–leucine peaks.

Basic amino acids, UR-40 resin. As with analysis of the acidic and neutral amino acids, reducing the sodium ion concentration will generally delay elution of the amino acid peaks. Decreasing the concentration below 0.4N has little effect upon the separation of the ethanola-mine–ammonia peaks, but it does reduce the resolution of the lysine and 1-methylhistidine peaks by a valley-to-peak height ratio of 0.45.

Basic amino acids, PA-35 resin. An increase from 0.36 to 0.38N sodium ion concentration has little, if any, effect on the resolution of the ethanolamine–ammonia peaks. Separation improves by 0.45 be-tween lysine and 1-methylhistidine.

1.4.5   Effect of Citrate Concentration

1.4.5.1   Hydrolysate procedures

Four-hour methodology, UR-30 resin. Increasing the citrate con-centration improves the separation between the glycine–alanine peaks.

Two-hour methodology, UR-30 resin. Increasing the citrate con-centration to 0.10M has no effect on the resolution of threonine–serine and the serine–glutamic acid peaks. Cystine is eluted with proline and the separation between the glycine–alanine peaks is improved.

Single-column methodology, UR-30 resin. While maintaining the buffer pH at 3.28, the Na^+ concentration at 0.20N, the column tem-perature at 55.6°C, and the buffer flow rate at 130 ml per hour, varying the citrate concentration from 0.066M to 0.100M has no effect on the resolution of threonine and serine peaks. The separation of glycine from alanine is enhanced; however, the cystine peak is eluted sooner, incompletely separated from alanine.

Single-column methodology, UR-40 resin. While maintaining the column temperature at 48.0°C, the sodium ion concentration at 0.20N, the first buffer pH at 3.545, and the buffer flow rate at 60 ml per hour; increasing citrate concentration from 0.066M to 0.200M has no effect on the resolution of the threonine and serine peaks. The separation of glycine–alanine is enhanced considerably and the cystine peak is eluted from the column sooner.

1.4.5.2   Physiological procedures

Acidic and neutral amino acids, UR-30 resin, lithium citrate buffers. While maintaining the buffer pH at 2.80, the column temperature at 38.8°C, and the buffer flow rate at 70 ml per hour, varying the citrate concentration from 0.033 to 0.166M produces the same general effect as increasing the pH of the buffer. Hydroxyproline and aspartic acid are not resolved at a citrate concentration of 0.033M, and as the citrate concentration is increased, aspartic acid is eluted ahead of hydroxypro-line. At a citrate concentration of 0.166M, cystathionine and methionine are not separated.

Acidic and neutral amino acids, UR-40 resin. While maintaining the buffer pH at 3.175, the Na^+ concentration at 0.18N, the column temperature constant at 32.5°C, and the buffer flow rate at 60 ml per hour, varying the citrate concentration from 0.066M to 0.133M does not change the position of glutamic acid relative to that of proline or citrulline. Increasing the citrate concentration improves the separation ratio of aspartic acid–threonine by 0.40 (valley-to-peak height ratio). At the higher citrate concentration most of the peaks are eluted sooner, the glycine–alanine separation is considerably improved, and the proline–glutamic acid–citrulline peaks are compressed.

1.4.6 Effect of Temperature

1.4.6.1 Hydrolysate procedures

Four-hour methodology, UR-30 resin. A decrease in temperature of only a few tenths of a degree will improve the separation of the threonine–serine peaks. An increase in temperature improves the separation of tyrosine–phenylalanine. There is, however, a loss of resolution between the isoleucine–leucine peaks.

Two-hour methodology, UR-30 resin. A decrease in column temperature improves the separation of threonine–serine and serine–glutamic acid peaks. Cystine is delayed and may be eluted with glycine, but a temperature decrease does improve the separation between the glycine–alanine and the isoleucine–leucine peaks. An increase in temperature will improve the separation between tyrosine–phenylalanine.

Single-column methodology, UR-30 resin. Increasing the column temperature from 52 to 55.6°C, while keeping the buffer pH, concentration, and flow rate constant, generally increases the elution rates of the amino acids. Ornithine and lysine are little affected by temperature; however, arginine is eluted sooner and the separation of threonine from serine and tyrosine from phenylalanine is improved.

Single-column methodology, UR-40 resin. Increasing the column temperature from 48.0°C to 50.0°C, while keeping the buffer pH, buffer concentration, and flow rate constant, generally increases the elution rates of the amino acid peaks. The separation of threonine–serine is reduced (valley-to-peak height ratio) about 0.01 and the loss to serine–glutamic acid is about 0.17. Cystine is little affected by temperature; however, the separation of glycine from alanine and tyrosine from phenylalanine is improved to 0.14 and 0.08 respectively. Of the basic amino acids, arginine is eluted sooner with an increase in temperature.

1.4.6.2 Physiological procedures

Acidic and neutral amino acids UR-30 resin, sodium citrate buffers. As column temperature increases, the resolution of α-aminoadipic acid and α-amino-n-butyric acid improves; however, the separation between the valine and cystine peak decreases. The higher temperature also improves the separation of aspartic acid–threonine and tyrosine–phenylalanine. Cystine is eluted more quickly from the column with an increase in temperature, as is glutamic acid. The triplet peaks, proline–glutamic acid–citrulline, are compressed, resulting in a loss of resolution.

Acidic and neutral amino acids, UR-30 resin, lithium citrate buffers. An increase in column temperature generally elutes the amino acid peaks sooner. At a temperature of 30°C, the separation of proline and α-aminoadipic acid is incomplete. An increase in temperature results in a loss of resolution of the cystathionine–methionine peaks; however, the separation between citrulline and α-aminobutyric acid, and proline–α-aminoadipic acid peaks is enhanced.

Acidic and neutral amino acids, UR-40 resin. Increasing the column temperature approximately 1.5°C, while maintaining the buffer pH, sodium ion and citrate concentration, and buffer flow rate constant, generally elutes the amino acid peaks sooner. Aspartic acid–threonine separation is improved to approximately 0.12. However, glutamic acid is eluted more quickly so that it is incompletely separated from proline. Cystine is eluted 2 minutes sooner and the proline–glutamic acid–citrulline is eluted in a narrower band which reduces resolution. Increasing column temperature greatly improves the separation of tyrosine–phenylalanine; however, there is a slight loss of separation between the isoleucine–leucine peaks.

Basic amino acids, UR-40 resin. Those basic amino acid peaks which are affected by slight variations in column temperature are arginine, carnosine, and tryptophan. An increase in temperature elutes all peaks from the column sooner. However, an increase of 3°C will improve the separation between ethanolamine–ammonia (at the sacrifice of lysine and 1-methylhistidine). γ-Aminobutyric acid also is eluted sooner than some of the other peaks and the separation of hydroxylysine and allohydroxylysine is enhanced. Critical areas of separation in this analysis are γ-amino-*n*-butyric acid from the ornithine peak, ethanolamine–ammonia, and the resolution of lysine and 1-methylhistidine.

Basic amino acid, PA-35 resin. An increase in temperature will cause arginine and carnosine to be eluted from the column sooner. If the temperature is too high, tryptophan will be eluted with anserine.

1.5 Evaluation Analyses
1.5.1 *Resolution*

The principal manual methods used for quantitating the amino acid peaks in amino acid chromatography are valid only when the symmetry of the peak approaches a Gaussian shape. Any deviation from this symmetry, due to the overlapping or the slight skewing of peaks, is a cause for error. In general, the accuracy and precision attained by evaluating the amount of material in the column effluent is dependent on peak shape and peak area. Many factors will affect the optimum shape of a peak and these are discussed in Section 1.4. An indication of the resolving power or separation of chromatographic peaks can be expressed as a valley-to-peak height ratio.

Where two peaks of similar peak height are juxtaposed so that the valley between them does not reach base line, the valley-to-peak height ratio is a:b where "a" is the absorbance read at the minimum of the valley and "b" is the absorbance read at the maximum of the lower of the two peaks. The absorbance values of the peaks and valleys must be corrected to zero base line on the chromatogram. To obtain optimum results, it is recommended in quantitating the amino acid peaks that the valley-to-peak height ratio not exceed 0.40.

1.5.2 Determination of Sample Concentration

The amount of each amino acid in a sample is usually determined by measuring the area enclosed by its corresponding peak on a chromatogram.

The integration of the area under the chromatographic peak is usually closely approximated by measuring the height of the peak and multiplying this value by the width at half the height. The height (H) of the peak in net absorbance is determined from the recorder chart. Using a multipoint recorder, the number of dots on each curve above the half-height of the total peak height can provide the width (W) of the peak in terms of time. Then, $H \times W/C$ is equivalent to a constant (C) in micromoles from the calibration run using amino acids of known concentration.

Correlation analysis of seven procedures for determining the areas of chromatographic curves has been shown by Mefferd.[141]

Harris and Habgood have indicated that precision in determining the area of a chromatographic peak by the height-width method is enhanced if the absolute values for the height and width are large.[142] General considerations on the integration of peaks was demonstrated by Tempé.[143] An experimental evaluation of the indeterminate errors in the measurement of chromatographic peaks is extensively presented by Ball, Harris and Habgood.[144-145] The usual sources of error were: error in placement of the base line; error in measuring the height of the peak from the base line; error in positioning the mark at one-half the peak height; and finally, the error involved in measuring the peak width. It was found that for peaks of a given height the errors decrease as the width increases up to a limiting value of 3–4 cm width. To reduce errors further, peaks should be as tall as the chart paper allows.

Therefore, to maximize accuracy the peak should be in the range of 2–10 for the ratio of height to width at half-height. The errors due to irregular travel of the recorder charts were determined by Gerling.[146]

In a new system of automatic amino acid analysis, Mondino[147] presents data using peak height only as a method of calculating the amount of material in the column effluent. The use of peak height alone as a quantitative measure of chromatographic peaks involves only two operations: drawing the base line and measuring the height of the peak from the base line. The error expected from this measurement is almost independent of peak shape; for a peak that is approximately 10 cm in height, the error would be about 0.13%. If experimental conditions (sample applications, flow rates, etc.) can be controlled, then measurement of height alone will give the best precision of manual methods.

Schroeder *et al.*[148] describes the construction and use of a "nomogram" which permits the calculation of the amino acid analysis within 5 minutes. In assessing the applicability of the nomogram calculations and those made by the usual $H \times W/C$ method, the results have been within $\pm 5\%$ in the 0–0.3-μmole range; in four out of five runs the results have agreed within $\pm 2\%$.

Electronic and mechanical integrator systems for computing the area under the amino acid peaks do exist, but we have found that calculation of concentration based on peak height only is adequate for most applications. Adding an internal standard to the unknown sample before analysis would indicate any deviation from the normal expected values (mechanical losses, chemical variations, equipment changes). Norleucine—or better, homocitrulline[139]—is suitable to use in the acidic and neutral chromatographic procedures eluted just after leucine. L-2-Aminoguanido propionic acid,* for the short basic hydrolysate column, is eluted between ammonia and arginine.[149]

1.6   Recommended Methodology

The procedures outlined below present those methods we use in our laboratory. Although many excellent methods are available, these are the ones with which we have had the most experience.

1.6.1   *Sample Preparation*

1.6.1.1   Hydrolysis

The procedure as outlined by Moore and Stein[137] for the hydrolysis of a sample avoids losses of amino acid due to artifacts and oxidation by air. The following steps are recommended:

---

*Norleucine and L-2-amino-3-guanidinopropionic acid are available from California Corporation for Biochemical Research.

(1) Measure 2 mg of protein sample, or the liquid equivalent, into a 16 × 125-mm Pyrex test tube without lip.

(2) Pipet into the tube (if sample is a solid) 0.5 ml of distilled water and 0.5 ml of concentrated hydrochloric acid, reagent grade. If the sample is a liquid, add an equal volume of concentrated hydrochloric acid.

(3) Attach a small glass rod temporarily to the lip of the tube to serve as a handle, and with an oxygen-gas flame draw the neck of the tube down to a capillary opening 1–3 mm, about 1.5 cm long.

(4) Remove the glass handle and cool the lower half of the tube in a dry-ice alcohol bath.

(5) After the sample-mixture has solidified, connect a vacuum line to the tube and evacuate to at least 0.05 mm mercury pressure.

(6) After the tube is evacuated, seal it off at the capillary constriction with the fine point of the oxygen flame.

(7) Place the sealed tube in a forced-draft oven regulated at $110°C \pm 1°C$ for 22 hours.

(8) After removing the tube from the oven and allowing it to cool to room temperature, open the top of the tube with a scratch from a file.

(9) Remove the hydrochloric acid from the sample by placing the sample tube, inclined at 15–20° from the horizontal, inside a desiccator that contains a shallow dish of sodium hydroxide pellets. The desiccator is then continuously evacuated through use of a water aspirator.

(10) To effect solution and adjust the sample to the correct pH for the ion-exchange columns, 2.0 ml of pH 2.2 (0.20N) sodium citrate buffer can be added to the sample.

(11) A volume of 400 μl will represent 0.4 mg of sample or approximately 0.1 μmole of each amino acid.

1.6.1.2 Deproteinization

Sulfosalicylic Acid

(1) Collect 1.0 ml of whole blood.

(2) Transfer it to a 5-ml conical glass centrifuge tube and allow to clot.

(3) Ring the top of the clot with a small glass stirring rod to free the clot from the walls of the tube.

(4) Centrifuge the sample at 2500–3500 rpm (standard 10-inch rotor) for 15 minutes.

(5) Then withdraw 500 μl of plasma with a micropipet and transfer it quantitatively to a clean centrifuge tube.

(6) Add 500 μl of 3% aqueous solution of sulfosalicylic acid (15 mg) to 500 μl plasma and mix thoroughly on a suitable shaker.

(7) After mixing, pack the precipitated proteins by centrifugation at 2500–3500 rpm.

(8) Pipet 0.5 ml of the clear supernatant containing 250 μl of the original plasma onto the ion-exchange column (appropriate for 1-mV recorder range sensitivity and a 6.6-mm cuvette).

Centrifugation

(1) Collect 1.0 ml of whole blood and transfer to a 5-ml conical centrifuge tube.

(2) Add 1.0 ml of pH 2.2 (0.20N) sodium citrate buffer and allow to stand in the refrigerator 30 minutes.

(3) Spin sample down in a high-speed centrifuge for 30 minutes at 18,000 $\times$ g or for 20 minutes at 78,000 $\times$ g.

(4) Clear supernatant, 0.5 ml, is equivalent to 0.25 ml of original plasma and can be applied directly to the column. Using a 0.5-mV recorder expansion and a 12-mm cuvette, 25% of the sample load indicated above can be used on the column.

1.6.1.3 Urine samples

For the analysis of urine it is necessary to remove the ammonia from the sample, particularly if the basic amino acids are to be determined. If the urine is to be kept longer than 2 days, it should be placed in a plastic bottle and stored in a freezer. About 1 ml of toluene is added per 24-hour collection to prevent decomposition. Removal of all ammonia from the sample is omitted if basic amino acids or ammonia are not chromatographed. The only treatment required in this case is adjustment of the sample to pH 2.0, and centrifugation to remove colloids.

To remove ammonia:

(1) Add 2N NaOH to 0.5 ml of urine in a beaker until the pH is 11.5–12.0 (pH indicator paper). Note the volume of NaOH required.

(2) Place sample in a vacuum desiccator over concentrated H_2SO_4 and evacuate continuously with a water aspirator for 6 hours to remove ammonia.

(3) Adjust the sample pH to 2.0–2.2 with 6N HCl.

(4) Filter for required time (10–30 minutes), depending upon colloid load or centrifuge at 2500–3500 rpm, 10-inch diameter rotor, using a filterfuge tube.
(5) Transfer the sample and washings to a 1.0-ml volumetric flask and adjust to volume with pH 2.2, 0.2N sodium citrate buffer.
(6) A 0.2-ml aliquot is equivalent to 0.1 ml of original urine.

1.6.1.4   Preparation of other samples

The procedures and methodologies found in *Amino Acid Determination Methods and Techniques* by S. Blackburn (New York: Marcel Dekker Inc.) should be followed for the preparation of amino acid samples from tissue extracts, foods, plant extracts, and other proteinaceous materials.

A general procedure for all samples before column application is a dilution of the sample with pH 2.2, 0.2N sodium citrate buffer. (Dilute sample so that the minimum-maximum applied is 0.20–1.50 ml; this range is for the ease and accuracy of chromatographic operation.)

1.6.2   Buffer Preparation

Many buffers are required as column eluants for the various special methodologies which follow. These buffers must be prepared with distilled or deionized water of known high purity.

Those methodologies developed earlier used Brij-35 as a recommended buffer component. Although it did help to make gas soluble within the column, it had the disadvantage upon deterioration of causing amino acid peaks to move (*e.g.*, glycine was eluted from the column sooner). Replacement of the ground or pulverized resin with spherical resin and the use of Micule filters has minimized the gas problems.

A preservative, usually caprylic acid (octanoic acid) or pentachlorophenol, is added to a buffer solution to prevent mold formation. Such inhibitors are not totally effective against the wide range of mold spores. Therefore, to assure retardation of mold growth, bulk portions of buffer should be kept under refrigeration and only a one-to-two-day operating supply left at room temperature.

Mold inhibitors:

(1) Caprylic acid: add 0.1 ml per liter of buffer
(2) Pentachlorophenol: stock solution—50 mg in 10 ml of 95% ethanol; add 0.1 ml per liter of buffer.

Thiodiglycol (TG) is added to the buffers to minimize the conversion of small amounts of methionine to the methionine sulfoxides during addition of the sample to the column and during analysis.

General outline of buffer preparation:

(1) Weigh the sodium citrate and dissolve in approximately 3.5 liters of water in a precalibrated glass or polyethylene container.

(2) Continue stirring and add concentrated hydrochloric acid, mold inhibitor, and TG (if required). Fill to the 4-liter mark with water.

(3) Mix the buffer solution thoroughly and record temperature.

(4) Standardize the pH meter with reference solution.

(5) Check the pH of the buffer. Before a reading is taken, the temperatures of the buffer and reference solutions should be within 0.5 degree of each other.

(6) If pH correction is necessary, the addition of approximately 0.4 ml concentrated HCl or 0.2 ml of 50% NaOH, for either the 0.35 or 0.38 normalities, will respectively lower or raise the buffer pH approximately 0.01 pH units. A 0.2N solution requires half of these amounts.

pH 2.2 Sample diluent. In sample preparation pH 2.2 (0.2N) sodium citrate buffer is used as a diluting solution. Table IV (see below) lists the constituents of this reagent.

0.3 N NaOH regeneration reagent. Weigh and dissolve 12.0 g of sodium hydroxide pellets in water sufficient to make 1 liter.

1.6.3 Ninhydrin Reagent

This reagent contains ninhydrin, stannous chloride, methyl Cellosolve, and 4N sodium acetate buffer. The methyl Cellosolve must be peroxide-free, because the presence of peroxides will quantitatively destroy the reducing capability of the stannous chloride to produce hydrindantin. Check for peroxides with the following test:

Mix approximately 4 ml of methyl Cellosolve with an equal volume of a 4% aqueous solution of potassium iodide (KI). The mixture will be colorless if peroxides are absent. Development of a yellow color indicates the presence of peroxides, and the methyl Cellosolve should be discarded.

Preparation of Ninhydrin Reagent. It is recommended that the ninhydrin reservoir be painted black, except for a ½-inch clear strip from the neck to the bottom of the bottle. This unpainted portion of the bottle is used for viewing the mixing of the reagents during preparation and for observing the liquid level before operation. A strip of cardboard or tape can be placed over the unpainted portion of the bottle to prevent leakage of light into the reagent while in use.

(1) Add 1 liter of 4N sodium acetate buffer and 3 liters of methyl Cellosolve to the ninhydrin reservoir. (Caution: undiluted methyl Cellosolve is toxic and should only be used in a hood.) Put a stirring bar in the bottle and place the bottle on a magnetic stirrer. Push the stopper assembly into place.

(2) Attach a nitrogen line to the long tube in the stopper assembly and adjust the nitrogen to deliver a slow, steady stream of bubbles to the mixture (approximately 500 cm^3 per minute). Turn on the magnetic stirrer.

(3) Lift the stopper assembly partially out of the bottle and slowly add 80.0 g of ninhydrin to the solution. Continue mixing until the reagent is dissolved. (A small powder funnel placed in the bottle will aid in adding the ninhydrin.) It may be necessary to move the bottle around on the magnetic stirrer to bring the stirring bar to the edge of the bottle where the ninhydrin collects.

(4) Add 1.600 g of stannous chloride to the mixture. Continue stirring and bubbling the nitrogen. If any of the reagent adheres to the bottle or stopper assembly, it may be washed into the mixture with a small quantity of water from a wash bottle. Remove the powder funnel.

(5) Push the stopper assembly firmly into place.

(6) Close the stopcock on the outlet side of the bottle and, finally, the stopcock on the inlet side of the bottle.

One liter of 4N sodium acetate buffer for use in the ninhydrin reagent is prepared as follows:

(1) Slowly add 544 g of NaOAc · 3H$_2$O to about 800 ml of deionized or distilled water in a graduated 1-liter beaker.

(2) Stir until dissolved, heating on a steam or water bath or on a hot plate to facilitate solution.

(3) Cool to room temperature and add 100 ml of glacial acetic acid.

(4) Dilute to the 1-liter volume mark with distilled water.

(5) The final pH should be 5.50 ± 0.02. For pH adjustment, 0.5 g of sodium hydroxide will change the solution about 0.02 pH units. To lower the pH, the use of acetic acid is required.

Occasional difficulties have been experienced in attempts at successful preparation of the ninhydrin reagent described above. The problems usually encountered have been attributed to the methyl Cellosolve reagent. Not only is it important that the reagent be peroxide-free

(<3 ppm) since the presence of peroxide inhibits the color development reaction, but occasional batches of methyl Cellosolve have turned cloudy when subjected to air, particularly if air is introduced during stirring of the reagents. Ninhydrin reagent prepared with this methyl Cellosolve has routinely formed a white precipitate that coats the inside walls of the coil in the reaction bath and the colorimeter cuvettes. This material is unlike a hydrindantin precipitate in that it is insoluble in methyl Cellosolve but soluble in dilute acid. To minimize these effects, a Micule filter can be placed on the pressure side of the pump to remove contaminants.

1.6.4   Hydrolysate Procedure

This procedure permits the analysis of those 20 amino acids outlined in Section 1.3.1.1. The detailed procedures outlined below are for the Type UR-30 and UR-40 resins. The Type UR-30 procedure requires a 56-cm resin column for analysis of the acidic and neutral amino acids and a 5.5-cm column of Type PA-35 resin for analysis of the basic amino acids. The total analysis time is approximately 4 hours. An analysis in 2 hours is also possible by overlapping the short- and long-column runs in a carefully timed sequence of steps. A single-column procedure is possible using the same long-column of UR-30 and introducing a third buffer to elute the basic amino acids after the acidic and neutral amino acids have been resolved.

Resin: Type UR-30.* A sulfonated styrene copolymer resin, nominally 7.25% cross-linked. The spherical particles have a mean diameter of 23 ± 6 μm.

Reagents: Composition of the sodium citrate buffers is presented in Table II. Preparation of the ninhydrin reagent is described in Section 1.6.3.

1.6.4.1   Preparation of ion-exchange column

The following is a general procedure for the preparation and packing of the resin column. It can be followed for the specific methodologies outlined below with only slight modifications of buffer flow rate, column temperature, and pH and ionic strength of the buffers.

The hydrolysate procedure requires a 69 × 0.9-cm glass chromatographic column for analysis of the acidic and neutral amino acids, and a 23 × 0.9-cm column for the basic amino acids. The following procedure is recommended:

The column in which the resin is to be packed must be clean. For this purpose, a piece of absorbent tissue can be soaked in a suitable

---

*Beckman Custom Research Resin, Beckman-Spinco, Palo Alto, California.

Table II
Sodium Citrate Buffers for Protein Hydrolysate Analyses

	4-Hour procedure			2-Hour procedure		
	Acidics and Neutrals (UR-30)		Basics (PA-35)	Acidics and Neutrals (UR-30)		Basics (PA-35)
pH	3.25 ± 0.01	4.30 ± 0.02	5.25 ± 0.01	3.488 ± 0.005	4.404 ± 0.010	5.359 ± 0.010
Sodium concentration, N	0.20	0.20	0.35	0.20	0.20	0.35
Citrate concentration, M	0.066	0.066	0.117	0.066	0.066	0.117
Sodium citrate · $2H_2O$, g	78.4	78.4	137.3	78.4	78.4	137.3
Concentrated HCl, ml	50.3	33.1	26.2	46.8	32.3	25.6
Thiodiglycol (TG), * ml	40.0	40.0	–	40.0	40.0	–
Pentachlorophenol, † ml	0.4	0.4	0.4	0.4	0.4	0.4
Final volume, liters	4	4	4	4	4	4

	Single-column (UR-30)			Single-column (UR-40)		
pH	3.28 ± 0.01	4.30 ± 0.01	6.71 ± 0.01	3.545 ± 0.01	4.250 ± 0.02	6.71 ± 0.01
Sodium concentration, N	0.20	0.20	1.00	0.20	0.20	1.00
Citrate concentration, M	0.066	0.066	0.333	0.200	0.066	0.333
Sodium citrate · $2H_2O$, g	78.43	78.43	392.15	78.43	78.43	392.15
Citric acid, g	–	–	–	112.0	–	–
Concentrated HCl, ml	49.3	33.0	5.0	12.0	33.5	5.0
Thiodiglycol (TG), * ml	40.0	–	–	20	20	–
Final volume, liters	4	4	4	4	4	4

*Pierce Chemical Company, 25% solution (purified).
†Stock solution, 50 mg in pentachlorophenol in 10 ml of 95% ethanol.

detergent, such as Triton X-100 or Brij-35,* and pushed through the column with a rod. Care must be taken not to scratch the glass column wall. It is also recommended that the bottom column fitting be cleaned with Ajax or a similar cleanser. Deionized water is then used to rinse both the column and fitting.

A stainless steel resin support screen is placed at the bottom of the column and pushed up into position by the bottom column fitting. To facilitate movement, the fitting O-ring should be moistened. A slight upward bowing will be apparent due to an oversized screen diameter. Hold the column jacket down to enable the bottom fitting to be fully inserted. At this point, the screen should lie horizontally within the column walls. If repositioning is necessary, remove the bottom fitting and gently push the screen out with a long rod, again being careful not to scratch the walls of the column. Remove grease caused by handling from the screen and fitting by pouring 5–10 ml of ethanol through the column. Deionized water can now be used for a final rinsing of the completed column assembly. Fill the column with water and allow to drain to within ½ cm of the screen surface. Upon applying a stopper to the bottom fitting, the column is ready for packing.

The resin, without pretreatment, is slurried with two column volumes of pH 3.2 (0.2N) sodium citrate buffer, which does not contain any detergent, Brij-35, or Triton X-100.

Dissolved air is expelled from the resin slurry by deaeration. This eliminates possible air pockets and liquid channeling in the packed bed. Place the resin beaker in a vacuum desiccator (without plate and desiccant) and aspirate, stirring continually, using a water aspirator. Continue to aspirate until the mixture is free of air bubbles (10–15 minutes). Another alternative is to place the resin slurry in a vacuum filtration flask. The top of the flask is stoppered and while maintaining a vacuum, the flask is frequently swirled to stir the resin slurry and to aid in the removal of any trapped gas bubbles. This procedure normally requires 15–20 minutes.

The resin column is packed in sections to ensure uniformity of packing and to minimize the entrapment of air. Before each section is poured, stir the resin slurry gently with a glass or Teflon stirring rod. Pour the first section into a column, stoppered at the bottom, and fill to within 2 cm from the top. Allow this section to set for 5 minutes or more to form a 2–4-cm packed bed of resin on the screen before attaching the top column fitting.

---

*Rohm and Haas Company, Philadelphia, Pennsylvania, supplies Triton X-100; Brij-35 is distributed by Pierce Chemical Company, P. O. Box 117, Rockford, Illinois and Atlas Powder Company, Wilmington, Delaware.

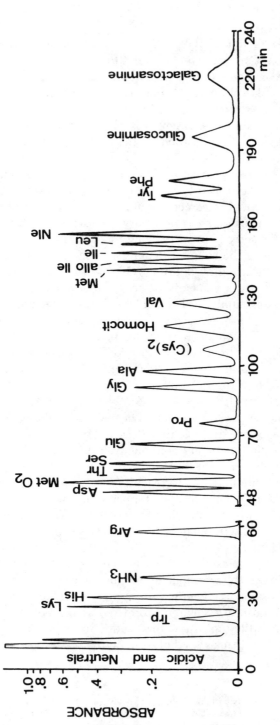

Figure 1. Analysis of an amino acid calibration mixture containing 0.25 μmole of each amino acid. The basic amino acids were separated on a short 5.0-cm column using spherical resin type PA-35. The acidic and neutral amino acids were separated on the 56-cm column of spherical resin type UR-30.

The resin columns are packed at a buffer flow rate of 68 ml per hour and a column temperature of 55°C. When the resin is packed, excess buffer is aspirated and the column again filled with resin slurry, with care taken not to disturb the previously packed section. Several additional packings are necessary to pack the long column to 56 cm and the short column to 5.5 cm. Each column is regenerated with 0.2N sodium hydroxide followed by equilibration with the appropriate starting buffer.

1.6.4.2 Chromatographic procedure

Acidic and Neutral Amino Acids. A buffer flow rate of 68 ml per hour is combined with a ninhydrin flow rate of 34 ml per hour. Elution of the amino acids is started at a column temperature of 55.2°C (column jacket outlet). The first buffer is pH 3.25 ± 0.01 (0.2N) sodium citrate with a change to 4.30 ± 0.02 (0.2N) at 85 minutes after the start of the analysis. The operating backpressure is approximately 130 psi and the analysis time is 180 minutes through phenylalanine.

Basic Amino Acids. Using the Type PA-35 resin, elution of the basic amino acids is started at the same buffer, ninhydrin flow rates, and column temperature as outlined above. A pH 5.25 ± 0.01 (0.35N) sodium citrate buffer is used throughout the analysis. Duration of the run is 50 minutes through arginine. (See Figure 1 for analysis of the acidic, neutral, and basic amino acids.)

1.6.4.3 *Two-hour methodology*

It is possible to analyze the total acidic, neutral, and basic amino acids in about 2 hours by overlapping the short- and long-column analyses. When the basic amino acids are being analyzed on the short column, the long column is started (to the drain position) before the basic analysis is complete.

Basic Amino Acids. A buffer flow rate of 70 ml per hour and a ninhydrin flow of 35 ml per hour are used throughout the analysis. The column temperature is 53.4°C ± 0.05°C. The buffer pH is 5.359 ± 0.01 (0.35N) sodium citrate. Duration of analysis for tryptophan, lysine, histidine, ammonia, and arginine is 48 minutes.

Acidic and Neutral Amino Acids. These are analyzed using the same buffer and ninhydrin flow rates and the same column operating temperature as above. The first buffer is 3.488 ± 0.005 (0.2N) sodium citrate and a change is made to a second buffer, pH 4.404 ± 0.010 (0.2N) at 30 minutes after the start of the run. The analysis time is 115 minutes. It should be noted that cystine is eluted between proline and glycine which is a deviation from the standard 4-hour methodology. See Table II for preparation of the buffers.

Resin: Type UR-40.[*] This sulfonated styrene copolymer resin is nominally 7.25% cross-linked; the spherical particles have a mean diameter of 14 ± 3 μm.

Reagents: The composition of the sodium citrate buffers is presented in Table II. Preparation of the ninhydrin reagent is described in Section 1.6.3.

1.6.4.4   Preparation of Ion-Exchange Columns

A 33.5×0.9-cm chromatographic column was used for the analysis of the acidic, neutral, and basic amino acids for the one-column procedure. The resin is packed to a height of 26 cm by the procedure outlined under the hydrolysate methodology using the UR-30 resin. The pH 3.545 (0.2N Na$^+$, 0.20M citrate) was used to pack the column.

The chromatographic procedure is outlined in Section 1.6.6.

1.6.5   *Physiological Procedure*

The physiological methodologies permit the analysis of amino acids and related compounds normally present in physiological fluids. A list of the amino acids detected and quantitated by this procedure is presented in Section 1.3.1.2. The detailed procedures outlined below are for the Type UR-30 and UR-40 resins.

1.6.5.1   Type UR-30 resin

Analysis of Acidic and Neutral Amino Acids Using Sodium Citrate Buffers. The same 56-cm resin column that was described under hydrolysate procedures can be used. If it is necessary to pack the resin column, the procedures outlined under the hydrolysate procedure, Section 1.6.4, can be followed. The starting buffer (Table III) pH 3.2 (0.2N) can be used (without detergent) at a flow rate of 50 ml per hour and a column operating temperature of 32.3°C. After applying the sample to the column, the analysis is started using a buffer flow rate of 50 ml per hour and a ninhydrin flow of 25 ml per hour. The starting column temperature of 32.3°C and a pH 3.20 ± 0.01 (0.2N) sodium citrate buffer is used. After 90 minutes the column temperature is switched to 62.0°C (on a temperature gradient of 55–60 minutes). At 158 minutes after starting the analysis, a change is made to a pH 4.26 ± 0.02 (0.2N) sodium citrate buffer. The normal operating column pressure is 180 psi (32.3°C). The analysis time is 320 minutes and is shown in Figure 2.

Analysis of basic amino acids. This analysis uses Type PA-35 resin packed to a column height of 23 cm. The same buffer and ninhydrin

---

[*]Sondell Scientific Instruments, Palo Alto, California.

flow rates are used as required for the analysis of the acidic and neutral amino acids above. Also the same 32.3°C and 62.0°C temperatures are used. The analysis is started with the pH 4.25 ± 0.01 (0.38N) sodium citrate buffer (Table III) and at 185 minutes is changed to pH 5.36 ± 0.01 (0.35N) buffer. Also at 185 minutes the temperature change is made, using the same 55–60-minute gradient. The normal operating pressure is 210 psi (32.3°C). The analysis time is 345 minutes as shown in Figure 2. A complete analysis of acidic, neutral and basic amino acids require a little over 11 hours.

Acidic and Neutral Amino Acids, Lithium Citrate Buffers. The separation of glutamine from asparagine without interfering with the rest of the chromatogram has always presented a problem. Other procedures using temperature programming[151] and enzymatic methods[152] have been described. The determination of asparagine alone in blood plasma was developed by Barry.[153] While Moore and Stein[2] mention the use of lithium citrate buffers for ion-exchange chromatography, a logical extension by Benson[150] finally produced the desirable chromatographic method.

It is believed that the differences in amino acid separation using the lithium cycle (from the usual sodium cycle) may be due to the hydrational effects on the amino acids. Heftmann[29] explains that the rate of ion exchange is primarily controlled by the diffusion of the ionic species from the resin and that the most hydrated ions are the least strongly bound. The hydrated lithium ion is 7.3–10.0 A° as compared to 5.6–7.9 A° for the sodium ion. Using Type UR-30 resin, analysis of the acidic and neutral amino acids by this physiological procedure requires 270 minutes through β-aminoisobutyric acid.

Preparation of Ion-Exchange Column. Before packing the column, the resin in the sodium form is converted to the lithium form. This is done by placing the resin on a Büchner funnel and washing it successively with ten-bed volumes of 2% lithium hydroxide, ten-bed volumes of deionized water, and finally, ten-bed volumes of the starting buffer. The resin is further equilibrated overnight in the starting buffer to ensure that the resin has been fully converted to the lithium form.

The 69 × 0.9-cm resin column is packed to a height of 56 cm using pH 2.80 (0.3N Li^+, 0.053M citrate) buffer at a flow rate of 70 ml per hour as outlined in Section 1.6.4. The column temperature is set at 38.8°C during the column packing procedure. After packing, the column is regenerated with 0.3N lithium hydroxide and equilibrated for 1 hour with starting buffer.

A buffer flow rate of 70 ml per hour, a ninhydrin flow rate of 35 ml per hour, and a column temperature of 38.8°C is maintained throughout the analysis. A buffer change is made at 137 minutes after the

Table III
Sodium and Lithium Citrate Buffers for Physiological Fluid Analysis

	11-Hour Procedure				6-Hour Procedure			
	Acidic and neutral (UR-30)		Basics (PA-35)		Acidic and neutral (UR-40)		Basics (UR-40)	
pH	3.20 ± 0.01	4.26 ± 0.02	4.25 ± 0.01	5.36 ± 0.01	3.175 ± 0.01	4.243 ± 0.01	4.148 ± 0.01	5.000 ± 0.01
Sodium concentration, N	0.20	0.20	0.38	0.35	0.18	0.18	0.40	0.40
Citrate concentration, M	0.066	0.066	0.127	0.117	0.133	0.060	0.133	0.133
Sodium citrate · $2H_2O$, g	78.4	78.4	149.0	137.3	70.56	70.56	156.9	156.9
Citric acid, g	—	—	—	—	61.6	—	—	—
Concentrated HCl, ml	51.1	33.5	61.0	25.6	35.0	29.0	65.0	60.0
Thiodiglycol (TG),° ml	40	40	—	—	20	20	—	—
Pentachloro-phenol†	0.4	0.4	0.4	0.4	—	—	—	—
Final volume, liters	4	4	4	4	4	4	4	4

Acidic and neutral analysis (UR-30) using lithium citrate buffers

pH	2.80 ± 0.01	4.16 ± 0.01
Lithium concentration, N	0.30	0.30
Citrate concentration, M	0.053	0.033
Lithium citrate · $4H_2O$, g	60.2	37.6
Lithium chloride, g	23.8	33.9
Thiodiglycol (TG),* ml	40	40
Concentrated HCl, ml	47.0	19.0
Final volume, liters	4	4

*Pierce Chemical Company, 25% solution (purified).
†Pentachlorophenol: stock solution, 50 mg in 10 ml of 95% ethanol.

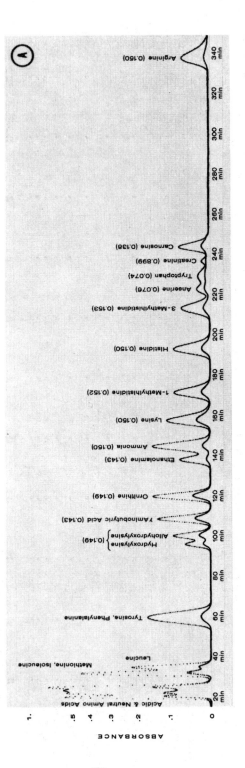

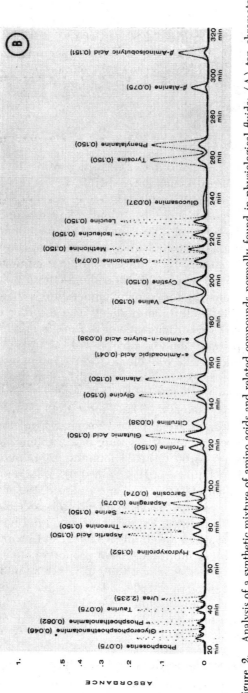

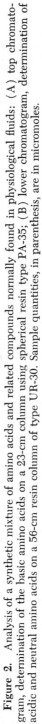

Figure 2. Analysis of a synthetic mixture of amino acids and related compounds normally found in physiological fluids: (A) top chromatogram, determination of the basic amino acids on a 23-cm column using spherical resin type PA-35; (B) lower chromatogram, determination of acidic and neutral amino acids on a 56-cm resin column of type UR-30. Sample quantities, in parenthesis, are in micromoles.

start of the analysis, from pH 2.80 (0.3N Li$^+$, 0.053M citrate) to a pH 4.16 (0.3N Li$^+$, 0.033M citrate). The operating backpressure is approximately 225 psi. Figure 3 shows the elution pattern of the amino acids using this procedure. Table III outlines the preparation of the lithium citrate buffers.

1.6.5.2 Type UR-40 resin

The resin column is packed to 26 cm according to the procedure outlined in Section 1.6.4, using the pH 3.175 (0.18N Na$^+$, 0.133M citrate) without detergent (Table III).

Acidic and Neutral Amino Acids. A buffer flow rate of 60 ml per hour and a ninhydrin flow rate of 30 ml per hour are maintained throughout the analysis. Elution is started at a column temperature of 32.5°C with the pH 3.175 (0.18N Na$^+$, 0.133M citrate) sodium citrate buffer. A buffer change to pH 4.243 (0.18N) sodium citrate buffer is made at 80 minutes. At 66 minutes, the column temperature is changed from 32.5 to 60°C using a 20-minute time gradient. The recorder chart speed is 15 cm per hour and a recorder range of 0–5 mV is used. Column backpressure is 225 psi and analysis time is 170 minutes. A single-length reaction coil is used (11.25 ml volume for a 7.5-minute color development reaction time). A 6.6-mm cuvette path length is used for the 570 and 440 nm wavelengths (Figure 4).

Basic Amino Acid Analysis. The same UR-40 resin column, buffer and ninhydrin flow rates are also used for analysis of the basic amino acids. Elution of the amino acids is started at a column temperature of 32.5°C and changed to 60°C at 33 minutes (20-minute temperature gradient). The first buffer is pH 4.148 (0.40N) sodium citrate; later changed to pH 5.000 (0.40N) sodium citrate at 190 minutes (Table III). The operating column backpressure is 225 psi (32.5°C) and the analysis time is 300 minutes. By using an 80-ml-per-hour buffer flow rate, a ninhydrin flow rate of 40 ml per hour, and changing the other operating conditions as indicated above proportionately, it is possible to complete the basic physiological analysis in 210 minutes. The operating backpressure in this case is 280 psi. It is possible to complete a complex physiological analysis (acidic, neutral and basic) in a little over 6 hours. Figure 5 illustrates the resolution of a synthetic mixture of basic amino acids.

1.6.6 *Accelerated Single-Column Hydrolysate Procedure*

To further extend the capability of the resins by accelerating the analysis of the acidic, neutral, and basic amino acids using the two-column procedures, the Types UR-30 and UR-40 resins will allow:

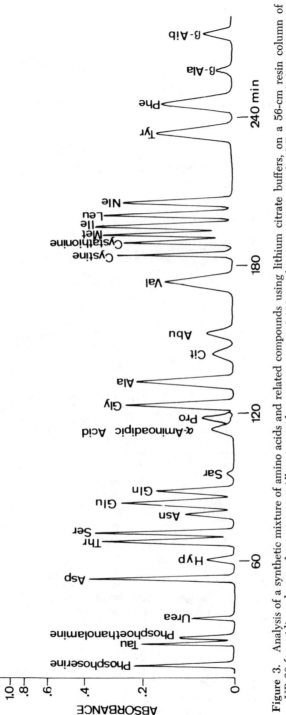

Figure 3. Analysis of a synthetic mixture of amino acids and related compounds using lithium citrate buffers, on a 56-cm resin column of type UR-30 for acidic and neutral components. All amino acids were present in 0.100-μmole quantities except the following and their respective amounts (in μmoles): phosphoserine, taurine, phosphoethanolamine, sarcosine, and cystathionine were present as 0.065; glutamine, 0.145; citrulline, 0.030; α-aminoadipic acid, 0.030; α-amino-n-butyric acid, 0.035; norleucine, 0.120; and β-aminoisobutyric acid, 0.130.

55

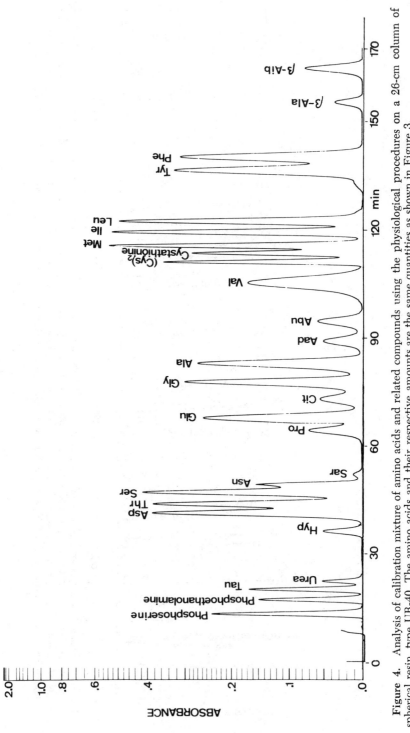

Figure 4. Analysis of calibration mixture of amino acids and related compounds using the physiological procedures on a 26-cm column of spherical resin, type UR-40. The amino acids and their respective amounts are the same quantities as shown in Figure 3.

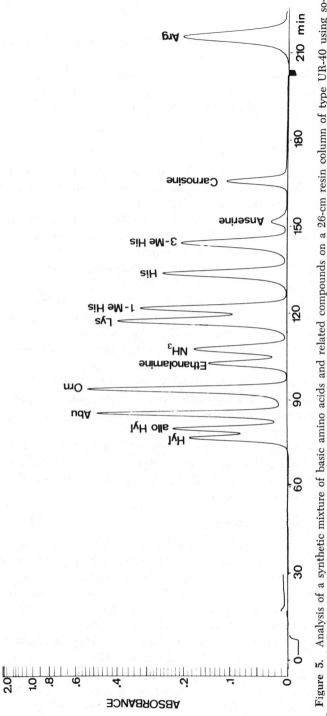

Figure 5. Analysis of a synthetic mixture of basic amino acids and related compounds on a 26-cm resin column of type UR-40 using sodium citrate buffers. All amino acids were present in 0.100-μmole quantities except anserine, 0.060, and carnosine, 0.110 μmoles.

(1) The accelerated analysis of acidic, neutral, and basic amino acids on a single column
(2) Operation at lower column backpressures over previously reported resins under these operating conditions
(3) Improved resolution of amino acid peaks.

1.6.6.1   Type UR-30 resin

The resin column is packed to a height of 56 cm. Analysis is started at a buffer flow rate of 130 ml per hour and a ninhydrin flow of 65 ml per hour. The first buffer is pH 3.28 ± 0.02 (0.20N) sodium citrate, and is later replaced by a pH 4.30 ± 0.02 (0.20N) buffer 37 minutes after the start of the analysis. The starting column temperature is 55.6°C, which is changed to 64.8°C at 42 minutes (using a 55–60-minute temperature gradient). At 54 minutes, a third buffer change is made, pH 6.71 (1.0N) sodium citrate, to elute the basic amino acids. The final operating column pressure is 425 psi. As shown in Figure 6, the analysis time is 150 minutes. The buffer composition is shown in Table II.

Discussion.   The 23 amino acid peaks which are often of interest in samples of free amino acids are usually determined by a simple hydrolysate procedure. When these amino acids are present in as complex a sample as serum, urine, or tissue homogenates, a more elaborate procedure is required for complete analysis (physiological procedures). The one-column system offers the advantage of eliminating column regeneration and equilibration.

Analysis of serum by this procedure will elute citrulline with proline. Asparagine and glutamine, also present in serum, are eluted with serine. To quantitate these amino acids, they must be converted from the amide by hydrolysis techniques.

1.6.6.2   Type UR-40 resin

Again using the 26-cm, Type UR-40 resin column previously described for the more complex physiological chromatographic procedures outlined in Section 1.6.5, a single-column procedure is used to chromatograph the acidic, neutral, and basic amino acids normally found in a protein or peptide hydrolysate. Using a sodium citrate, three-buffer, step-change system, a complete analysis requires 170 minutes. The method outlined below uses a buffer system which permits the complete analysis to be run at a single temperature. However, it is possible to make a hydrolysate type analysis by following the physiological methodology outlined above for the acidic and neutral amino acids and then introducing a third buffer, similar to the one used below, to elute the basic amino acids. This procedure not only will permit the analysis of

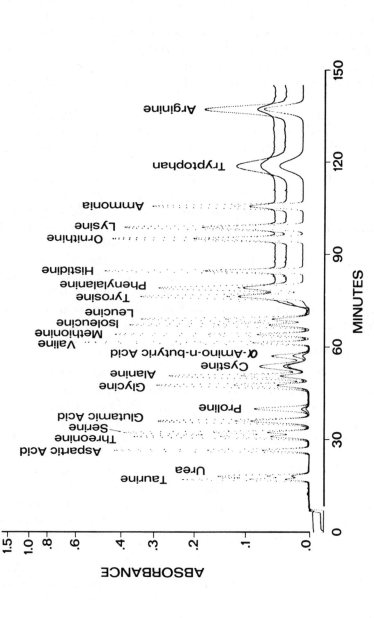

Figure 6. Analysis of a synthetic mixture of amino acids and related compounds using spherical resin type UR-30. Amino acids were present in 0.125-μmole quantities except (in μmoles): taurine, 0.085; urea, 0.300; α-amino-n-butyric acid, 0.028; and tryptophan, 0.216.

59

those amino acids normally found in simple hydrolysate samples, but will also allow the unusual or unknown amino acids in a sample to be chromatographed with minimum interference and peak overlapping.

A column temperature of 48.0°C is used throughout the analysis. The buffer flow rate is 60 ml per hour and the ninhydrin flow rate is 30 ml per hour. Elution is started with the pH 3.545 (0.20 Na$^+$, 0.20M citrate) sodium citrate buffer with a change to the second buffer, pH 4.250 (0.20N), timed to take effect just after alanine. The third buffer, pH 6.176 (1.0N) sodium citrate, is added to take effect just after phenylalanine and elutes the basic amino acids histidine, lysine, ammonia, and arginine.

Results. The elution sequence of the amino acids is the same as shown in Figure 6, using the longer column, Type UR-30 resin. If cysteic acid, taurine, and urea are present in the sample, they will be eluted early in the analysis. Using the Type UR-40 resin column (26 cm), the valley-to-peak height ratio of threonine–serine is 0.40 and of serine–glutamic acid, 0.23. Cystine is centered between proline and glycine and the valley-to-peak height ratio of the glycine–alanine peaks are 0.10. The separation of tyrosine–phenylalanine is 0.16. In this system, histidine is also eluted before lysine, followed by ammonia and arginine. Table II shows the composition of the buffers.

1.6.7 Specific Amino Acid Patterns

Clinical interest in diseases associated with abnormalities in amino acid metabolism has increased the desirability of having more rapid techniques for chromatographing only the amino acids diagnostically related to specific metabolic disorders. Techniques have been developed for chromatographing only certain areas of interest from a total pattern, using a 23-cm Type PA-35 resin column also used for analysis of basic amino acids. Modification of those procedures previously described (Scriver[69,81]) allows the investigator to divide the simple hydrolysate amino acid procedure into three groups of amino acids. All of these procedures use the same buffer flow rate of 90 ml per hour, a ninhydrin flow rate of 45 ml per hour, a column operating temperature of 55.0°C, and a single-length reaction coil. There are three procedures.

Procedure A. This procedure allows aspartic acid, glutamic acid, proline, glycine, alanine, cystine, and valine to be analyzed in less than 60 minutes. A sodium citrate pH 3.270 (0.20N) is used (see Figure 6).

Procedure B. This procedure uses a pH 3.165 (0.35N) sodium citrate buffer for the chromatography of the amino acids associated with branched-chain ketoaciduria, homocystinuria, hyperprolinemia, and others. The amino acids of interest are: proline, valine, methionine, alloisoleucine, isoleucine, and leucine. These amino acids are chromatographed in about 70 minutes (Figure 7).

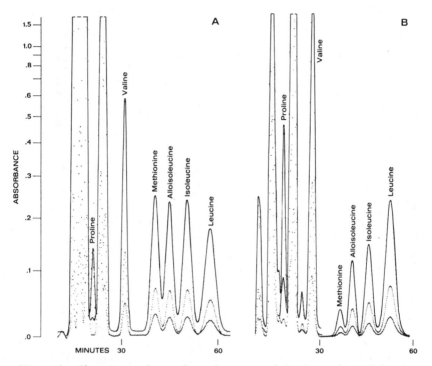

Figure 7. Chromatographic analysis of amino acids by "Procedure B" for those related to specific aminoacidopathies on a 23-cm resin column of type PA-35 resin: (A) synthetic mixture of amino acids at 0.025-μmole level using a 4–5-mV recorder range; and (B) 0.20 ml normal human blood plasma.

Procedure C. For the chromatography of the amino acids associated with phenylketonuria, tyrosinosis and histidinemia, a pH 6.46 (0.35N) sodium citrate buffer is used. Tyrosine can be analyzed in 19 minutes, phenylalanine in 21 minutes, and histidine in 30 minutes (Figure 8).

Many of the diseases showing amino acid abnormalities have primary changes in one of the groups. Having a methodology available to concentrate on a specific area of a chromatogram permits a significant increase in the number of analyses in a day. Buffer compositions are presented in Table IV.

Procedure BP. This accelerated procedure resolves all of those basic amino acids normally present in a physiological fluid. It should find its application in quantitation of those amino acids related to Wilson's disease, citrullinemia, psoriasis, and dietary levels. The analysis time is 240 minutes. A buffer flow rate of 90 ml per hour and a ninhydrin flow rate of 45 ml per hour is used. The analysis is started with the pH 4.263 (0.38N) sodium citrate buffer at a column temperature of 33.0°C. A buffer change is made at 125 minutes to pH 5.360 (0.35N) sodium citrate buffer and a temperature change to 55.0°C at the same time.

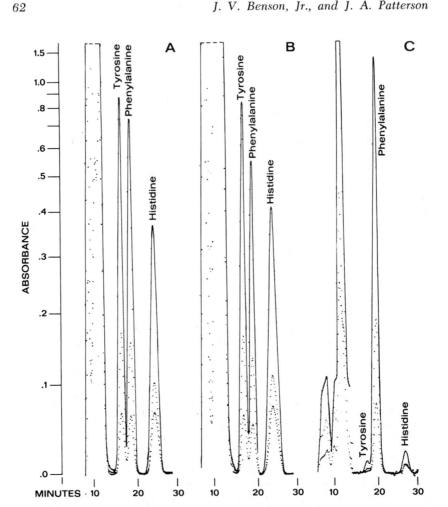

Figure 8. Analysis of amino acids using "Procedure C" for specific amino acid patterns (4–5-mV recorder range) on type PA-35 resin: (A) synthetic mixture of amino acids at 0.020-μmole load level; (B) 0.40 ml normal human blood plasma; and (C) 0.05 ml blood plasma of patient with phenylketonuria.

1.6.8 Selenium Compounds

The importance of selenium and organic selenium derivatives in biochemical reactions makes it desirable to provide an accelerated quantitative method for the analyses of selenocystine and selenomethionine.[158-161]

Analysis of selenocystine requires 45 minutes and selenomethionine 90 minutes, using one buffer (Table V), one temperature, one buffer flow rate and Type PA-35 resin. The resin column is prepared as described in Section 1.6.4, using a buffer flow rate of 70 ml per hour and a column temperature of 30°C. The final column height is 23 cm.

Table IV
Sodium Citrate Buffers for Amino Acid Patterns (see Section 1.6)

Procedure	A	B	C	BP		Sample diluent
Buffer pH	3.27 ± 0.01	3.165 ± 0.02	6.46 ± 0.02	4.26 ± 0.02	5.36 ± 0.02	2.2 ± 0.03
Sodium concentration, N	0.20	0.35	0.35	0.38	0.35	0.2
Sodium citrate · 2H₂O, g	78.4	137.3	137.3	149.0	137.3	19.6
Concentrated HCl, ml	49.2	87.4	3.6	61.0	25.6	16.5
Pentachlorophenol, ml	0.4	0.4	0.4	0.4	0.4	0.1
Thiodiglycol (TG), ml	40.0	40.0	—	—	—	10.0
Final volume, liters	4	4	4	4	4	1

Table V

Sodium Citrate Buffer for Analysis of Seleno Amino Acids

pH	*3.23 ± 0.02*
Sodium concentration, N	0.45
Sodium citrate · 2H$_2$O, g	176.6
Concentrated hydrochloric acid, ml	112.6
Thiodiglycol (TG), ml	20
Final volume, liters	4

Figure 9 illustrates the resolution of selenocystine and selenomethionine from other amino acids in the sample. The analysis is performed at a buffer flow rate of 70 ml per hour and a ninhydrin flow rate of 35 ml per hour. The column temperature remains at 30°C throughout the analysis. The column operating backpressure is 330 psi. A 0.5-mV recorder is used and the sample contains 0.1 μmoles of each amino acid.

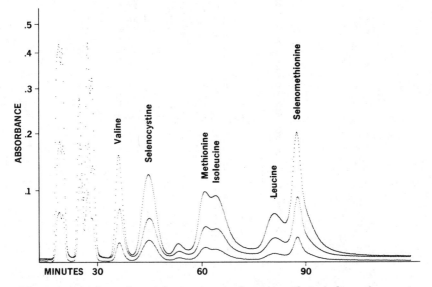

Figure 9.   Analysis of a synthetic mixture of amino acids including selenocystine and selenomethionine on a 23-cm resin column of type PA-35.

1.6.9   *Peptide Chromatographic Procedures*

The final item in this discussion deals with procedures followed in the separation of peptides. Classically, the separation of peptide mixtures has been made on low cross-linked resins, particularly for mixtures containing large peptides. Peptides of more than 100 residues have been separated on columns of Dowex 1-X2, at low flow rates, by Piez.[154] In a study of sulfonated cation-exchange resins for use in chromatographing long-chain peptides, it was concluded by Moore and Stein[5] that the 2% cross-linked resin was preferred to the 4 and 8% cross-linked resins. Use of the lower cross-linked material eluted the peptides in relatively narrow bands. Using a low cross-linked resin does have the disadvantage of undergoing considerable bed compression and occlusion with concentrated gradients of buffers. As the resin bed compresses, this is usually accompanied by a high-column backpressure, which generally requires that the column be repacked before the next analysis.

Table VI
Pyridine Acetate buffers for Peptide Analysis

pH	3.50 ± 0.02	5.00 ± 0.02
Pyridine concentration, M	0.10	2.0
Pyridine,* ml	32.2	645
Glacial acetic acid, ml	320	573
Final volume, liters	4	4

*Baker and Adamson, purified; treated with ninhydrin and redistilled.

Jones[100] has reported that peptides as large as 30 residues can be successfully chromatographed on an 8% cross-linked pulverized sulfonated cation-exchange resin, Type 15A.* The use of higher cross-linked resin minimized column-bed compression during the changing of the buffer or eluent gradient. However, a pulverized resin exposes the internal matrix of the resin which contains some linear material. This linear material is leached from the resin by the buffers and occasionally results in blockage of the liquid tubing in the flow system. Due to the development of a highly cross-linked spherical cation-exchange resin (Type PA-35) originally produced for the chromatography of amino acids, these problems have been minimized.[103]

Dowex 1-X2, a strong base anion-exchange resin, has also been used for the chromatography of peptides[134, 155] and chromatographic studies on lysozyme have utilized a carboxylic acid ion-exchange resin, Amberlite 1RC-50.[156] These three resins compliment each other when used for the separation of peptide mixtures because of their different chromatographic properties.

Resin. Type PA-35. A sulfonated styrene copolymer resin, nominally 7.5% crosslinked. Spherical mean diameter of the beads is 16 ± 6 μm.

Reagents. The composition of the pyridine acetate eluents is presented in Table VI. The ninhydrin reagent is prepared as outlined in Section 1.6.3.

---

*Beckman Instruments, Inc., Palo Alto, California.

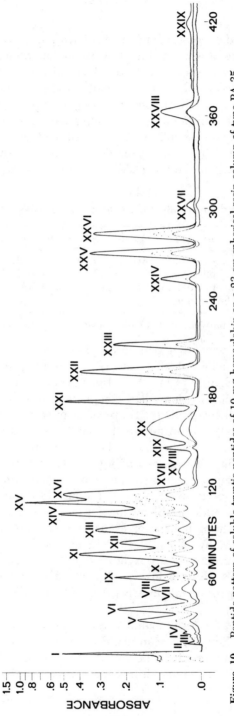

Figure 10. Peptide pattern of soluble tryptic peptides of 10 mg hemoglobin on a 23-cm spherical resin column of type PA-35.

Sample preparation. The tryptic peptides of hemoglobin can be prepared according to the methods outlined by Jones.[100] The lyophilized peptide mixture is dissolved in 0.3M pyridine and adjusted to a pH 2.2 with concentrated HCl. Other techniques in enzymatic hydrolysis are presented by Smyth[36] and Blackburn.[34]

Preparation of resin column. The resin-column packing procedure is described in Section 1.6.4. However, before packing the column, the resin is washed on a Büchner funnel with 4% sodium hydroxide, deionized water, 2M acetic acid, deionized water, 2M pyridine, deionized water, and starting eluent. About ten volumes of each reagent is used per volume of resin. The resin suspended in five-bed volumes of buffer is packed into the column at 50°C at a flow rate of 90 ml per hour to a height of 23 cm.

Chromatographic procedure. The column is equilibrated with five-bed volumes of the starting buffer, 0.1M pyridine pH 3.5, and is maintained at 50°C throughout the analysis. The first portion of the analysis is made using a linear gradient of pyridine–acetic acid from pH 3.5, 0.1M pyridine to pH 5.0, 2.0M pyridine. This gradient is produced by placing 259 g of the starting buffer in the first mixing-chamber (pH 3.5, 0.1M) and 259 g of the limiting buffer (pH 5.0, 2.0M) in a second connecting mixing-chamber of the same internal diameter. The device is similar to the simple gradient device described by Block and Ling in their Figure 6b.[157] The buffer line is connected from the first mixing-chamber to the buffer pump. (It is recommended that a valve or three-way stopcock be connected into the line to allow a third buffer to be added to the system.) A stirring device (magnetic stirrer) is necessary to mix the starting buffer as the limiting buffer is fed into the first mixing-chamber. Both mixing chambers must be at the same level.

A buffer change is made at 330 minutes so that as the gradient nears completion, only limiting buffer is pumped into the column to elute those components more tightly bound to the resin. This analysis time is 430 minutes. Figure 10 shows the peptide pattern of soluble tryptic peptides of hemoglobin.

REFERENCES

1. Stein, W. H., and S. Moore.   Cold Spring Harbor Symp. Quant. Biol. **14**, 179 (1949).
2. Moore, S., and W. H. Stein.   J. Biol. Chem. **192**, 663 (1951).
3. Stein, W. H.   Proc. Soc. Exptl. Biol. Med. **78**, 705 (1951).
4. Stein, W. H.   J. Biol. Chem. **201**, 45 (1953).
5. Moore, S., and W. H. Stein.   J. Biol. Chem. **211**, 893 (1954).
6. Stein, W. H., and S. Moore.   J. Biol. Chem. **211**, 915 (1954).

68 J. V. Benson, Jr., and J. A. Patterson

7. Tallen, H. H., S. Moore, and W. H. Stein. J. Biol. Chem. 211, 927 (1954).
8. Tallen, H. H., W. H. Stein, and S. Moore. J. Biol. Chem. 206, 825 (1954).
9. Stein, W. H., A. G. Bearn, and S. Moore. J. Clin. Invest. 33, 410 (1954).
10. Stein, W. H., A. C. Paladini, C. H. W. Hirs, and S. Moore. J. Am. Chem. Soc. 76, 2848 (1954).
11. Moore, S., and W. H. Stein. J. Biol. Chem. 211, 907 (1954).
12. Moore, S., D. H. Spackman, and W. H. Stein. Anal. Chem. 30, 1185 (1958).
13. Spackman, D. H., W. H. Stein, and S. Moore. Anal. Chem. 30, 1190 (1958).
14. Walton, H. F. Ann. Rev. Phys. Chem. 10, 123 (1959).
15. Strain, H. H. Anal. Chem. 32, 3R (1960).
16. Kunin, R. Anal. Chem. 32, 67R (1960).
17. Kunin, R. Ann. Rev. Phys. Chem. 12, 381 (1961).
18. Heftmann, E. Anal. Chem. 34, 13R (1962).
19. Kunin, R., and F. X. McGarvey. Anal. Chem. 34, 48R (1962).
20. Kunin, R., and F. X. McGarvey. Anal. Chem. 34, 101R (1962).
21. Heftmann, E. Anal. Chem. 36, 14R (1964).
22. Walton, H. F. Anal. Chem. 36, 51R (1964).
23. Kunin, R. Anal. Chem. 36, 142R (1964).
24. Hamilton, P. B. Advan. Chromatogr. 2, 3 (1966).
25. Walton, H. F. Anal. Chem. 40, 51R (1968).
26. Helfferich, F. Ion Exchange (New York: McGraw-Hill, 1962).
27. Giddings, J. C. Dynamics of Chromatography, Part 1, (New York: Marcel Dekker, Inc. 1965).
28. Glueckauf, E. Ion Exchange and its Applications (London: Chemical Society Industry, 1955).
29. Heftmann, E. Chromatography (New York: Reinhold Publishing Corp., 1961).
30. Samuelson, O. Ion Exchange Separations in Analytical Chemistry (New York: John Wiley and Sons, Inc., 1963).
31. Kingsley, G. R. Anal. Chem. 37, 20R (1965).
32. D'Eustachio, A. J. Anal. Chem. 40, 19R (1968).
33. Zweig, G. Anal. Chem. 40, 490R (1968).
34. Blackburn, S. Amino Acid Determination: Methods and Techniques (New York: Marcel Dekker, Inc., 1968).
35. Edman, P. Arch. Biochem. 22, 475 (1949).
36. Smyth, D. G. Methods in Enzymology, Vol. 11, Ed. by C. H. W. Hirs (New York: Academic Press, 1967), p 214.
37. Benisek, W. F., and R. D. Cole. Biochem. Biophys. Res. Commun. 20, 655 (1965).
38. Young, J. D., E. Benjamin, M. Shimizu, and C. Y. Leung. Biochemistry 5, 1481 (1966).

39. Wang, S. S., and F. H. Carpenter. J. Biol. Chem. **240**, 1619 (1965).
40. Lai, C. Y., P. Hoffee, and B. L. Horecker. Arch. Biochem. Biophys. **112**, 567 (1965).
41. Akabori, S., K. Ohiro, and K. Narita. Bull. Chem. Soc. Japan **25**, 214 (1952).
41a. Niederwieser, A., and H. C. Curtius. Z. Klin. Chem. Klin. Biochem. **7**, 404 (1969).
41b. Hamilton, P. B., in *Handbook of Biochemistry: Selected Data for Molecular Biology*, Ed. by H. A. Sober (Cleveland, Ohio: The Chemical Rubber Co., 1968), p B-43.
42. Berlow, S., R. Arends, and C. Harries. Lancet **85**, 241 (1965).
43. Ghadimi, H., M. W. Partington, and A. Hunter. New Eng. J. Med. **265**, 221 (1961).
44. La Du, B. N., R. R. Howell, G. A. Jacoby *et al.* Pediatrics **32**, 216 (1963).
45. Frimpter, G. W., A. Haymovitz, and M. Horwith. New Eng. J. Med. **268**, 333 (1963).
46. Perry, T. L., S. Hansen, H. Bar, and L. MacDougall. Science **152**, 776 (1966).
47. Gerritsen, T., J. G. Vaughn, and H. A. Waisman. Biochem. Biophys. Res. Commun. **9**, 493 (1962).
48. Dancis, J., M. Levitz, S. Miller, and R. G. Westall. Brit. Med. J. **1**, 91 (1959).
49. Norton, P. M., E. Roitman, S. E. Snyderman, and L. E. Holt, Jr. Lancet **1**, 26 (1962).
50. Lonsdale, D., and D. H. Barber. New Eng. J. Med. **271**, 1338 (1964).
51. Guthrie, R., and S. Whitney. *Phenylketonuria* (Washington, D.C.: U. S. Govt. Printing Office, Children's Bureau Publication No. 419, 1964).
52. Sacks, W. J. Appl. Physiol. **17**, 985 (1962).
53. Berlet, H. H. *Progress on Brain Research*, Vol. 16, Ed. by W. A. Himevich and J. P. Schade (Amsterdam: Elsevier, 1965), pp 184–215.
54. Jervis, G. A., R. J. Block, D. Bolling, and E. Kanze. J. Biol Chem. **134**, 105 (1940).
55. Lorincz, A. B., and R. E. Kuttner. Nebraska State Med. J. **50**, 609 (1965).
56. Dent, C. E., and G. A. Rose. Quart. J. Med. **20**, 205 (1951).
57. Rosenberg, L. E. Science **154**, 1341 (1966).
58. Crawhall, J. C., and C. J. Thompson. Science **147**, 1459 (1965).
59. Bell, G. E., D. C. Slivka, and J. R. Huston. J. Lab. Clin. Med. **71**, 113 (1968).
60. Schafer, I. A., C. R. Scriver, and M. L. Efron. New Eng. J. Med. **267**, 51 (1962).
61. Gerritsen, T., E. Kaveggia, and H. A. Waisman. Pediatrics **36**, 882 (1965).
62. Salzer, J. S., and M. E. Balis. Cancer Res. **28**, 595 (1968).

63. Peisakhov, B. I.   Klin. Med. **43**, 83 (1964).
64. Ryan, W. L., and A. B. Lorincz.   Nebraska State Med. J. **49**, 321 (1964).
65. Saunders, S. J., G. O. Barbezat, W. Wittmann, and J. D. L. Hansen. Am. J. Clin. Nutr. **20**, 760 (1967).
66. Block, W. D., and M. H. Westhoff.   Metabolism **15**, 46 (1966).
67. Armstrong, M. D., and K. N. Yates.   Am. J. Obstet. Gynecol. **88**, 381 (1964).
68. Brown, R. R.   Science **157**, 432 (1967).
69. Scriver, C. R., and E. Davies.   Pediatrics **36**, 592 (1965).
70. McCance, R. A., A. B. Morrison, and C. E. Dent.   Lancet **1**, 131 (1955).
71. Dingman, H. F., and S. W. Wright.   J. Mental Deficiency Res. **8**, 77 (1964).
72. Perry, T. L. and R. T. Jones.   J. Clin. Invest. **40**, 1363 (1961).
73. Dickinson, J. C., and P. B. Hamilton.   J. Neurochem. **13**, 1179 (1966).
74. Soupart, P.   *Free Amino Acids of Blood and Urine in the Human, In Amino Acid Pools.* Ed. by J. T. Holden (New York: Elsevier, 1962).
75. King, J. S., Jr.   Clin. Chim. Acta **9**, 441 (1964).
76. Dickinson, J. C., H. Rosenblum, and P. B. Hamilton.   Pediatrics **36**, 2 (1965).
77. Takao, T., *et al.*   Nature **217**, 365 (1968).
78. Agar, W. T., F. J. R. Hird, and G. S. Sedhu.   Biochim. Biophys. Acta **14**, 80 (1954).
79. Lewis, O. A. M.   Nature **5029**, 1239 (1966).
80. Shih, V. E., M. L. Efron, and G. L. Mechanic.   Anal. Biochem. **20**, 299 (1967).
81. Benson, J. V., Jr., J. Cormick, and J. A. Patterson.   Anal. Biochem. **18**, 481 (1967).
82. Piez, K. A., and L. Morris.   Anal. Biochem. **1**, 189 (1960).
83. Miller, E. J., and K. A. Piez.   Anal. Biochem. **16**, 320 (1966).
84. Thomson, A. R., and B. J. Miles.   Nature **203**, 483 (1964).
85. Hamilton, P. B.   Anal. Chem. **35**, 2055 (1963).
86. Dus, K., S. Lindroth, R. Pabst, and R. M. Smith.   Anal. Biochem. **14**, 41 (1966).
87. Kunin, R.   *Ion Exchange Resins* (New York: John Wiley and Sons, Inc., 1958).
88. Benson, J. V., Jr., and J. A. Patterson.   Anal. Biochem. **13**, 265 (1965).
89. Senō, M., and T. Yamabe.   Bull. Chem. Soc. Japan **34**, 1021 (1961).
90. Cohn, E. J., and J. T. Edsall.   *Proteins, Amino Acids, and Peptides* (New York: Reinhold Publishing Corp., 1943).
91. Partridge, S. M.   Brit. Med. Bull. **10**, 241 (1954).
92. Ohms, J. I., J. Zec, J. V. Benson, Jr., and J. A. Patterson,   Anal. Biochem. **20**, 51 (1967).
93. Green, J. G.   *Nat. Cancer Inst. Monograph No. 21*, 447 (1966).

94. Anderson, N. G., J. G. Green, M. L. Barber, and S. F. C. Ladd. Anal. Biochem. **6**, 153 (1963).
95. Martin, A. J. P., and R. L. M. Synge. Biochem. J. **35**, 1358 (1941).
96. Haller, W. Nature **206**, 693 (1965).
97. Doremus, R. H. "Ion Exchange in Glasses," *Ion Exchange* Ed. by J. A. Marinsky (New York: Marcel Dekker, 1969).
98. Thompson, A. R. Biochem. J. **61**, 253 (1955).
99. Schroeder, W. A., R. T. Jones, J. Cormick and K. McCalla. Anal. Chem. **34**, 1570 (1962).
100. Jones, R. T. Cold Spring Harbor Symp. Quant. Biol. **29**, 297 (1964).
101. Margoliash, E., and E. L. Smith. Nature **192**, 1121 (1961).
102. Rudloff, V., and G. Braunitzer. Z. Physiol. Chem. **323**, 129 (1961).
103. Benson, J. V., Jr., R. T. Jones, J. Cormick, and J. A. Patterson. Anal. Biochem. **16**, 91 (1966).
104. Moore, S., and W. H. Stein. J. Biol. Chem. **176**, 367 (1948).
105. Jacobs, S. Microchem. J. **9**, 387 (1965).
106. Jacob, S. Analyst **85**, 257 (1960).
107. Frame, E. G., J. A. Russell, and A. E. Wilhelmi. J. Biol. Chem. **149**, 255 (1943).
108. Matsushita, S., and N. Iwami. Mem. Res. Inst. Food Sci. **28**, 75 (1967).
109. Yemm, E. W., and E. C. Cocking. Analyst **80**, 209 (1955).
110. Singh, C. J. Chromatog. **18**, 194 (1965).
111. Ruhemann, S. J. Chem. Soc. **99**, 792 (1911).
112. Troll, W., and R. K. Cannan. J. Biol. Chem. **200**, 803 (1953).
113. Schilling, E. D., P. I. Burchill, and R. A. Clayton. Anal. Biochem. **5**, 1 (1963).
114. Wainer, A. J. Chromatog. **26**, 48 (1967).
115. Zacharius, R. M., and W. L. Porter. J. Chromatog. **30**, 190 (1967).
116. Cadavid, N. G., and A. C. Paladini. Anal. Biochem. **9**, 170 (1964).
117. Rosen, H. Arch. Biochem. Biophys. **67**, 10 (1957).
118. Knight, W. S. Anal. Biochem. **22**, 539 (1968).
119. Thomas, A. J., R. A. Evans, J. A. D. Siriwardene, and A. J. Robins. Biochem. J. **99**, 5C (1966).
120. Kirschenbaum, D. M. Anal. Biochem. **12**, 189 (1965).
121. Moore, S. Personal communication.
122. Pope, C. G., and M. F. Stevens. Biochem. J. **33**, 1070 (1939).
123. Satake, K., A. Matsuo, and T. Take. J. Biochem. (Tokyo) **58**, 90 (1965).
124. Hupe, K. P., and E. Bayer. J. Gas Chromatog. **5**, 197 (1967).
125. U. S. Patent approved for J. A. Patterson, Sondell Research and Development Company.
126. Anderson, N. G. Personal communication.
127. Blackburn, S. *Amino Acid Determination* (New York: Marcel Dekker, 1968), p 124.
128. Hamilton, P. B. Anal. Chem. **30**, 914 (1958).

129. Spackman, D. H.   Federation Proc. **23**, 371 (1964).
130. Benson, J. V., Jr., and J. A. Patterson.   Anal. Chem. **37**, 1108 (1965).
131. Horvath, C., and S. R. Lipsky.   J. Chromatog. Sci. **7**, 109 (1969).
132. Avrameas, S., and T. Ternynck.   J. Biol. Chem. **242**, 1651 (1967).
133. Kaplan, M. E., and E. A. Kabat.   J. Exptl. Med. **123**, 1061 (1966).
134. Schroeder, W. A., and B. Robberson.   Anal. Chem. **37**, 1583 (1965).
135. Siegel, A., and E. T. Degens.   Science **151**, 1098 (1966).
136. Catravas, G. N.   Anal. Chem. **36**, 1146 (1964).
137. Moore, S., and W. H. Stein.   *Methods in Enzymology*, Vol. 6, Ed. by S. P. Colowick and N. O. Kaplan (New York: Academic Press, 1963), p 819.
138. DeWolfe, M. S., S. Baskurt, and W. A. Cochrane.   Clin. Biochem. **1**, 75 (1967).
139. Gerritsen, T., M. L. Rehberg, and H. A. Waisman.   Anal. Biochem. **11**, 460 (1965).
140. Prescott, B. A., and H. Waelsch.   J. Biol. Chem. **167**, 855 (1947).
141. Mefferd, R. B., R. M. Summers, and J. D. Crayton.   J. Chromatog. **35**, 469 (1968).
142. Harris, W. E., and H. W. Habgood.   *Programmed Temperature Gas Chromatography* (New York: John Wiley and Sons, Inc., 1966), pp 203–207.
143. Tempé, J.   "Analyse des acides par chromatographie—considérations générales sur l'intégration des pics," J. Chromatog. **24**, 169 (1966).
144. Ball, D. L., W. E. Harris, and H. W. Habgood.   Anal. Chem. **40**, 129 (1968).
145. Ball, D. L., W. E. Harris, and H. W. Habgood.   Separ. Sci. **2**, 81 (1967).
146. Gerling, G. W., A. R. Gigg, and M. R. Heley.   J. Chromatog. **31**, 525 (1967).
147. Mondino, A.   J. Chromatog. **30**, 100 (1967).
148. Schroeder, W. A., W. R. Holmquist, and J. R. Shelton.   Anal. Chem. **38**, 1281 (1966).
149. Walsh, K. A., and J. R. Brown.   Biochim. Biophys. Acta **58**, 596 (1962).
150. Benson, J. V., Jr., M. J. Gordon, and J. A. Patterson.   Anal. Biochem. **18**, 228 (1967).
151. Oreskes, I., F. Cantor, and S. Kupfer.   Anal. Chem. **37**, 1720 (1965).
152. Ramadan, A. Mohyi El-Din, and D. M. Greenberg.   Anal. Biochem. **6**, 144 (1963).
153. Barry, J. M.   Nature **171**, 1123 (1953).
154. Piez, K. A., and H. A. Saroff, in *Chromatograpy*, Ed. by E. Heftmann (New York: Reinhold Publishing Corp., 1961), p 347.
155. Chernoff, A. I., and N. Pettit, Jr.   Biochim. Biophys. Acta **97**, 47 (1965).
156. Tallan, H. H., and W. H. Stein.   J. Biol. Chem. **200**, 507 (1953).
157. Block, R. M., and N. Ling.   Anal. Chem. **26**, 1543 (1954).

158. Jauregui-Adell, J.   Advan. Protein Chem. **21**, 387 (1966).
159. Rosenfeld, I., and O. A. Beath.   *Selenium* (New York: Academic Press, 1964).
160. Schwarz, K., and C. M. Foltz.   J. Am. Chem. Soc. **79**, 3292 (1957).
161. Muth, O. H.   *Selenium in Biomedicine* (Westport, Connecticut: The AVI Publishing Company, Inc., 1967).

Chapter 2

Gas Chromatography of Amino Acids

by J. R. Coulter and C. S. Hann

2.1 Introduction

The word *protein* (Greek *proteios*—of the first rank) was first used by Berzelius in 1838 and from then on work on protein structure and function has been in the forefront of biochemical research. The isolation of the amino acids spanned 132 years, from 1806 (cystine) to 1938 (hydroxylysine), and throughout this long period of research effort many elegant and unusual methods of separation were developed.

Because the first step in understanding a protein structure is a quantitative estimate of the 20 or so amino acids in a hydrolysate, methods for achieving this have received considerable attention by many of the great names in protein chemistry. Emil Fischer[38] was the first to use an analytical technique bearing any resemblance to modern gas-chromatographic methods. He converted the amino acids to their ethyl esters and separated the esters by fractional distillation. Chromatographic methods were first used in 1941 when Martin and Synge[86] separated the N-acetyl derivatives of proline, valine, phenylalanine, isoleucine, and norleucine on a silica gel column eluted with chloroform. By 1949 Moore and Stein[119] had entered the field with elution from a starch column, and in 1954, when they developed a liquid column separation using Dowex 50 as an ion-exchange adsorbent, they set the pattern for all subsequent quantitative amino acid analyses.[90] In 1958, together with Spackman, they produced an automated version of their method and since then practically all quantitative analysis of amino acid mixtures has been done on machines of this type[91] (see Chapter 1 in this book).

Separation in this system depends upon the amphoteric nature of amino acids which attach to and detach from the resin under controlled gradients of pH, ionic strength and temperature. Detection is achieved by spectrophotometry subsequent to reaction of the effluent with ninhydrin and quantitation by the measurement of peak areas by hand or electronic methods. Quantitation of 10^{-9} mole in approximately 1 hour (Technicon Analyzer TMS 1) is the best so far attained, and it is against this performance that any gas-chromatographic method must compete. Any method which is either faster or more sensitive (or both) will become the method of choice because the number of proteins yet to be analyzed is vast. Also, many of them are labile or hard to purify and are therefore available in small amounts only. We believe that GC* has already shown a sufficient superiority over resin analysis to displace the latter for many applications.

Gas chromatography has four advantages over resin analysis:

(1) Gas chromatographs fitted with flame-ionization detectors will quantitate as little as 10^{-12} mole, and mass spectrometer detectors, 10^{-17} mole. While both these detectors could be adapted to monitor the resin effluent, they are already an integral part of the gas chromatograph.

(2) Analysis time has been reduced to less than half an hour and may be reduced to less than 10 minutes.

(3) Instrumentation cost may be reduced by as much as 50%.

(4) The gas chromatograph can readily be used for other substances should it no longer be required for amino acid analysis.

In the past decade there have been many attempts at amino acid analysis by GC—some exhaustive, some rudimentary. Taken together, these publications not only bear witness to the theoretical superiority of amino acid analysis by GC but also suggest that the development and adoption of a universal method is imminent. Only time will show whether any of the more promising methods already described will become widespread.

The amino acids are not a homologous series, with a cline in physical and chemical properties, but a heterogeneous assortment of compounds aliphatic and cyclic, and bearing (besides the α-amino and carboxyl groups common to all members) such diverse functional groups as secondary carboxyl and amino, imino, hydroxyl, sulfhydryl, imidazole, guanidine, and indole. These polar groups confer a high boiling point on the amino acids. At temperatures at which they are stable, no separation by gas chromatography has been attempted. Their vapor pressures, however, have been measured and they have been sublimed.[53] It seems possible that in the future with very sensitive detectors it may be practical to separate and quantitate the native amino

---

*See abbreviations, Section 2.6. For GC rather than GLC, see p. 113.

acids by gas chromatography. At present these polar groups must be masked or removed; the conversion to the derivative must be rapid and precise; the yield, if not 100%, must at least be a constant percentage at all times. It is also desirable that the derivative-making reaction be specific so that contaminating substances are not rendered volatile, thus interfering with the chromatogram. Furthermore, if the inherent sensitivity of GC is to be fully exploited, the derivation procedure must be capable of handling less than 10^{-9} mole of amino acid. This amount of material can often no longer be seen, so that transfer from one vessel to another with possible loss must be avoided. Derivative preparation should, therefore, be carried out in a single vessel. It is then necessary that the chosen reaction conditions be suitable for all amino acids. It is convenient if mixtures can be analyzed in a single gas-chromatographic run.

Several factors have mitigated against the acceptance of gas chromatography for amino acid analysis. Throughout the literature one finds many methods which, while obviously satisfactory in the hands of their authors, have been difficult or even impossible to reproduce elsewhere. Intrinsic to a good chromatographic technique is the adoption and testing of arbitrary values for a number of interrelated variables—supports, liquid phases, temperatures, etc. When the added variables of choice of derivative and method of synthesis are taken into account, it is not surprising to find that the literature contains a number of papers sharing few points of real contact but each claiming some improvement or advantage. For this reason confidence in GC as a practical method for amino acid determination has been undermined. These same features make a review of the literature difficult because the multidimensional matrix prescribed by all the variables is by no means fully explored. Explanations offered in one paper often cannot be validated by reference to another. For example, derivatives made by process A are analyzed by one worker on column type X and an arginine peak is not shown; another worker makes derivatives by process B, analyzes them on column Y, and shows an arginine peak. It is often impossible to say whether the first worker failed to make the hoped-for arginine derivative, made a different derivative, destroyed his derivative in the column, or adsorbed it irreversibly to the support or liquid phase.

Rather than confuse the reader by launching into a detailed survey of the literature, paper by paper or method by method, we propose to deal first with methods of historical interest, of limited or special application, or of recent development not yet fully explored. The TMS-derivatives are then introduced in a separate section. These derivatives, while showing great promise, have not yet achieved the status of the acyl esters. Methods utilizing the acyl esters have been shown to be quantitative and reproducible, and applicable to a wide range of amino acids. The section discussing these esters is therefore dealt with as one would approach the problems of selecting a practical method—choice of ester, of acyl group, of support, and of liquid phase. Within each of these categories we shall generalize the chemistry and chromatography as much as possible. In this way we hope that those who are looking for a method will be able to find it and at the same time to understand something of its shortcomings, advantages and pitfalls. Those who wish to follow a particular author's method are referred to the original papers or to one or another of several recent reviews.[11, 36, 132]

2.2   Methods of Historical or Exploratory Interest, or of Special or Limited Application

2.2.1   *Ninhydrin Oxidation*

The earliest gas-chromatographic method for analysis of amino acids (1956) involved decarboxylation and deamination with ninhydrin.[64] An aldehyde with one less carbon atom resulted and these aldehydes were then separated by gas chromatography.

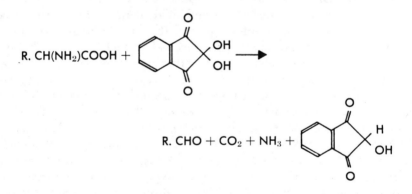

Certain amino acids had been determined by this reaction in 1947 when the aldehydes had been fractionated by steam distillation.[128] Limited investigation of this method ensued[4, 6, 7, 65] and attempts were also made to shortcut the procedure by reacting the amino acids with ninhydrin in the chromatographic column at elevated temperatures.[148, 149] Diphenylmethane and *p*-dimethylamino benzaldehyde have also been used as decarboxylating agents, but yields of between 40 and 80% only have been achieved.[7, 33]

The ninhydrin method is not applicable to all amino acids and does not appear to have been used since 1960. Glycine becomes formaldehyde which polymerizes, while histidine, arginine, tryptophan, cysteine, and aspartic and glutamic acids also appear to be unsuited to this method.[7] Both silicone[7, 64, 148] and polyester[4, 149] liquid phases have been used. Decarboxylation with N-bromosuccinimide has also been tried[121] but both a nitrile and the aldehyde with one less carbon atom are formed. The distribution between these two end-products varies with the amino acid.

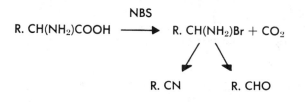

2.2.2 α-Chloroamino Acid Methyl Esters

The α-monoamino monocarboxylic acids have been converted with hydrochloric and nitric acid to the α-chloro- acids, which were then esterified with diazomethane. Quantitation of eight amino acids on both silicone and polyethyleneglycol phases was achieved.[88]

$$R. CH(NH_2)COOH + HCl + HNO_3 \longrightarrow R. CH(Cl)COOH + H_2O + N_2O$$

$$R. CH(Cl)COOH + CH_2N_2 \longrightarrow R. CH(Cl)COOCH_3 + N_2$$

2.2.3 2,4-Dinitrophenyl Amino Acid Methyl Esters

These DNP-amino acids have been separated on silicone[67, 69, 79, 96] liquid phases and also on polyesters[79, 80] after esterification with diazomethane, or boron trifluoride in methanol.[96]

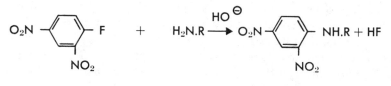

1-fluoro-2,4-dinitrobenzene   DNP-amino acid

These derivatives are of special interest for several reasons. The presence of the DNP-group allows the use of electron-capture detection with sensitivity down to 3×10^{-16} mole/sec[79] and it also introduces specificity. But, although DNP-derivatives have been used for N-terminus and sequence determination in peptides, scope will be limited for the use of these derivatives in determining the sequence of peptides difficult to purify or available in small amounts. And they are not useful generally, because amino acids with additional functional groups are not resolved without further treatment. The hydroxyl groups of threonine, serine, and hydroxyproline have been treated with trimethylsilylating reagents[67] but tryptophan and tyrosine do not appear to have been chromatographed satisfactorily (see Discussion in Reference 96).

2.2.4   *Phenylthiohydantoin Amino Acids*

phenyl isothiocyanate

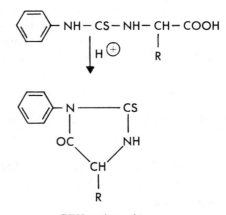

PTH-amino acid

These derivatives have also been used for N-terminus and peptide-sequence determination.[34] They were found to chromatograph well on silicone liquid phases but serine, threonine, asparagine, glutamine, and basic amino acids presented certain problems.[96] The second carboxyl groups of aspartic and glutamic acids were first esterified with boron trifluoride in methanol. We can expect GC of these derivatives, like the DNP-derivatives, to find limited application.

2.2.5   **N, N-***Dimethyl Amino Acid Methyl Esters*

The reductive condensation of amino groups of a peptide with aldehydes to give an N, N-dialkyl peptide was investigated in 1950.[16] Using aqueous formaldehyde for the condensation, the subsequent hydrolysis of the peptide

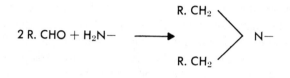

$$2 \text{ R. CHO} + \text{H}_2\text{N}- \longrightarrow \begin{array}{c} \text{R. CH}_2 \\ \\ \text{R. CH}_2 \end{array} \Big\rangle \text{N}-$$

was found to liberate the undamaged N, N-dimethyl amino acid. Because of the possibility of monoalkylation occurring, the reaction with the higher aldehydes was not sufficiently general to use with all amino acids. However, the methyl esters of the N, N-dimethyl amino acids were found to be very volatile.[12] Although it does not appear to have been used for that purpose, this simple and specific reaction could be the method of choice for estimating the N-terminus of peptides by GC.

2.2.6 α-Hydroxy Amino Acids

The Van Slyke reaction between an α-amino acid and nitrous acid has been used to prepare the α-hydroxy acids which then have been esterified with diazomethane.[129]

$$\text{HNO}_2 + \text{R. NH}_2 \longrightarrow \text{R. OH} + \text{H}_2\text{O} + \text{N}_2$$

(as nitrite
+ sulfuric
acid)

Valine, leucine, isoleucine, threonine, methionine, phenylalanine, histidine, and arginine were separated and quantitation to ±10% was claimed.

2.2.7 N-Diethylphosphate Amino Acid Methyl Esters

Reaction of the methyl esters of amino acids with diethylchlorophosphate in the presence of triethylamine (TEA) has been found to produce volatile derivatives suitable for use with the alkali-flame detector.[35]

$$\begin{array}{c} \text{C}_2\text{H}_5\text{O} \\ \\ \text{C}_2\text{H}_5\text{O} \end{array} \!\! \text{P} \!\! \begin{array}{c} \text{O} \\ \\ \text{Cl} \end{array} + \text{H}_2\text{N. R} \xrightarrow{\text{TEA}} \begin{array}{c} \text{C}_2\text{H}_5\text{O} \\ \\ \text{C}_2\text{H}_5\text{O} \end{array} \!\! \text{P} \!\! \begin{array}{c} \text{O} \\ \\ \text{NH. R} \end{array} + \text{HCl}$$

This apparatus is essentially a flame-ionization detector containing a bead of alkali metal within the flame. It shows greatly increased sensitivity for compounds containing halogens and phosphorus,[2, 50, 75] and it can detect as little as 10^{-12} mole amino acid as diethylphosphate amino methyl esters.[35] We have achieved this sensitivity with an unmodified flame-ionization detector and there are models already several orders of magnitude more sensitive than this. However, the alkali-flame detector possesses an advantage in that it is far more specific for these phosphorus-containing amino acids. So far, amino acids phosphorylated in the α-amino position only have been reported.[35]

2.2.8 *Alkylidine and Alkyl Amino Acid Esters*

The methyl ester hydrochlorides of amino acids have been reacted with isobutyraldehyde and sodium sulfite in sodium carbonate solution according to the following equation:[31]

$$
\begin{array}{c}
\text{R. CH. COOCH}_3 \\
| \\
\text{NH}_2 \cdot \text{HCl}
\end{array}
+ \text{R'CHO} \cdot \text{NaHSO}_3
\xrightarrow{\text{Na}_2\text{CO}_3}
\begin{array}{c}
\text{R. CH. COOCH}_3 \\
| \\
\text{N} = \text{CHR'}
\end{array}
$$

amino acid methyl ester   alkylidine amino acid ester

The alkylidine amino acid esters so formed were chromatographed as such or reduced with zinc dust in methanolic HCl to the alkyl ester

$$
\begin{array}{c}
\text{R. CH. COOCH}_3 \\
| \\
\text{HN. CH}_2\text{.R'}
\end{array}
$$

These derivatives were separated on capillary columns coated with Carbowax 1540. This method would appear to be of limited application, for yields of between 50 and 70% only were achieved and both the arginine and histidine derivatives were decomposed.

2.2.9 *Iodine-containing Amino Acids**

Because of the importance of these amino acids to medical science and their low levels of normal occurrence in the bloodstream, considerable interest has arisen in exploiting the essential sensitivity of GC in their separation. Six amino acids are of clinical importance: mono-iodotyrosine, diiodotyrosine, diiodothyronine, 3,3',5'- and 3,5,3'-tri-iodothyronines, and 3,5,3',5'-tetraiodothyronine. These are generally shortened to MIT, DIT, T_2, T_3, Reverse T_3, and T_4. Amino acid T_4 is

---

*See also Section 2.7.7.

the thyroid hormone thyroxine, whereas T_3 has similar but greater physiological activity; Reverse T_3 has an action antagonistic to T_3 and T_4. A difference of opinion has been raised concerning the ratios of concentrations of these substances both in health and disease[97,133,134] and this has provided a stimulus for better assay methods. The presence of iodine in all these amino acids allows use of the sensitive electron-capture detector, but the problem has been to render these large polar molecules (thyroxine, mol wt 777) volatile and to resolve them on columns at moderate temperatures. Most work has been done with artificial mixtures of the iodine amino acids.

The silyl derivatives have been prepared[1,3,112] by the reaction of the amino acids with BSA.* Other silylating agents, TMCS and HMDS, were investigated[112] and found not to give 100% conversion to the TMS-derivative. All these workers achieved satisfactory separations on silicone columns but analyses of naturally occurring mixtures were not reported; one group[3] does mention difficulties in their extraction from blood. Quantitation down to 6×10^{-12} mole for MIT using electron-capture detection is shown in Reference 3 and down to $6 \cdot 10^{-11}$ mole using flame-ionization detection in Reference 112. These sensitivities are of the right order of magnitude to analyze one or several ml of whole blood. Trimethylsilylation suffers from being fairly nonspecific so that many contaminating compounds will also be derivatized and appear in a chromatogram using flame-ionization detection. The electron-capture detector should overcome this defect. Iodinated contrast media which may interfere with the chemical assay of iodinated amino acids have already been separated from the latter as TMS-derivatives by GC.[3]

Methyl esters of iodinated amino acids have also been synthesized using methanol–HCl[70,102] or thionyl chloride–methanol[70,122] and the esters have then been acylated with trifluoroacetic anhydride[102] or pivalic anhydride.[70,122] The methyl trifluoroacetyl derivatives resolved well on SE-30 with flame-ionization detection but the amounts analyzed $(1 \times 10^{-8}$ mole) could be obtained only from a liter of blood. Only the N,O-dipivalyl methyl esters have been used for assay of blood; satisfactory results have been obtained with 5 ml of serum, OV-17 as liquid phase, and electron-capture detection. Amounts of T_3 as small as 1×10^{-12} mole have been chromatographed, the amount theoretically obtained from about 0.02 ml whole blood. Although the acetyl group has not been used for acylation, we see no reason why the N,O-acetyl methyl esters should not be just as easy to prepare, as stable,

---

*See Section 2.6 for abbreviations.

and perhaps easier to chromatograph. Acetic acid is more volatile than pivalic acid and we would expect the acetyl derivatives to be more volatile than the pivalyl.

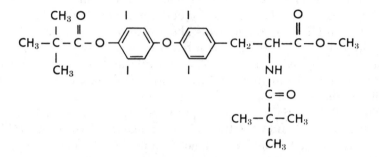

N,O-dipivalyl methyl ester of T_4.

DIT is also known to occur in some corals where it may be accompanied by the corresponding dibromotyrosine. GC should provide an easy method for distinguishing these two and for their assay.

2.2.10   Sulfo- and Selenoamino Acids

2.2.10.1   Sulfur amino acids

Methionine and cysteine have been chromatographed by a number of methods as have some of the methionine breakdown products. There has been some doubt, however, as to which breakdown products have been chromatographed, and cystine and cysteic acid appear to have caused difficulty for most workers. This is particularly true of the N-acyl esters of cystine and cysteic acid. The latter is probably too polar by reason of the -SO$_3$H group while the former is both large (mol wt 240) and difficult to esterify.[41] The TMS-derivatives of cysteine, cystine, cysteic acid, and a number of other nonprotein sulfur-containing amino acids and breakdown products (S-methyl cysteine, taurine, homocystine, djenkolic acid and ethionine,[111] and cysteinesulfinic acid[18]) also have been investigated. Both BSA[18, 111] and BSTFA*[111] have been used and the percentage conversion has differed. BSTFA[111] is most suitable for all the amino acids listed above except S-methyl cysteine for which BSA is found to give greater yield.

With proteins the problem of cysteic acid can be avoided by conversion of the cysteine to carboxymethyl cysteine before hydrolysis of the protein. This method is already widely used in the preparation of proteins for amino acid resin analysis.[24] Carboxymethyl cysteine can

---

*See abbreviations, Section 2.6.

be esterified and acylated; in our hands N-acetylcarboxymethyl cysteine di-n-propyl ester is easily and quantitatively chromatographed, appearing between glutamic acid and tyrosine on Carbowax-coated columns.

Confusion existed regarding the chromatography of methionine, and methionine sulfone and sulfoxide[72, 78, 150] until these were made the subject of a special study using the N-TFA-methyl esters.[125] Methionine sulfoxide was converted to N-TFA-methionine methyl ester during the process of derivative formation, and therefore methionine and its sulfoxide eluted together. The conversion, however, was not total because only 52% of the sulfoxide appeared in the chromatogram as the native amino acid, the rest having been destroyed. The experimenters[125] failed to obtain any peak for the sulfone and supposed it, too, to have been destroyed. Tests from our laboratory, however, have obtained a peak for N-acetylmethionine sulfone n-propyl ester, although quantitation of this substance has not been attempted.

2.2.10.2 Selenium amino acids

Selenium (Se) is a trace element essential in very small amounts and it is thought by some to play its part incorporated into amino acids and not in the inorganic form. It is known that selenium can replace sulfur in methionine and cysteine and that these selenium amino acids can function in proteins. Whether some Se-containing proteins have functions not shared by the sulfur analogs is not known. The Se-amino acids have been gas-chromatographed as their TMS-derivatives, prepared by reaction with BSA.[18] These derivatives appeared to be unstable. The N-acetyl-n-propyl esters of Se-methionine and Se-carboxymethyl cysteine chromatograph well. In each case the Se-analog has a longer retention time. Under standard conditions (Figure 5) the methionine derivative showed a retention time of 893 seconds compared with the Se-analog with a retention time of 955 seconds (phenylalanine: 971 seconds). Carboxymethyl cysteine and Se-carboxymethyl cysteine showed retention times of 1,201 and 1,252 seconds, respectively. (Authors' unpublished results.)

2.2.11 *Resolution of Stereoisomers**

The specific rotation of plane-polarized light by enantiomorphs of many biologically active molecules, including the amino acids, is frequently small; polarimetric methods are not always sufficiently sensitive to detect trace amounts of one isomer contaminating a large excess of its antipode. This analysis may be important for several reasons. When making biologically active peptides the effects of isomeric contamina-

---

*See also Section 3.6 in this volume.

tion are cumulative; 1% D-amino acid in each amino acid built into a peptide of 10 units will result in 10% inactive material. Another area of considerable interest relates to the detection of extraterrestrial life. A unique feature of macromolecules of biological origin is their ability to distinguish and selectively incorporate particular optical isomers. The detection of a preponderance of one optical isomer on a distant planet may be taken to indicate the existence of life.[59] Although optical isomers will not separate easily on the ordinary liquid phases, it is not surprising that considerable attention has been paid to GC as a rapid means for their separation and identification. Two approaches have been used: (a) use of optically active liquid phases and (b) introduction of a second asymmetric center into the compounds being separated. Both have been successful.

2.2.11.1   Use of optically active liquid phases

This method was initially unsuccessful when tried with several racemic mixtures, not amino acids.[51, 73, 74] Very slight resolution was achieved in some cases. Using capillary columns and a liquid phase of N-TFA-L-isoleucine lauryl ester,[48] N-TFA-alanine esters of four alcohols were investigated and partially resolved. Proceeding from this work the same authors described the almost complete resolution of N-TFA-(±)-alanine t-butyl ester on a liquid phase of N-TFA-L-valyl-L-valine cyclohexyl ester in a packed column only 2 m in length.[46] They speculated that the N-TFA-t-butyl esters of other (±) amino acids also should be separable on the same column. Using capillary columns, 18 pairs of enantiomers were resolved using N-TFA-isoleucine lauryl ester and N-TFA-L-phenylalanine cyclohexyl ester as liquid phases.[48] It has been shown that racemization of the N-TFA-amino acid esters does not occur under the conditions of chromatography.[101] Analysis by this method, therefore, gives an accurate measure of the proportion of each isomer present in a mixture. Possible mechanisms of separation are discussed in References 37, 48, 120, and 135, although the last two papers deal particularly with diastereoisomeric separation. It is thought that hydrogen bridging and possibly electronic interactions occur between the liquid phase and the isomers undergoing separation. In the presence of an optically active liquid phase one enantiomer will form these loose diastereoisomers slightly more readily than the other and will therefore be retained slightly longer in the column. In using L-isoleucine lauryl ester as a liquid phase it was found that N-TFA-L-amino acid isopropyl esters were retained longer than the corresponding D-esters, and that using D-isoleucine lauryl ester as a liquid phase reversed the elution order.[48] Functional groups in close proximity to the asymmetric center confer bigger differences in retention times than functional groups further away. Optical purity has also been determined by NMR in optically active solvents.[17, 95] Both NMR and GC on optically active stationary phases possess the advantage of measuring with precision the concentration of enantiomers in a racemic mixture. In either method, contamination of the active solvent with the opposite enantiomer will affect the resolution but not the estimation of absolute quantities in racemic mixtures.

When the logarithm of retention volume is plotted against the number of carbon atoms for a homologous series, each isomeric configuration gives a straight line.[19, 37, 45, 49] GC may find a place as a means of rapidly establishing the absolute configuration of an unknown compound.

2.2.11.2 Diastereoisomers

A method more thoroughly explored than the one above is the introduction of a second center of asymmetry into compounds in racemic mixtures. It has been observed in a number of systems that when LL, LD, DL, and DD diastereoisomers are resolved on an inactive stationary phase, the LL and DD compounds elute together before the LD and DL isomers which also elute together.[45, 100, 127] For unequivocal resolution of a racemic mixture it therefore follows that a pure enantiomorph must be used to introduce the second asymmetric center. Separation of diastereoisomeric dipeptides on Sephadex[146] has led to the speculation that DL compounds preferentially form ring structures while LL and DD forms tend to remain as open chains. Diastereoisomers have also been used since the time of Pasteur to separate optical isomers by crystallization. This method has been recently employed to obtain optically pure alcohols for amino acid esterification.[61] Proximity of the two asymmetric centers and functional groups aids resolution by GC and it seems likely that, as with the optically active phases mentioned above, hydrogen bridging and electronic effects between the diastereoisomer and the inactive liquid phase determine the resolution.[120, 135] Polar phases such as Carbowax have been found to give better resolution than nonpolar phases.[101, 135]

The N-TFA-D-2-butyl esters have been separated on a capillary column coated with Carbowax 1540 or Ucon LB550.[100] The DD esters eluted before the LD esters, and the test was used to measure racemization of 11 amino acids during acid hydrolysis of bovine serum albumin. Unfortunately, the 2-butanol used in preparation of the esters was only 90% D configuration but the paper sets out the way in which accurate analysis of racemic mixtures can be achieved. The N-TFA-methyl esters of valyl–valine have also been used to investigate racemization during peptide synthesis by different methods.[143] Secondary alkanol esters of the N-TFA-amino acids have been used in a number of stereoisomer studies: valine and alanine;[19] configuration of amino acids in the antibiotic doricin;[20] alanine, valine, leucine, isoleucine, proline, phenylalanine, glutamic acid;[44] threonine, serine, hydroxyproline, cysteine, aspartic acid, tyrosine, and tryptophan (O-acetyl where appropriate);[99] alanine, valine, leucine, isoleucine, alloisoleucine (in this study methionine, hydroxyproline, lysine, and aspartic and glutamic acids with secondary functional centers gave irregular results);[45] and on-column racemization.[101] Dipeptides have been used to obtain a second asymmetric center in several studies: L-α-chloroisovaleryl amino acid methyl ester;[58] N-TFA-L-prolyl amino acid methyl ester[59, 60] (preparation of these diastereoisomers);[139] N-TFA-valyl-valine (racemization during peptide synthesis);[143] all dipeptide combinations of alanine, glycine, valine, leucine, isoleucine, methionine, and phenylalanine as N-TFA-methyl esters;[141] and N-TFA-S*-prolyl amides of asymmetric amines and amino acid esters.[135] L-Menthyl N-TFA-amino acids have also been investigated.[127]

Making a dipeptide to introduce the second asymmetric center leads to loss of volatility and it is unlikely that any of the less volatile amino acids can be analyzed this way. For this reason and because accurate quantitation of a racemic mixture by making a diastereoisomer demands that one isomer be optically pure, it appears that analysis of a racemate directly on optically active stationary phases will become increasingly more important.

2.2.12   Methods of Single, Limited or Recent Application

Here we propose to list those methods which fall outside the scope of other sections either because they do not have, or have not yet been found to have, general application.

N-Pentafluoropropionyl and N-heptafluorobutyryl derivatives were compared with the more usual trifluoroacetyl derivatives. Fourteen amino acid derivatives having lower retention times than the corresponding TFA-derivatives were made and separated.[98] The author considered their greater volatility and sensitivity to electron-capture detection to have possible advantages although no quantitative studies were done.

N-Carbobenzyloxy-trans-3-hydroxy-D-L-proline was chromatographed during a study on the occurrence of cis-trans-3-hydroxy-L-proline in Telomycin.[68] Several other proline isomers were separated as N-acetyl-n-butyl esters after nitrous acid had been used to remove the α-amino

---

*Refers to S-R designation of optically active compounds.

groups of other amino acids leaving the imino groups of the proline isomers intact.[92]

A recent reaction used for GC of amino acids is condensation with hexane-2,5-dione: [131]

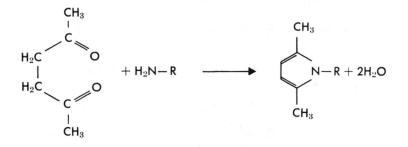

The condensation product was then esterified with methanol–HCl. Only a few amino acids are reported and no quantitative data.

2.3 Trimethylsilyl Amino Acids

The substitution of a proton with the trimethylsilyl (TMS-) group (in which context the term silylation is ordinarily used) has been successfully exploited for the preparation of volatile derivatives of such biologically important compounds as sugars,[124] purine and pyrimidine bases, nucleosides and nucleotides,[42] steroids,[126] amines,[63] and amino acids. A variety of reagents and a great number of methods have now become available for silylations, the selection of any particular one being determined by the compound in question and the scale on which the reaction is to be carried out. These methods have recently been reviewed.[94]

2.3.1 Silylations with TMCS, HMDS, and TMS-DEA*

Much of this work was pioneered by Rühlmann who investigated trans-silylation reactions of TMS-amino acids and amines[104] and noted the instability of TMS-derivatives in aqueous, alcoholic and ethereal solutions. The lability of one silylated group in comparison to that of another in a fully silylated amino acid varies markedly. Using both amino acids and model compounds—hexanoic acid (COOH), 1-octanol (OH), and n-octylamine (NH₂)—it was found that the Si-N bond was more labile than the Si-O bond of a hydroxyl group, which in turn was more labile than the Si-O bond of a carboxyl group.[87] The reverse order

---

*See abbreviations, Section 2.6.

represents the ease with which TMS groups may be introduced, with carboxyl groups the easiest to silylate and also the most stable. N-TMS-Imidazole will silylate -OH groups but not basic -NH$_2$ groups.[63] The extent to which guanidine can be silylated is unknown but one might expect this group to be the most difficult to convert and also the most labile.

The silylation of an amino acid may be thought to proceed in two stages:

R. CH. COOH   TMS-   R. CH. COO Si(CH$_3$)$_3$ TMS-   R. CH. COO Si(CH$_3$)$_3$
| $\longrightarrow$ | $\longrightarrow$ |
NH$_2$ NH$_2$ NH. Si(CH$_3$)$_3$

Esterification of the carboxyl group occurs more readily and can be achieved under mild conditions with HMDS,[9, 10] but basic amino and other functions such as -OH and -SH require stronger conditions. HMDS, either alone or with TMCS, does not give satisfactory yields.[114] Reaction of amino acid sodium salts with TMCS[103] or with TMS-DEA[87, 105, 114] has been found to be better.

R. CH. COOH
| + 2 (CH$_3$)$_3$— Si — N — (C$_2$H$_5$)$_2$
NH$_2$ TMS - DEA

$\longrightarrow$ R. CH. COO-Si(CH$_3$)$_3$ + 2 HN—(C$_2$H$_5$)$_2$
                                     |
                                     NH
                                     |
                                     Si(CH$_3$)$_3$

OH- and SH- groups, as well as the imidazole of histidine were silylated.[105] The best yields of several TMS-amino acids were 75%.[114] TMS-Derivatives of alanine, valine, leucine, isoleucine, glutamic acid and phenylalanine were separated by GC using a silicone-oil liquid phase.[106] N-TMS-Amino acid ethyl esters were also chromatographed, as well as free bases of the TMS-esters prepared by reaction of the amino acid with HMDS or by stripping off the labile N-TMS group with ammonia.[107] Peaks were claimed for arginine, histidine and lysine but no retention times were reported.

2.3.2   Silylations with Silylamides

The introduction of BSA, which was stated to be 50 times more potent as a silyl donor than any of the monosubstituted amides, permitted rapid and quantitative silylation of amides, ureas, amino acids, hindered phenols, carboxylic acids, and enols.[77] NMR and IR evidence favors the structure I rather than the isomer II, with rapid intramolecular ex-

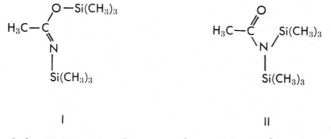

change of the TMS-groups between the oxygen and nitrogen atoms. The reactivity can be explained by competition of one TMS-group for the O- or N-position, weakening the bond of the other (see Discussion in Chapter 4 of Reference 94). Both silyl groups can be exchanged where equilibrium favors transfer as in carboxylic hydroxyl groups.[77] More recently BSTFA has been introduced as a silylating reagent but firm evidence for its structure is lacking. Its formula was stated to be related to II.[116] Other silylamides (N-TMS-N-methyl acetamide and formamide) have also been used for silylation of amino acids.[8]

Using BSA, 22 amino acid derivatives were prepared and gave single chromatographic peaks on SE-30 columns.[77] Arginine showed decomposition, and glycine and alanine were swamped by the peak of one of the reaction products (mono-N-TMS-acetamide). When BSTFA was used to silylate 18 amino acids by carrying out the reaction in acetonitrile at 125°C, the corresponding reaction product was stated to elute ahead of alanine and glycine. The position of the latter was not shown in the accompanying chromatogram.[116] Cystine, glycine and glutamic acid showed variability with the time of derivatization, the latter two giving rise to two chromatographic peaks when silylated at higher temperatures (150°C).

TMS-Asparagine and glutamine could be synthesized under special conditions and chromatographed (asparagine, 150°C for ½ hour; glutamine, 70°C for ½ hour). Relative molar response data for individual amino acids silylated for increasing periods of time showed high reproducibility in synthesis.

2.3.3   GC of TMS-Amino Acids*

Single chromatographic peaks for TMS-amino acids have been obtained on a number of silicone liquid phases (such as SE-30,[8, 77, 116]

---

*See also Section 2.7.6.

silicone oil,[106] SE-52, QF-1 and DC-200,[114] and DC-550[116]). Unrecognizable peaks were obtained on polyesters and polyglycols,[114] and it is safe to conclude that TMS-derivatives are insufficiently stable to chromatograph on these liquid phases.

One useful feature of the silylation reaction is that it is accomplished quantitatively in one step in a short time, but quantitative conversion without suitable gas-chromatographic columns for analysis is of limited use. In those papers where illustrative chromatograms have been presented, separations are confined to a few amino acids only.[8, 106, 116] We can only assume that resolutions so far achieved of the complete range of amino acids (or at least those expected in an acid hydrolysate) are unsatisfactory (however, see Section 2.7.6). Quality and speed of resolution are often improved by increasing the polarity of the liquid phase but for reasons already given polar liquid phases cannot be used with silyl derivatives. A further disadvantage of these derivatives lies in the ease with which likely contaminants of amino acid mixtures can be silylated and chromatographed under similar conditions. Esterification and acylation confers a measure of specificity for producing volatile amino acid derivatives, which allows accurate analysis of many impure samples. For these reasons we hesitate to predict the future of TMS-derivatives for general amino acid analysis.

2.4 Methods of General Application: Acylated Amino Acid Esters

In 1929 Cherbuliez et al.[21] measured the boiling points of certain N-acetyl amino acid ethyl esters under reduced pressure and later subjected these same compounds to fractional distillation.[22] Vacuum sublimation of N-TFA-amino acids and their methyl and ethyl esters revealed that the esters sublimed at a temperature some 40°C lower than when the carboxyl group was free.[138] N-TFA-Amino acid methyl esters were also fractionated by vacuum distillation.[137]

The first use of GC for separation of N-acylated amino acid esters appeared in 1959 when Youngs[147] successfully analyzed six amino acids as the N-acetyl-n-butyl esters. The great number of succeeding publications (Table I) attests to the popularity and potential of this class of derivatives for determination of amino acids. In this section production of derivatives, their stability, and gas chromatography are discussed separately so far as this is possible.

2.4.1 Esterification of Carboxyl Groups

2.4.1.1 With alcohols using acid catalysts

Amino acid carboxyl groups may be esterified by refluxing with the appropriate alcohol using an acid catalyst, preferably dry hydrogen

Table I

Amino Acid Derivatives Chromatographed after Esterification of the Carboxyl Group and Acylation of the Amino and Other Functional Groups

Ester group	Acyl group	Reference
methyl	formyl	82
methyl	acetyl	27, 78, 83
methyl	trifluoroacetyl	25, 27, 54, 55, 56, 57, 66, 78, 83, 84, 102, 125, 130
methyl	pivalyl	70, 122
ethyl	acetyl	83, 113
ethyl	trifluoroacetyl	83
n-propyl	acetyl	23, 52, 83, 113
isopropyl	acetyl	113
n-butyl	acetyl	72, 78, 83, 113, 147
n-butyl	trifluoroacetyl	39, 40, 41, 78, 83, 85, 115, 117, 150
n-butyl	pentafluoropropionyl	98
n-butyl	heptafluorobutyryl	98
n-amyl	acetyl	72, 83
isoamyl	acetyl	72
n-amyl	trifluoroacetyl	13, 14, 26, 27, 28, 29
benzyl	trifluoroacetyl	27
cyclohexyl	trifluoroacetyl	12

chloride because of the ease with which it can be removed. Hydrogen bromide has enjoyed less popularity.[72, 113] In this way amino acids have been methylated,[30] ethylated,[83] propylated,[23, 52] butylated,[41, 147] and amylated.[26, 27] The readiness with which the reaction proceeds depends on such factors as chain length of the alcohol, the amino acid being esterified, the amount of catalyst, temperature, and quality of reagents.

Methylation takes place readily in 30 minutes at room temperature[117] (using 1.25M HCl), but increasing time and/or temperature are required through the series methanol–pentanol. Lysine, histidine and cystine have proved especially difficult to esterify using butanol or pentanol with low concentrations (1.25M) of HCl. The problem was circumvented by first esterifying with methanolic HCl (room temperature for 30 minutes), followed by trans-esterification at 100°C for 150 minutes with butanol–HCl to yield the butyl ester.[117] Temperature, rather than HCl concentration, was the important factor in the trans-esterification reaction. At 90°C the conversion of methyl to butyl ester was significantly slower; increasing HCl concentration from 1.25M to 3.25M influenced neither yield (which approached 100%) nor rate of reaction. An alternative approach has been to dissolve the amino acids in trifluoroacetic acid. Lysine hydrochloride could then be esterified at 108°C with n-amyl alcohol[14, 29] through which dry HCl gas was bubbled continuously. The methyl ester of histidine has also been obtained using H_2SO_4 as the esterification catalyst.[84] When only small quantities (less than 1 mg) of amino acids are employed,[23] all common amino acids including lysine and histidine can be esterified in 20 minutes at 100°C using *n*-propanol made 8M with dry HCl. Larger amounts of lysine or histidine (much more than ever handled in a routine analysis) require a trans-esterification step.

No figure for the optimal concentration of HCl gas in alcohol can be given from the literature as this has ranged from about 1.25M[41, 150] to saturation[147] with many arbitrary intermediate values chosen by various investigators. In our own experience with propylations a high concentration of gas has been desirable (see also Section 2.7.2). Since the esterification reaction produces water, it is necessary that reagents (alcohol and HCl gas) be dry. Alcohols may be dried by refluxing with magnesium turnings[117] or, for the higher homologs, with calcium hydride.[23] Thorough desiccation of glassware and protection against atmospheric moisture should be observed. We have found dry HCl gas (Matheson, Coleman and Bell, >99% purity) may be used as supplied, or sufficiently dry HCl gas can be generated from fused ammonium chloride and concentrated H_2SO_4.[23] Water of reaction has been removed azeotropically with benzene from a refluxing solution of the amino acids in propanol–HCl.[52] Water "scavengers" (such as dibutoxypropane[150]) have also been used to increase ester yields.

2.4.1.2   *With dimethyl sulfite*

This procedure for methylation has not been widely used but esterification is both rapid and complete.[25, 125] The reaction is carried out in methanolic–HCl under reflux providing two alternative esterification pathways.

$$
\underset{\substack{| \\ H_2N . CH . COOH.}}{\overset{R}{}} \quad \xrightarrow[\text{CH}_3\text{OH/HCl}]{(\text{CH}_3\text{O})_2\text{SO}} \quad \underset{\substack{| \\ Cl^- H_3N^+ . CH . COOCH_3}}{\overset{R}{}}
$$

2.4.1.3   With thionyl chloride

Amino acid methyl esters have been produced by treatment with methanol in the presence of thionyl chloride.[56, 58, 59, 122] Using between 10 and 20% thionyl chloride at 40°C for 2 hours, reproducible yields ensued.[56] The reaction proceeds via the formation of an intermediate dimethyl sulfite (q.v.) which is the active donor of methyl groups.

2.4.1.4   With diazomethane

Treatment of amino acids with diazomethane has been used for small-scale esterifications. A notable feature of the method is that it has been successfully used on some amino acids which have already been N-acylated.[54, 69, 82, 88, 130] Although low yields have been reported,[130] most other workers believe the reaction to be quantitative.[82] Benzyl esters of amino acids have been prepared by an analogous reaction using phenyl diazomethane.[27]

2.4.1.5   With cation-exchange resins as catalysts

In the H^+-cycle, these resins were shown to catalyze esterification of oleic acid with n-butanol.[123] The kinetics of this reaction were studied.[81] In an analogous way amino acid esters have been prepared by refluxing with the appropriate alcohol in the presence of Dowex-50 or other strong cation exchangers.[85, 89, 92, 108] When resin-bound n-butyl esters were eluted with citrate buffer at pH 6.95, thin-layer chromatography revealed no unesterified amino acid.[85] Eleven butyl esters were produced in this manner but those difficult to esterify by other means were not included.

2.4.1.6   Sundry methods for esterification

Besides boron trifluoride[96] reference has been made to the use of boron trichloride and p-toluene sulfonic acid as catalysts for esterification.[78]

2.4.2   Amino Acid Ester Free-Base Production

Esterification in the presence of HCl (HBr) yields an amino acid ester hydrochloride (hydrobromide) which lacks sufficient volatility for GC. Neutralization of the salt by mild alkali, or better, by ion-exchange, gives a volatile free amine in the case of simple amino acids.

Free bases from amino acid methyl ester hydrochlorides were produced by treatment with NaOH[5] or an anion-exchange resin (Dowex-1, OH-form) suspended in anhydrous methanol.[93] (Moisture poses a serious risk of ester destruction by liberation of free OH-ions from the resin, creating intense hydrolytic foci.) The free bases were considered to be reasonably stable when stored.

Ester hydrochlorides of some amino acids have been chromatographed by including ammonia in the carrier gas.[109] The free bases, and chloride and acetate salts have also been chromatographed.[93] At elevated temperatures the acetate salts (the chloride salts more reluctantly) gave rise to gas-chromatographic peaks of similar retention time and area to those for the corresponding free base, indicating heat dissociation of the salt. With esters of lysine and arginine no such dissociation was apparent, while cysteine methyl ester gave a peak of good yield as the acetate salt but not as the free base. Tyrosine, tryptophan and histidine could not be chromatographed either as free base or acetate salt.

2.4.3   *Acylation of Amino and Other Functions*

Acylation reactions are generally carried out with acid anhydrides or acid chlorides, often in the presence of pyridine or quinoline or less often with other tertiary amines. The reaction with a symmetric anhydride produces the corresponding acid.

$$(R. CO)_2O + H_2N.R' \longrightarrow R. CO. NH. R' + R.COOH$$

Alkaline conditions are not obligatory because the reaction also proceeds under acidic conditions, particularly when primary amines are concerned. The tendency to amino diacyl formation is prevented by dilution of the anhydride. In addition to the primary amino (or imino) functions common to all amino acids, other groups conferring polarity which can be acylated include the -OH groups of serine, threonine, hydroxyproline, and tyrosine; the secondary $-NH_2$ groups of ornithine and lysine; the -SH group of cysteine; and the imidazole, guanidine and indole group of histidine, arginine and tryptophan, respectively. Under specified conditions it is possible to obtain diacyl derivatives of all these amino acids except arginine. Introduction of two acyl groups into the guanidine nucleus creates a triacyl species.

2.4.3.1   Formylation

In a single reference formylation of amino acids was carried out with formic acid in acetic anhydride[82] but details of the amino acids used were not given. After methylation with diazomethane, yields were reported to be 100%. Above 100°C, N-formylglutamic acid dimethyl ester was converted to methyl pyrrolidone carboxylate.

2.4.3.2   Acetylation

Amino acids or their esters (hydrochlorides) have been acetylated with acetic anhydride either alone[113] or in the presence of sodium acetate,[21, 22] acetic acid,[52] or pyridine.[23] With pyridine we have found that acetylation of the α-amino function of amino acid ester hydrochlorides occurs very rapidly at room temperature. Within 5 minutes acetylation of all secondary groups (except the guanidine and imidazole) is complete.

2.4.3.3   Trifluoroacetylation by TFA-transfers

In a long series of publications Weygand and co-workers reported on a variety of methods and reagents for trifluoroacetylation of amino acids. Transfer of the TFA-moiety from the phenyl[145] and methyl[137] esters of trifluoroacetic acid to the α-amino group of a number of amino acids and peptides occurred with high yields. This method was also used to acylate 14 resin-bound amino acid methyl esters[108] including serine, threonine, hydroxyproline and lysine. The occasional appearance of multiple gas-chromatographic peaks was consistent with incomplete acylation of -OH or ϵ-NH$_2$ functions, although in other hands[66] TFA-amino acid methyl esters (including the hydroxy amino acids and tryptophan but not arginine, histidine or cystine) were successfully chromatographed after a similar acylation procedure.

TFA-Imidazole also readily acylated some amino acids between 0 and 20°C but multiple peaks occurred with serine and threonine. All these methods appear to be of limited practical use in quantitative derivative production.

2.4.3.4 Trifluoroacetylation with acid anhydride

This is undoubtedly the best method for TFA-acylation but the conditions under which the reaction has been carried out vary widely (−10–+150°C for minutes to hours). Confusion has arisen regarding the facility with which amino acid secondary groups react and the stability of the resulting acyl derivative (see Section 2.4.5). The basic amino acids (lysine, ornithine, arginine and tryptophan) have often proved refractory. On reaction with trifluoroacetic anhydride in trifluoroacetic acid, ornithine, lysine and arginine were acylated in the α-NH$_2$ position only.[136] More effective acylation of amino acid hydrochlorides has been achieved with undiluted trifluoroacetic anhydride[25, 26, 85, 130] or with trifluoroacetic anhydride diluted with methylene chloride or ethyl acetate.[41, 84, 100, 150] *N, O*-bis-TFA-Serine and threonine and other bifunctional amino acids have been obtained in high yield by using a two- to fivefold molar excess of reagent;[84, 144] pyridine is unsuitable as a basic solvent, leaving a brown syrup on evaporation.[15] At room temperature in methylene chloride all amino acids except arginine are trifluoroacetylated in 2 hours. At 150°C the reaction is complete in 5 minutes.[115]

2.4.3.5 Acylation of arginine, histidine and tryptophan

Esterification of arginine and histidine with acid–alcohols yields the dihydrochloride salts involving both the guanidine and imidazole groups. When an acid hydrolysate is the source of amino acids, these dihydrochloride salts are present before esterification. Diazomethane for esterification does not therefore overcome the problem of salt formation discussed below. Salt formation reduces volatility and in the case of arginine opposes complete acylation. The important influence of salt formation on acylation, volatility and gas-chromatographic behavior appears to have been insufficiently appreciated and many confusing claims and counterclaims regarding these amino acids have been made.

Histidine, when trifluoroacetylated in the presence of triethylamine using trifluoroacetyl methyl ester, formed a triethyl ammonium salt;

desalting was achieved with Amberlite XE-64.[145] No indication was given whether the product was mono- or diacylated. The important paper of Makasumi and Saroff,[84] in which bulk preparations and properties of highly purified TFA-amino acid methyl esters were described, confirmed that the histidine derivative isolated and crystallized from ether was the mono-N-TFA-compound (III) unsuited to GC. (Trifluoroacetylation was carried out with trifluoroacetic anhydride in ethyl acetate.) It was evidently injected as the TFA-salt which was heat-dissociated and gave rise to a poor peak. However, a histidine derivative which could be prepared in solution (but not isolated) and was amenable to gas-chromatographic analysis was assumed to be the diacyl (N,N′-TFA) species (IV). This compound is very labile. However, see Section 2.7.3.

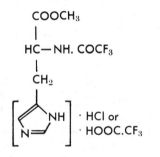

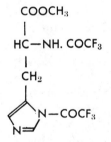

N-TFA-histidine methyl ester, HCl or TFA salt   N,N′-TFA-histidine methyl ester

We also reported that the propyl ester dihydrochloride of histidine could not be acetylated with pyridine–acetic anhydride to give a product suitable for GC.[23] When acetylated after neutralization or in the presence of anhydrous sodium carbonate, the product could be chromatographed on Chromosorb-G (High Performance) coated with 3% OV-17. The quantitative aspect of this procedure is questionable.

Although chromatographic peaks for arginine have been claimed after trifluoroacetylation at room temperature[56] it has generally proved necessary to acylate with trifluoroacetic anhydride between 100 and 150°C[25,29,115] or in the presence of anhydrous sodium carbonate[39] and dimethyl formamide.[150] This procedure[39] was not considered satisfactory as some conversion to ornithine occurred. When arginine methyl ester dihydrochloride was acylated with trifluoroacetic anhydride in ethyl acetate at room temperature, elemental analysis revealed an approximate empirical formula $C_{13}H_{13}O_5N_4F_9$, the sensitivity of the analysis not permitting distinction between the tri-TFA-compound (V) and the bis-TFA-trifluoroacetate salt (VI).[84]

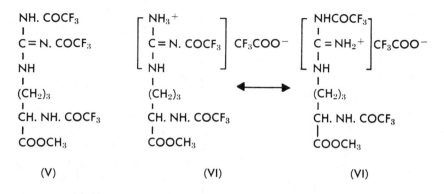

Stalling and Gehrke[115] discussed at some length the problems associated with acylation of arginine. In essence they concluded that trifluoroacetylation at room temperature yielded a guanidine salt (VI) insufficiently volatile for GC. If injected into flash-evaporators (especially metal) in the presence of excess trifluoroacetic anhydride, some degree of synthesis of the tri-TFA-compound (V) occurred at the high temperature (see also Reference 25) and a peak was sometimes observed. (Column packing and selection of liquid phase were important.) Varying decomposition in the flash heaters produced some ornithine. The fully acylated derivative of arginine showing no tendency to salt formation was quantitatively produced by heating with trifluoroacetic anhydride at 140–150°C.[29, 115] Under these conditions tryptophan was also fully acylated.[115] The less potent acylating properties of acetic anhydride do not permit acetylation of the guanidine hydrochloride group under similar conditions. However, we have produced an acetyl derivative suitable for GC by neutralizing the *n*-propyl ester hydrochloride with alkali carbonate, or better, by treatment with Dowex-1 (OH-form) in anhydrous *n*-propanol before acetylation. Analysis of the derivatives of arginine and histidine on Chromosorb-W (Johns-Manville, High Performance), 3% OV-17 is shown in Figure 1. No arginine peak could be eluted from Chromosorb-G (H.P.) with the same liquid phase.[23] These arginine and histidine derivatives would not elute from either support coated with Carbowax.

2.4.4 Conversion of Arginine to Ornithine and Histidine to Aspartic Acid

For mixtures containing a variety of unknown amino acids it is clearly desirable to have standard conditions for esterification and acylation. So far it has proved impossible to acetylate arginine and histidine under the same conditions as the other amino acids. We have chosen to solve this problem by conversion of arginine to ornithine and

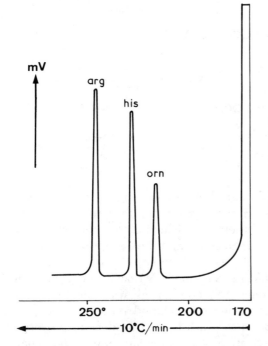

Figure 1. Separation of acetyl *n*-propyl derivatives of histidine and arginine (1.5×10^{-8} moles) by GC. Acetylation carried out after neutralization of the ester hydrochlorides with Na_2CO_3. Columns: glass, 40 cm $\times$ 3 mm i.d., 3% OV-17 on Chromosorb-W (H.P.). Carrier gas N_2, 30 ml per minute; 170–250° C at 10° C per minute.

histidine to aspartic acid. Each reaction is specific, making it applicable to complex mixtures of amino acids and allowing routine production of derivatives under standard conditions.

After activation of arginase with manganese sulfate at pH 7.8, the conversion of arginine to urea and ornithine was followed by urea pro-

Table II

Cleavage of Arginine to Urea (+ Ornithine) by Arginase

µg Arginine · HCl	Amino acids added (µmoles)	µg Urea produced (theoretical: 60 µg = 1 µmole)	% Conversion
211 (1 µmole)	none	58, 60	99%
211 (1 µmole)	7°	62, 59	100%
211 (1 µmole)	14	60	100%
211 (1 µmole)	21	57	95%
211 (1 µmole)	1 lysine	58	97%
211 (1 µmole)	2 lysine	60	100%
211 (1 µmole)	3 lysine	58	97%

In each case 20 µg enzyme was added. Urea was estimated colorimetrically using diacetylmonoxime–thiosemicarbazide reagent by relating to a standard graph.
° 1 µmole each of proline, threonine, serine, aspartic and glutamic acids, methionine and phenylalanine.

duction. Arginase (Worthington Biochemical) was activated by incubating 1 mg of the enzyme with 0.5 ml of 0.1M ammonium acetate pH 7.8 and 0.5 ml of 0.5M $MnSO_4$ at 37°C for 4 hours. The activated enzyme was diluted 1:10 with distilled water (brought to pH 9.5 with ammonia) before use and 0.2 ml of the dilution added to 1 μmole of arginine hydrochloride in 0.5 ml ammonium acetate at the same pH.[23] Experimental results given in Table II show quantitative enzymic cleavage even in the presence of large excesses of other amino acids.

Ozone causes fission of the C double bond of imidazoles;[76] histidine is thereby converted to aspartic acid. Our procedure is described in detail in Section 2.7.4.

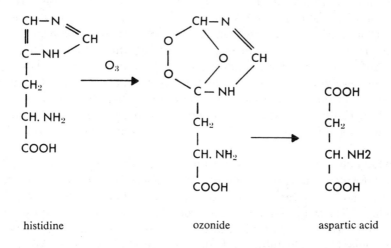

histidine   ozonide   aspartic acid

The quantitative aspects of this reaction were investigated by passing ozone (generated by high-voltage discharge in a stream of pure oxygen) over 10^{-7} mole of histidine in 0.1N HCl for 15 minutes before conversion to the N-acetyl-n-propyl derivative (Section 2.4.8). The aspartic acid ester so formed was dissolved in 100 μl of ethyl acetate and 1 μl (corresponding theoretically to 10^{-9} mole) subjected to gas-chromatographic assay. Percentage conversion of histidine to aspartic acid was calculated by reference to the peak area of an authentic sample of 10^{-9} mole of aspartic acid and was found to be 96%. For quantitation of mixtures containing both aspartic acid and histidine, ozonized and untreated samples are chromatographed. The molar difference in aspartic acid then measures the amount of histidine in the mixture.

2.4.5 Stability of Acylated Amino Acids

The tendency to deacylation of acyl-substituted O-, S-, secondary NH_2-, imidazole, and guanidine functions is much greater than for the primary NH$_2$ group, the TFA-derivatives being much more sensitive in this respect than the corresponding acetyl ones. These properties are of vital importance when choosing a liquid phase and support for GC. Bis-TFA-serine was shown to revert rapidly to the mono (N-) acyl species in aqueous media.[144] If exposed for a sufficient length of time $N \rightarrow O$ acyl shifts were thought to occur, the N-TFA-serine eventually reverting to the native amino acid. Bis-TFA-derivatives of hydroxyproline, serine, threonine, tyrosine, and cysteine were also decomposed to the N-TFA-compounds with methanol or water and even in some carefully dried solvents (benzene, ethylene dichloride and n-amyl trifluoroacetate).[28, 84, 99] Tyrosine was most labile, the β-O-TFA- of threonine and the indole nitrogen of tryptophan being most stable. GC of methanolic solutions of TFA-amino acid methyl esters showed long retention times of serine and tyrosine in accord with breakdown to the free OH-forms. Further evidence of the labile nature of O- and S-TFA-tyrosine and cysteine was afforded by their propensity to acylate isoleucine n-amyl ester. In this respect threonine, serine, and hydroxyproline were inactive.[28] It has also been found expedient to chromatograph bis-TFA-histidine-n-butyl ester by totally decomposing it with butanol to the mono-TFA-derivative on the gas-chromatographic column.[43] See also Section 2.7.3.

Although it was claimed that esterification with propanol–HCl could be performed after acetylation of amino acids with acetic anhydride–acetic acid,[52] this can only be true for those amino acids having the single α-NH$_2$ group. We have found total breakdown of O-acetyl serine, threonine, tyrosine, and ϵ-NH-acetyl lysine during esterification with propanol–HCl. (This also applied to TFA-methyl esters.[25]) We have also shown that unlike the bis-TFA-derivatives, bisacetyl amino acid n-propyl esters (except histidine) are relatively stable in the presence of alcohol (Table III).

Table III

Percentage Recoveries Estimated by GC of Acetyl Amino Acid n-Propyl Esters after Exposure to 20% Methanol in Ethyl Acetate; Zero Time Taken as 100%

Hours treated (20% methanol in ethyl acetate)	Thr	Ser	Tyr	Lys	Asp	Val
½	100%	100%	98%	98%	100%	100%
2	100%	101%	100%	98%	99%	98%
18	103%	98%	100%	97%	101%	97%
100% methanol						
3	102%	98%	96%	97%	99%	97%

Norleucine derivative was included as an internal standard. GC was as in Figure 4.

Thermal stability of both TFA- and acetyl amino acid esters is good (200°C+) under anhydrous conditions[28, 144] although metal flash injectors appear to cause decomposition of O-TFA-derivatives, particularly tyrosine and threonine.[41, 78] The corresponding acetyl compounds are much more stable.[78] Similar differences in stability are observed in the presence of polar liquid phases (see Section 2.4.7).

2.4.6 Choice of Acyl Amino Acid Esters for GC*

The adoption of a particular derivative for gas-chromatographic analysis is determined by a number of interrelated factors—simplicity of preparation, high yield, stability, and the volatility of derivatives and choice of columns and chromatographic conditions suitable for their resolution.

Several groups have compared volatilities of different acyl amino acids and their esters. Melting points[142] and more pertinently the sublimation temperatures[138] and vapor pressures[140] of a number of N-TFA-amino acids have been measured; under a vacuum of 0.02–0.06 Torr several N-TFA-amino acids sublimed between 70 and 80°C (tyrosine, tryptophan and lysine were exceptions) while the corresponding methyl esters sublimed some 40°C lower. Comparison of retention times or temperatures of different derivatives chromatographed under the same conditions is the most meaningful measurement relevant to gas-chromatographic studies. Increasing the chain length of the esterifying alkanol increases the retention time[83] and the N-TFA-amino acid methyl and n-butyl esters are more volatile than the corresponding N-acetyl compounds.[78, 83] (The derivative chosen must be volatile enough to elute within the time and temperature schedule dictated by the liquid phase; too great a volatility may lead to crowding of the early part of the chromatogram and certainly to mechanical losses if a final evaporative step is involved in derivative preparation.)

From quantitative measurements of the evaporative losses of N-TFA-alanine esters the conclusion was reached that only the n-amyl derivative could be handled with any certainty[27] although Gehrke selected the n-butyl esters.[78] Only by evaporating the excess reagents at low temperature have losses of the N-TFA-amino acid methyl esters been avoided.[30] Our preliminary observations showed the methyl, ethyl and n-propyl esters of acetylated amino acids to be suited to gas-chromatographic separation between 100 and 240°C. Esterification is rapid and in a simple two-step procedure for making the acetyl-n-propyl esters[23] no evaporative losses occurred (Table IV). For these reasons and also because of the ready separation they undergo on Carbowax-coated

---

*See also Sections 2.7.2 and 2.7.5.

Chromosorbs-G or -W (High Performance grades), we have adopted the acetyl-*n*-propyl esters (see Section 2.4.7).

Table IV

Percentage Recoveries Estimated by GC of Acetyl Amino Acid *n*-Propyl Esters after Controlled Evaporative Procedures

Evaporation minutes	°C	Ala	Val	Gly	Ser	Glu
5	30	96	99	94	98	100
20	30	73	81	80	97	99
5	90	32	56	44	100	100
20	90	<10	<10	<10	53	96

Norleucine derivative as internal standard was added to each tube after treatment. Evaporative gas flow: dry nitrogen, 1200 ml per minute passing through apparatus routinely used for derivative preparation (Figure 8, Section 2.4.8).

Acetylated *O*- and *S*- groups are comparatively stable. The acylating ability of acetic anhydride is insufficient to acetylate arginine ester dihydrochloride. Free base production is first necessary, either by desalting with anion-exchange resin or by neutralization (Section 2.4.3.5). The higher degree of volatility conferred by the TFA-group demands the use of esters of higher alkanols if evaporation losses are to be easily avoided. A protracted esterification time is required using butanol or pentanol, but on the credit side, complete trifluoroacetylation of the guanidine group can be achieved with trifluoroacetic anhydride at 150°C. TFA-Derivatives are less stable than their acetyl opposites and problems arising through on-column destruction of *O*- and *S*-TFA-derivatives are discussed in Section 2.4.7.

2.4.7 Supports, Liquid Phases and Techniques for Separations of Acyl Amino Acid Esters by GC*

2.4.7.1 Supports

The most widely used solid supports for packed columns have been the specially prepared and graded diatomaceous earths. The surfaces of these materials have been modified in many ways but two treatments in particular are generally believed to improve performance: washing in strong mineral acids (a.w. grades) to remove metallic ions and 'deactivation' of the surface with silanizing agents. The active sites on the surface of siliceous materials are of two types: (a) the proton-donating silanol (-Si-OH) group and (b) the proton acceptor siloxane

---

*See also Section 2.7.1.

group (-Si-O-Si-). The activity of the former can be blocked by reaction with HMDS or DMCS.

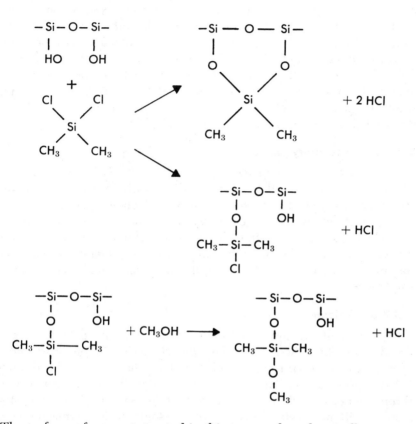

The surfaces of supports treated in this way are less chemically reactive than untreated supports. The ability of various supports to resolve the acetyl amino acid n-propyl esters without tailing could be used to place them in a sequence which roughly parallels the degree of surface deactivation. Using 0.5–1.0% Carbowax 6M we found we could arrange the following in increasing order of efficiency: Chromosorb-W (a.w.) < W (a.w., DMCS) < Gas Chrom Q (Applied Science) < Chromosorb-W or G (a.w. DMCS, High Performance grades).

Gehrke et al. showed that heat treatment of a.w. Chromosorb-G (unsilanized) between 450 and 600°C prior to coating gave improved resolution of TFA-amino acid-n-butyl derivatives. Without such treatment the peak for tyrosine was absent and that for serine much reduced. The improved performance was ascribed to removal of traces of residual water but some alteration of the surface activity could also be important.[43] Erratic elution of derivatives (principally of the basic amino acids) has often been blamed on the liquid phase alone,[40] but the support should not be exonerated altogether.[93] For example, we have been unable to elute arginine (triacetyl-*n*-propyl derivative) from Chromosorb-G (H.P.) coated with 1 or 3% OV-17 up to 300°C; yet the same derivative elutes from Chromosorb-W (H.P.) with 3% OV-17 at 250°C, the only difference being the support. We believe that interaction among the liquid phase, the support, and the materials undergoing separation may often play an important part in resolution and this we discuss below after consideration of liquid phases.

2.4.7.2 Stationary phases

The variety of liquid phases now available for GC is bewildering. Many have been applied to amino acid derivative separations. About 100 stationary phases including silicones, surfactants, polyesters, polyglycols, metallo-organics, and sundry others have been evaluated in several papers.[13, 14, 27, 29, 40] In endeavoring to sift out the salient features we have discarded details pertaining to column dimensions, temperature programs, carrier gas flows, and the particular liquid-phase loading employed.

2.4.7.3 Silicones

As a general principle resolution improves as the polarity of the liquid phase increases. As a class the silicones are the least polar phases. It is not surprising to find, therefore, that silicones have often failed to give adequate separations of the aliphatic amino acids where chromatograms tend to be crowded. Leucine and isoleucine, for example, often coincide.[27] To improve resolution, isothermal or very slowly programmed temperatures have been employed. This makes for slower analyses. SE-30 is one of the least polar, XE-60 one of the most polar liquid phases in this category. TFA-Amino acid methyl esters have been separated on a mixed stationary phase (2.5% on deactivated Diatoport S) of XE-60, QF-1 and MS-200 in proportions 46%, 27%, and 27% by weight.[30] Combinations of temperature programming and isothermal operation separated 22 amino acids, including arginine, and resolution was excellent. Separation of 12 acetyl-amino acid-*n*-propyl esters on 1% XE-60 is shown in Figure 2.

Catalytic breakdown of labile groups (especially *O*-TFA-serine, tyrosine and hydroxyproline) is not a problem on any of the silicones.[13, 14, 27, 29] Silicone phases appear to be obligatory for quantitative GC of arginine, histidine and cystine. This applies to the acetyl- as well as the TFA- and TMS-derivatives.

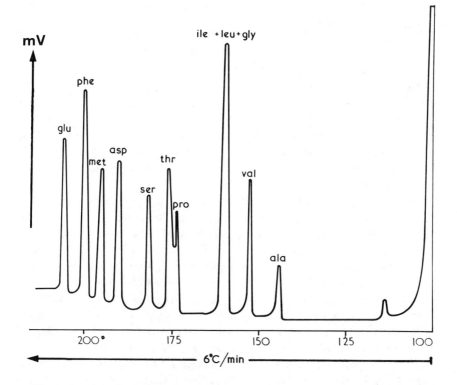

Figure 2. GC of 12 *N*-acetyl-amino acid *n*-propyl esters on XE-60. Columns: glass, 106 cm × 3 mm i.d., 1% XE-60 on Chromosorb-G (H.P.); carrier N_2, 30 ml per minute; 100–220° C at 6° C per minute.

2.4.7.4 Polyesters

These liquid phases, intermediate between the silicones and polyglycols in polarity, have enjoyed considerable popularity and moderate success. Coatings employed are usually between 0.75 and 2% w/w, but sometimes higher. Of nine polyesters tested[27] BDS was most polar and NPGS* least. Choice among the polyesters is dictated not only by resolution but also by the ease with which these phases catalytically decompose secondarily acylated amino acids. Sensitivity of these to on-column destruction roughly parallels their stability to water or methanol, with serine, cysteine, and histidine being most sensitive, threonine more resistant. *N-O*-bis-TFA-Serine amyl ester failed to elute from BDS, NPGSb, NPGG, PPSb and Hi-Eff-8B; the corresponding

---

*See abbreviations, Section 2.6.

threonine derivative did not elute from the last two.[27] TFA-Cysteine and hydroxyproline were also destroyed on a number of polyesters (Hi-Eff-8B, DEGA, DEGS, PEGS, NPGG and PPSb).[13] The column support in all the above-mentioned studies was Silocel C-22 (L. Light & Co.) and liquid coatings of 5%.

At variance with these findings are those of Gehrke who has obtained chromatographic peaks for bis-TFA-serine, threonine, cysteine, and hydroxyproline *n*-butyl esters on NPGSb and DEGS.[41] Since it is improbable that the different ester groups are related to the anomalies, reconciliation can best be made by noting the different supports (Chromosorbs-G or -W) and lower liquid-phase loadings (0.5% NPGSb; 0.75% DEGS–0.25% EGSS-X).

When polyesters of the NPG series were examined, maximum column efficiency for separation of 17 amino acids was obtained when the carbon number of the dicarboxylic acid was 10, *i.e.*, NPGSb.[43] However, EGA seems to have increased thermal stability and is perhaps the most useful polyester for separation of TFA-*n*-butyl amino acid esters.[43, 118] Peak shifts of these derivatives were observed as the EGA coating on Chromosorb-W was increased from 0.5–2.0%.[118] Cysteine–methionine separation improved, while glycine–isoleucine, proline–threonine, and phenylalanine–aspartic acid merged as the level of liquid phase increased. An optimum coating of 0.65% was adopted.

The only polyester phase we have examined for GC for acetyl amino acid methyl, ethyl and *n*-propyl esters was DEGS, which failed to give adequate separations (Figure 3).

2.4.7.5   Polyglycols

There are isolated reports of GC of TFA-amino acids on Carbowaxes (1% 20M and 1540 on Diatoport-S[56] capillary columns).[57] Peaks for the recalcitrant amino acids, including arginine and histidine, and linearity of dose response were claimed.[56] The long retention times of serine and hydroxyproline suggest deacylation; diminutive serine and threonine peaks eluted from capillary columns also suggest deacylation.[57] Cysteine, hydroxyproline,[13] serine, and threonine[27] TFA-*n*-amyl esters were not recovered from Silocel C-22 coated with 5% PEG.

Our own experiments indicate on-column destruction of TFA-*n*-propyl esters of threonine and serine using 0.7% PEG-6M on Chromosorb-W (H.P.). The greater stability of acetyl amino acid esters permits gas-chromatographic separation of all common amino acids expected in a protein hydrolysate on polyglycols, with the exception of arginine and histidine.[23, 52, 78, 113] Resolution of acetyl amino acid methyl, ethyl and *n*-propyl esters on 0.7% PEG-6M and 0.05% TCEPE together is shown in Figures 4 and 5.

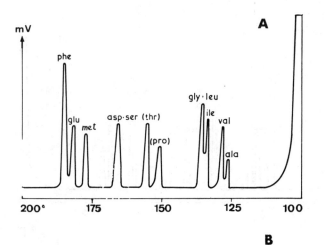

Figure 3. GC of (A) methyl, (B) ethyl, and (C) *n*-propyl esters of *N*-acetyl amino acids on a polyester phase. Columns: glass, 106 cm × 3 mm i.d., 0.7% DEGS on Chromosorb-W (H.P.); carrier N_2, 30 ml per minute; 100–220° C at 6° C per minute; qualitative run only. Unequivocal assignment of proline and threonine peaks was not made.

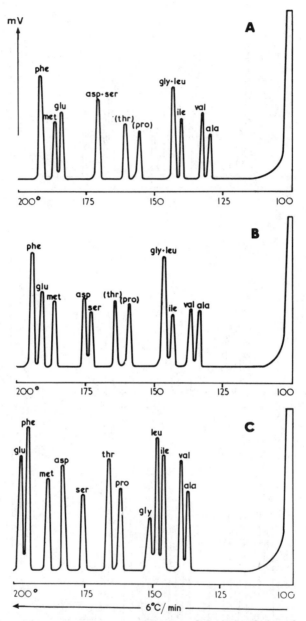

Figure 4. Separation of (A) methyl, (B) ethyl, and (C) *n*-propyl esters of *N*-acetyl amino acids on a poly-glycol liquid phase. Instrument: Perkin-Elmer Model 881. Columns: dual glass, standard analytical columns 106 cm × 3 mm i.d., 0.7% PEG 6M and 0.05% TCEPE on 100/120 mesh Chromosorb-W (H.P.); carrier N_2, 30 ml/mm; flash injector 250° C; 100–240° C at 6° per minute; (A) and (B) qualitative only, (C) 1×10^{-9} moles of each amino acid.

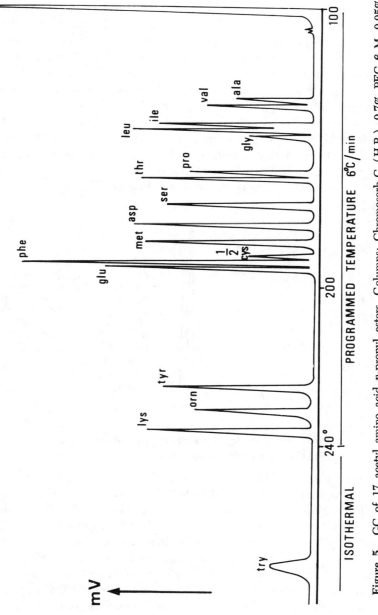

Figure 5. GC of 17 acetyl amino acid *n*-propyl esters. Columns: Chromosorb-G (H.P.), 0.7% PEG 6 M, 0.05% TCEPE; GC as in Figure 4. Chromosorb-W (H.P.) may be used with analogous results. [Reprinted, by permission, from J. R. Coulter and C. S. Hann, Journal of Chromatography 36, 42 (1968). © Elsevier Publishing Co.]

2.4.7.6 Mixed stationary phases

When single liquid phases are unable to achieve the necessary resolution of a mixture, it is often possible to find an adequate compromise
by mixing two liquid phases together. The same resolution is obtained if
supports, each coated with a single liquid phase, are blended together in
the appropriate proportions or packed sequentially into columns.[23, 29, 41]
In each case a mean resolution between the two phases ensues. We
have occasionally observed relative peak shifts within the N-acetyl-n-
propyl isoleucine–leucine–glycine trio on Chromosorb-G or -W (H.P.)
with identical liquid-phase loadings. The two extremes (illustrated in
Figure 6) show resolution of isoleucine–leucine (a) and leucine–glycine
(b). By mixing such batches a compromise is reached with adequate
separation of all three peaks. See also Section 2.7.2.

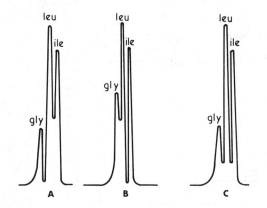

Figure 6. Relative shifts of
peaks observed in batches of
packing coated under like conditions using standard columns
as in Figure 4. (A) resolution
of gly/leu; (B) resolution of
ile/leu. By blending packings,
acceptable resolution of all three
results (C).

2.4.7.7 Interaction among support, liquid phase, and derivative, and effect on resolution

Although partition between the gas and liquid phases in gas-liquid
chromatography is often thought to depend on volatility and solubility
in the liquid phase, it was recognized at the outset (1952)[71] that hydrogen bonding between solvent and solute played an important part. In
comparing primary, secondary and tertiary amines on paraffins and
"Lubrol-MO" (a polyethylene oxide–long-chain alcohol condensate),
James and Martin found a correlation between retention volume and
ability to form hydrogen bonds. The longer retention of trimethylamine
on Lubrol was explained by assuming that the methyl groups would be
sufficiently 'active' to form hydrogen bonds. This effect on retention
was not seen with higher homologs (cf., glycine). We believe that
similar interactions occur between liquid phase and support, and support and the materials undergoing partition. Throughout this paper,

therefore, we have chosen to use the general term gas chromatography (GC) rather than gas–liquid chromatography (GLC). The use of the latter term implies that gas-solid chromatography depends upon entirely different mechanisms. That the surface of the support is important in GLC is evident from a number of observations.

Increasing the liquid loading often alters the order of elution[118] whereas if interaction with the liquid phase only were involved, one would expect all solutes to be retained proportionately longer. We have found that triacetyl-*n*-propyl arginine will elute from Chromosorb-W (H.P.) but not from -G (H.P.) when both are coated with the same liquid phase. For other basic amino acids the elution patterns were identical. It should also be noted that when polar phases have been used, the best separations have been achieved with very low loadings (0.7% PEG,[23] 0.65% EGA,[118] 0.325% EGA[43]). Use of 0.7% PEG-6M on Chromosorb-W (H.P.) 100-120 mesh implies a layer of monomolecular proportions.* It is surprising that such a polar phase will distribute itself on a silanized surface. However, it has been shown generally that unsilanized supports do not give such crisp resolution. We have also found that the addition of a small amount (0.05%) of the highly polar TCEPE improves resolution. Furthermore, although TCEPE should vaporize rapidly and bleed from the column above 180°C, columns with this small amount can be taken to 240°C repeatedly without excessive bleed. It is assumed either that the TCEPE is bonded in some way to the surface or that its presence during coating with PEG alters the way in which the PEG distributes itself on the surface. The latter explanation seems likely for the polarity of the depositing solvent is also important.[23] Depositing the liquid phases from 20% methanol in chloroform produces a better coating than from chloroform alone. For these reasons we believe that interactions between the liquid phase and the support are real and that they determine the distribution of liquid phase on the support and ultimately some of the forces involved in resolution.

Mention has already been made of the hydrogen bonding which occurs between optical isomers and optically active stationary phases and between diastereoisomers and polar nonoptically active liquid phases. Further evidence of the role of hydrogen bonding in resolution is provided by two examples. (a) Glycine, although the smallest and most volatile of the amino acids,[140] generally elutes after alanine and valine (and often after leucine and isoleucine as well) on a polar liquid phase and the delay in the elution of glycine is a measure of this polarity.[27] On nonpolar liquid phases glycine usually elutes earlier. (b) Serine and threonine elute in the reverse order of their molecular

---

*See also Section 2.7.1.

weights, with the difference in retention times and the position of elution relative to leucine acting as a function of liquid-phase polarity.[27] It follows from this concept of liquid-phase function that one could not expect isoleucine and leucine to separate well on an inert phase and that, since polar phases must be used, the effect of this polarity on resolution is bound to be complex until all the possible bond sites can be "explained."

2.4.8   *Quantitation of Acetyl Amino Acid n-Propyl Esters*

Esterification may be followed by thin-layer chromatography while quenching of reactivity towards ninhydrin indicates acylation of the α-amino group. Although useful in the preliminary phase of an investigation, neither technique is satisfactory for determining percentage conversions of small amounts of amino acid. Comparison of GC peak areas of amino acids carried through all the manipulations of a microsynthetic procedure with those of highly purified reference standards permits accurate measurement of the derivation procedure.[41, 84] It is much harder to design experiments to prove on-column stability and that the peak areas recorded do, therefore, represent a known amount of amino acid. Most workers have been content to assume that a well-formed peak measures all the amino acid applied to the column. A partial answer to this problem has been suggested by Blau[11] who proposes that peak areas from very nonpolar phases (SE-30) should be compared with areas from the chosen liquid phase. It is necessary to include an internal standard to compensate for handling losses or injection variability.

The molar response of flame-ionization detectors varies with each amino acid but linearity of the response in normal working ranges allows quantitation of unknowns by direct proportion. Computation of molar ratios (for example, in a peptide hydrolysate) does not require the use of an internal standard. One must be included in both the reference standard and the unknown in the same concentration if absolute amounts of each amino acid are to be calculated. (See Discussion in Reference 41.)

2.4.8.1   Synthesis of reference standards of
acetyl amino acid *n*-propyl esters

We have synthesized in bulk the acetyl-*n*-propyl derivatives of all common amino acids except histidine. The amino acids (2–10 g) were refluxed with *n*-propanol (60 ml containing 2.5M HCl) and dry benzene (20 ml) in a Dean-Stark apparatus, removing water of reaction azeotropically. The propyl ester hydrochlorides were dried in a rotary evaporator. Acetylation was carried out with a three- to fivefold molar excess of acetic anhydride in pyridine (1:2) for 1 hour after which the excess reagent was again removed with a rotary evaporator. The derivative was extracted into ethyl acetate (100–300 ml) and insoluble pyridine hydrochloride and pyridine acetate filtered off. The filtrate was washed briefly with 0.1M aqueous HCl to remove remnants of pyridine and its salts. (This step was omitted for those amino acids having acetylated secondary groups; these were repeatedly extracted into ethyl acetate and filtered and dried.)

After evaporation of ethyl acetate the syrups were distilled, where appropriate, under vacuum in a specially designed cold-finger apparatus (Figure 7). Threonine, ornithine, lysine, and histidine were

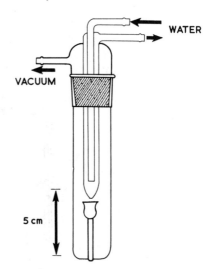

WATER

VACUUM

Figure 7. Cold finger condenser for purification of acetyl amino acid *n*-propyl esters. The impure derivatives in the bottom were distilled, condensed, and collected in the thistle cup.

5 cm

crystallized from petroleum ether (bp 100–120°C)–ethyl acetate. Purities were checked by IR, elemental analysis, or NMR. The histidine derivative was recovered only in low yield (<10%) and an integrated NMR spectrum indicated a mixture of both mono- and diacetylated forms. Stock solutions in ethyl acetate containing 1, 2, 5, 10, and 20×10^{-9} moles per μl were prepared for use as reference standards. Acetyl mercaptoethanol (which elutes conveniently between threonine and serine), 1.5 μg/μl, was included in each solution as an internal standard.

2.4.8.2 Microsynthetic procedure

Small samples were synthesized in an apparatus designed to obviate manual transfers[23] (Figure 8). A known amount of amino acid solution (0.5–2 μmoles) was placed in the tube and dried at 100°C in a stream of dry nitrogen. After drying, the gas flow was stopped and *n*-propanol

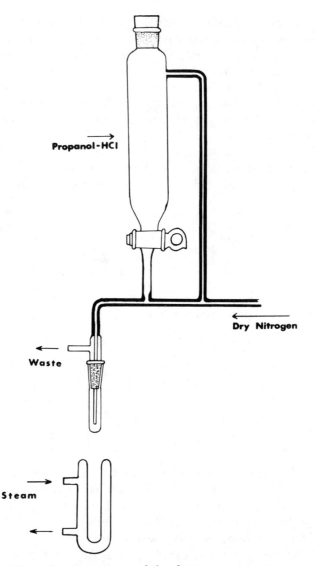

Figure 8. Apparatus used for derivative preparation on microscale. Amino acid solution is placed in the tube and propanol–HCl in the funnel. Dry nitrogen is passed through the apparatus as required, and the tube is heated by raising the steam jacket. [Reprinted, by permission, from J. R. Coulter and C. S. Hann, Journal of Chromatography 36, 42 (1968). © Elsevier Publishing Co.]

containing 8M HCl was run in (0.2–0.4 ml). The tube was incubated at 100°C for 10 minutes by raising the closed steam jacket. After evaporation with dry nitrogen (350 ml/minute, passed through Union Carbide Molecular Sieve type 4A to lower water content to 2 ppm or less) the propylation step was repeated. Acetylation was achieved in pyridine–acetic anhydride for 5 minutes at room temperature, whereupon the excess reagents were carefully evaporated with nitrogen. The tube was warmed during the evaporation, but as soon as the last of the acylation mixture disappeared the nitrogen stream was shut off. The derivatives were dissolved in ethyl acetate which contained internal standard (acetyl mercaptoethanol, 1.5 $\mu g/\mu l$), and 1-μl samples were injected.

Percentage yields were calculated from peak areas obtained in the microprocedure compared with those of the reference standard after equating internal-standard peaks. Columns and conditions used for these studies were as in Figure 4. Peak areas were measured using an Infotronics Digital Integrator model CRS-100 or by planimetry. Yields of all amino acids that could be chromatographed on these columns were between 96.0% (lysine) and 102%. As already described arginine and histidine were quantitatively converted to ornithine and aspartic acid, respectively, while cystine was reduced and carboxymethylated.

2.4.9 GC Analyses of Protein and Peptide Hydrolysates

There is a wide gap between the number of papers describing gas-chromatographic methods and those describing applications. Qualitative analyses or those restricted to the measurement of only a few amino acids of peptide or protein hydrolysates were made quite early.[25, 69, 88, 147] The most complete data, however, are those of Gehrke whose gas-chromatographic estimations of amino acids of the hydroly-

Table V

Gas-Chromatographic Analysis (in Moles $\times$ 10^{-9}) of Amino Acids in Hydrolysates of Three Dipeptides

Leucyl-serine		*Glycyl-tyrosine*		*Glycyl-lysine*	
leucine	3.6	glycine	5.0	glycine	7.0
serine	3.4	tyrosine	4.7	lysine	6.7

sates of bovine serum albumin, kappa casein, soybean protein,[41] and ribonuclease[43] are in excellent agreement with ion-exchange results. The accompanying Tables V, VI and VII show some experimental results obtained by GC of the acetyl-amino acid *n*-propyl esters of several peptide hydrolysates. The same molar ratios of fibrinopeptide hydrolysates (Table VI) were obtained by ion-exchange.[62] Bradykinin (Table VII) was produced synthetically and its synthesis monitored by GC.

We have also measured the free amino acids in serum after deproteinization and cleanup on Dowex-50, a procedure which would seem to be essential for all samples of biological origin likely to be contami-

Table VI

Gas-Chromatographic Analysis (in Moles $\times$ 10^{-9}) of Peptides Released
from *Anser* and *Cairina* Fibrinogens during Coagulation
(Figures in Parentheses Are Molar Ratios)

	Anser Peptide A-1	*Cairina* Peptide C-1
Ala	3.09(0.9)	3.25(1.0)
Arg*	6.43(1.9)	2.80(0.9)
Asp (+Asn)	14.66(4.4)	15.50(4.9)
Glu (+Gln)	13.12(4.0)	15.13(4.8)
Gly	3.75(1.1)	0
Phe	3.16(1.0)	0
Pro	7.17(2.2)	6.26(2.0)
Ser	6.77(2.0)	5.72(1.8)
Thr	3.43(1.0)	3.90(1.2)
Tyr	1.84(0.6)	3.50(1.1)
Val	0	2.85(0.9)

*Determined as ornithine after enzymic conversion. Data taken in part from Hann.[62]

nated with a large excess of protein. Peptides eluted from paper may
also be concentrated by adsorption to small columns of Dowex-50 (H^+
form) and quantitatively eluted with triethylamine, water, acetone
(24:24:7) before hydrolysis. In experiments in which amino acids were
added to exhaustively dialyzed serum, recoveries were between 90 and
102%. Analyses could be performed on 0.125 ml of serum using the
techniques outlined above for making the acetyl-*n*-propyl esters.

2.5 Conclusion

At present, GC has established a clear lead as the most accurate and
sensitive method for quantitative analysis of racemic mixtures of amino
acids. The intrinsic sensitivity of GC has also been demonstrated in its
value in analysis of iodine amino acids in blood. The method has not
been widely adopted, and there seems no good reason for this. Selenium
amino acids are in this same category although the acetyl-*n*-propyl derivatives are easily prepared and chromatographed.

Table VII

Amino Acid Content (in Moles $\times$ 10^{-9}) of a Hydrolysate of Synthetic
Bradykinin (Molar Ratios Are Shown in Parentheses; GC as in Figure 4)

	Found	*Theoretical*
Gly	2.48(1.1)	(1)
Pro	7.19(3.2)	(3)
Ser	2.06(0.9)	(1)
Phe	4.34(1.9)	(2)
Arg	not determined	(1)

Analysis of peptide and protein hydrolysates is the most commonly encountered problem in amino acid assay. The TFA-methyl,[30] butyl,[41, 43, 117] and amyl[13, 14, 26, 27, 28, 29] esters and the acetyl-*n*-propyl esters[23, 52] have all been successfully applied to the full range of amino acids. Problems with arginine and histidine (lability of acyl guanidine and imidazole) have been overcome by using less polar liquid phases both in the column used to resolve the other amino acids[30] and in a separate column used only for arginine and histidine.[43] Alternatively, arginine and histidine have been converted to ornithine and aspartic acids.[23] The former procedures require longer one-column analysis time while the latter requires longer derivatization.

In terms of speed, sensitivity and cost, analysis by any of these gas-chromatographic methods compares favorably with the existing automated resin analyzers. In terms of convenience and ease of operation, the automatic resin analyzer has clear advantages. Wider use of GC will depend largely on the development of more automated methods. It is not difficult to envisage a machine into which a number of peptide or protein samples are placed to be treated in turn by a number of reagents and then flushed one at a time onto a chromatographic column. Upon elution the peaks would be recognized and integrated and printed out as so many nanomoles of amino acid X. Saroff (personal communication) is already moving in this direction with the use of gas-phase derivatization. Reagents to make the TFA-methyl esters are passed in the carrier gas over the sample which is then evaporated in a hot gas stream and delivered onto the column. It should be pointed out that methylation with methanolic HCl may also effect alcoholysis of peptides and proteins, so avoiding the long and tedious hydrolysis step.[11] Saroff has also reduced the elution time of less volatile amino acids and abolished temperature programming by using a column divided into three sections, each maintained isothermally at decreasing temperatures.[84] Valves between each section shunt amino acids either onto the succeeding section or directly to a detector. Such a procedure should simplify automation, speed up separation, and greatly increase column life.

Until such machines are commercially available, those who have gained experience as gas chromatographers or who require the increased sensitivity, speed or lower cost which the method offers will continue to analyze amino acids by GC. Others will wait a little longer.

ACKNOWLEDGMENTS

The authors wish to thank the National Health and Medical Research Council of Australia for the purchase of a gas chromatograph, chart recorder, and digital integrator. In addition, gratitude is extended to Miss J. Mitton, who by untiring secretarial and technical assistance made the preparation of this chapter much easier, to Drs. E. Hackett, S. Mander, and L. Mander for helpful discussion and criticism, and to Mrs. J. Guiney for typing the manuscript.

2.6   Abbreviations Used in Text

2.6.1   *General*

GC	gas chromatography; used in preference to GLC for the reasons discussed in Section 2.4.7.7
GLC	gas-liquid chromatography
i.d.	inner diameter
ir	infrared
NMR	nuclear magnetic resonance

2.6.2   *Reagents and Groups*

BSA	bis(trimethylsilyl) acetamide
BSTFA	bis(trimethylsilyl) trifluoroacetamide
DEA	diethylamine
DNP-	dinitrophenyl-
HMDS	hexamethyldisilazane
PTH	phenyl thiohydantoin
TEA	triethylamine
TFA-	trifluoroacetyl-
TMCS	trimethylchlorosilane
TMS	trimethylsilyl-
TMS-DEA	trimethylsilyl-diethylamine

2.6.3   *Gas-Chromatographic Terms and Liquid Phases*

a.w.	acid-washed
BDS	butane-1,4-diol succinate
DEGA	diethyleneglycol adipate
DEGS	diethyleneglycol succinate
EGA	ethyleneglycol adipate
EGSS-X	ethyleneglycol succinate silicone polymer
Hi-Eff-8B	cyclohexanedimethanol succinate (Applied Science)
H.P.	high-performance (relating to deactivated supports)
MS-200	methyl siloxane (Hopkins and Williams)
NPGG	neopentylglycol glutarate
NPGS	neopentylglycol succinate
NPGSb	neopentylglycol sebacate
OV-17	methyl phenyl silicone
PEG	polyethyleneglycol (Carbowax) usually followed by molecular weight

PEGS	polyethyleneglycol succinate
PPSb	polypropylene sebacate
QF-1	trifluoropropyl methyl silicone gum
SE-30	methyl silicone
TCEPE	tetracyanoethylpentaerythritol
Ucon LB550	polyalkyleneglycol (Union Carbide)
XE-60	cyanoethyl methyl silicone gum

2.7 Addendum

2.7.1 *Supports and Liquid Phases*

Since the above chapter was written, several papers have discussed the interaction of support, liquid phase, and derivative. In the absence of a detailed understanding of these interactions, a standardized coating procedure giving absolutely reproducible chromatograms is still not possible.

Because longer conditioning of columns improved resolution,[159] these workers believed that loss of water from the support was important. Accordingly, experiments in which the support (a.w. Chromosorb W) was heated under various time-temperature conditions were conducted. As adsorbed water would be lost during normal conditioning and coated silanized supports show similar improvements in resolution if conditioned slowly, it seems unlikely that this is the whole explanation. Silanized supports should not adsorb water. Heating the support probably causes modification of the surface affecting the distribution of liquid phase. The method used in drying the support during coating also affects resolution. It has been shown that fluidized packings show improvement over cake-dried packings if the liquid phase is viscous.[157] In the former technique a stream of preheated dry nitrogen is made to pass upward through the damp packing so that the coated support appears to be gently boiling. However, even the most careful standardization of coating technique still produces batch variability.[156] Different batches of the same support may possess slightly different properties. It is not surprising therefore that unexplained departures from customary resolution should continue to irritate the expert and discourage the beginner.

2.7.2 *Acylated Amino Acid Esters*

Our observation[23] that isoleucine is difficult to esterify if the concentration of HCl in alkanol (in this case butanol[161]) is too low has been verified. Also confirmed is the destruction of tryptophan during esterification and the consequent need for standardizing HCl concentration, time and temperature if tryptophan is to be quantitated. A concentration of 3M was found to be satisfactory for making the *n*-butyl esters and avoided the necessity for transesterification.

Islam and Darbre, continuing their meticulous investigations, have perfected the separation of N-TFA-amino acid methyl esters (except histidine) on a mixed silicone stationary phase of XE-60–QF-1–MS200, 100cS (46:27:27) on Diatoport-S using combinations of temperature programming and isothermal operation.[156] Highly purified radioactive N-TFA-amino acid-*n*-butyl esters have been prepared by chromatography on silica gel.[163] When these radioactive derivatives were injected onto a column packed with XE-60–QF-1–MS200, 100cS (46:27:27) coated onto Celite 560, not all the radioactivity was associated with the eluted amino acid peak.[151] A fair per cent of the activity appeared spread throughout the remainder of the chromatogram, principally after the corresponding peak. The TFA-methyl ester of phenylalanine was an exception; 13.1% of the activity appeared before the elution of the derivative and this activity was not associated with a second peak. The phenylalanine methyl ester is more volatile than the corresponding TFA-derivative, making these observations consistent with slow, continuous degradation on the column. It should be stressed that this observation applies only to this particular liquid phase and support. It is unfortunate that comparable studies with other derivatives, supports and liquid phases are lacking.

2.7.3   Histidine

Bis-TFA-histidine-*n*-butyl ester has been protected from partial deacylation by injection in an excess of trifluoroacetic anhydride. On OV-liquid phases, however, this derivative eluted with aspartic acid.[160] These workers claim that the bis-TFA-derivative can be readily converted to the mono-TFA-derivative by evaporation of the excess trifluoroacetic anhydride in a stream of nitrogen. After injection, this less volatile product is retained in the upper part of the column until it is reconverted to bis-TFA-histidine ester by injecting trifluoroacetic anhydride. By careful timing of this injection the resynthesized volatile ester can be eluted in a vacant part of the chromatogram. It is surprising, considering the method of preparation of the monoacyl species, that the diacyl derivative is sufficiently stable in the stream of hot carrier gas.

Histidine (along with arginine, tryptophan and cystine) has been determined as the TMS-derivative on columns of Chromosorb G (H.P.) coated with a mixed phase of 3% OV-17 and 1.5% OV-22.[160]

2.7.4   Ozonolysis of Histidine

Several workers in private communications have reported difficulties with the quantitative conversion of histidine to aspartic acid by ozonolysis. This reaction is pH-, temperature- and concentration-dependent. To ensure 100% conversion at all times the following modification of the published method[23] is used.

The sample, say 0.1 ml in 0.1N HCl, is placed in the tube in which it will be derivatized. This tube is connected to an apparatus similar to that of Figure 8 but without the funnel. The gas stream from the ozonizer is connected (oxygen flow 40 ml per minute, 0.8–1.0% O_3) and the bottom 1 cm of the tube immersed in liquid nitrogen. The blue ozone is seen to condense together with some oxygen. After 5 minutes the gas flow is shut off and the tube raised just above the surface of the nitrogen so that a fractional distillation occurs. As the last ozone evaporates the inlet and outlet are sealed. The apparatus stands at room temperature for 30 minutes and then is heated, still sealed, at 100°C for 5 minutes. The sample is then dried in a stream of nitrogen and derivatized in the usual way.

2.7.5 Combined GC-Mass Spectroscopy

Mass spectrographic data on the N-TFA-butyl esters have been reported.[154] Unique fragmentation pathways of individual amino acids allowed positive identification of poorly resolved peaks. This work showed the usefulness of the mass spectrograph as a very sensitive, albeit expensive, identifier-detector.

2.7.6 TMS Amino Acids

Gehrke *et al.*[153] have recently synthesized and resolved suitable TMS-derivatives of all 20 common amino acids. BSTFA was used in equal proportion with acetonitrile in a 30-molar excess. Fourteen amino acids were derivatized at 135°C for 15 minutes and analyzed on a column of 3% OV-7 and 1.5% OV-22 coated on Chromosorb G (H.P.) Glutamic acid, arginine, lysine, histidine, tryptophan and cystine were resolved on the same column after silylation at 135°C for 4 hours. Each chromatographic run lasted nearly 70 minutes. Glycine with three TMS groups was poorly resolved from isoleucine and proline making impossible its quantitative analysis in mixtures in which it is present in excess (*e.g.*, urine). Although silylation is a simple one-step procedure, the generality of this reaction is still a source of complication. Urine required cation and anion-exchange clean-up before derivatization.

2.7.7 Iodoamino Acids

The TMS-derivatives of iodoamino acids have been successfully resolved on short columns of OV-17 with a linear detector response curve in the range 10^{-6}–10^{-9} moles.[152] The extraction of iodoamino acids from serum has been considerably improved using Sephadex LH-20. However, fully to exploit the sensitivity of the electron capture detector and so analyze very small amounts of serum (<1 ml) it was necessary to extract the derivatives from the mixture in which they were synthesized. A second LH-20 chromatography step achieved this separation but recovery was only about 50%.[162]

A recent review of gas chromatography of amino acids, its historical development and applications, has appeared in Italian.[158]

REFERENCES

1. Alexander, N. M., and R. Scheig. Anal. Biochem. **22**, 187 (1968).
2. Aue, W. A., C. W. Gehrke, R. C. Tindle, D. L. Stalling, and C. D. Ruyle. J. Gas Chromatog. **5**, 381 (1967).
3. Backer, E. T., and V. J. Pileggi. J. Chromatog. **36**, 351 (1968).
4. Baraud, J. Bull. Soc. Chim. France **1**, 785 (1960).
5. Bayer, E., K. H. Reuther, and F. Born. Angew. Chem. **69**, 640 (1957).
6. Bier, M., and P. Teitelbaum. Federation Proc. **17**, 191 (1958).
7. Bier, M., and P. Teitelbaum. Ann. N.Y. Acad. Sci. **72**, 641 (1959).
8. Birkofer, L., and M. Donike. J. Chromatog. **26**, 270 (1967).
9. Birkofer, L., and A. Ritter. Chem. Ber. **93**, 424 (1960).
10. Birkofer, L., and A. Ritter. Angew. Chem. Intern. Ed. Engl. **4**, 417 (1965).
11. Blau, K., in *Biomedical Applications of Gas Chromatography*. Ed. by H. A. Szymanski (New York: Plenum Publishing Corp., 1968).
12. Blau, K., and A. Darbre. Biochem. J. **88**, 8P (1963).
13. Blau, K., and A. Darbre. J. Chromatog. **17**, 445 (1965).
14. Blau, K., and A. Darbre. J. Chromatog. **26**, 35 (1967).
15. Bourne, E. J., C. E. M. Tatlow, and J. C. Tatlow. J. Chem. Soc. **1950**, 1367.
16. Bowman, R. E. J. Chem. Soc. **1950**, 1349.
17. Burlingame, T. G., and W. H. Pirkle. J. Am. Chem. Soc. **88**, 4294 (1966).
18. Caldwell, K. A., and A. L. Tappel. J. Chromatog. **32**, 635 (1968).
19. Charles, R., G. Fischer, and E. Gil-Av. Israel J. Chem. Proc. **1**, 234 (1963).
20. Charles-Sigler, R., and E. Gil-Av. Tetrahedron Letters **35**, 4231 (1966).
21. Cherbuliez, E., and P. Plattner. Helv. Chim. Acta **12**, 317 (1929).
22. Cherbuliez, E., P. Plattner, and S. Ariel. Helv. Chim. Acta **13**, 1390 (1930).
23. Coulter, J. R., and C. S. Hann. J. Chromatog. **36**, 42 (1968).
24. Crestfield, A. M., S. Moore, and W. H. Stein. J. Biol. Chem. **238**, 622 (1963).
25. Cruickshank, P. A., and J. C. Sheehan. Anal. Chem. **36**, 1191 (1964).
26. Darbre, A., and K. Blau. Biochem. J. **88**, 8P (1963).

27. Darbre, A., and K. Blau. J. Chromatog. **17**, 31 (1965a).
28. Darbre, A., and K. Blau. Biochim. Biophys. Acta **100**, 298 (1965b).
29. Darbre, A., and K. Blau. J. Chromatog. **29**, 49 (1967).
31. Davis, J. W., Jr., and A. Furst. Anal. Chem. **40**, 1910 (1968).
32. Dorlet, C. Bull. Soc. Chim. Belg. **72**, 560 (1963).
33. Dose, K. Nature **179**, 734 (1957).
34. Edman, P. Acta Chem. Scand. **4**, 283 (1950).
35. Ertinghausen, G., C. W. Gehrke, and W. A. Aue. Separ. Sci. **2**, 681 (1967).
36. Fales, H. M., and J. J. Pisano, in *Biomedical Applications of Gas Chromatography*. Ed. by H. A. Szymanski (New York: Plenum Publishing Corp., 1964), p 39.
37. Feibush, B., and E. Gil-Av. J. Gas Chromatog. **5**, 257 (1967).
38. Fischer, E. Ber. Deut. Chem. Ges. **39**, 530 (1906).
39. Gehrke, C. W., W. M. Lamkin, D. L. Stalling, and F. Shahrokhi. Biochem. Biophys. Res. Commun. **19**, 328 (1965).
40. Gehrke, C. W., and F. Shahrokhi. Anal. Biochem. **15**, 97 (1966).
41. Gehrke, C. W., and D. L. Stalling. Separ. Sci. **2**, 101 (1967).
42. Gehrke, C. W., D. L. Stalling, and C. D. Ruyle. Biochem. Biophys. Res. Commun. **28**, 869 (1967).
43. Gehrke, C. W., R. W. Zumwalt, and L. L. Wall. J. Chromatog. **37**, 398 (1968).
44. Gil-Av, E., R. Charles, and G. Fischer. J. Chromatog. **17**, 408 (1965).
45. Gil-Av, E., R. Charles-Sigler, G. Fischer, and D. Nurok. J. Gas Chromatog. **4**, 51 (1966).
46. Gil-Av, E., and B. Feibush. Tetrahedron Letters **35**, 3345 (1967).
47. Gil-Av, E., B. Feibush, and R. Charles-Sigler. Tetrahedron Letters **10**, 1009 (1966).
48. Gil-Av, E., B. Feibush, and R. Charles-Sigler, in *Gas Chromatography*. Ed. by A. B. Littlewood (London: Bartholomew Press, Sixth Intern. Symp. 1966), p 227.
49. Gil-Av, E., and D. Nurok. Proc. Chem. Soc. **1962**, 146.
50. Giuffrida, L. J. Assoc. Office Agr. Chemists **47**, 293 (1964).
51. Goldberg, G., and W. A. Ross. Chem. Ind. (London) **1962**, 657.
52. Graff, J., J. P. Wein, and M. Winitz. Federation Proc. **22**, 244 (1963).
53. Gross, D., and G. Grodsky. J. Am. Chem. Soc. **77**, 1678 (1955).
54. Greer, M., and C. M. Williams. Anal. Biochem. **19**, 40 (1967).
55. Hagen, P., and W. Black. Federation Proc. **23**, 371 (1964).
56. Hagen, P. B., and W. Black. Can. J. Biochem. **43**, 309 (1965).
57. Halász, I., and K. Bünnig. Z. Anal. Chem. **211**, 1 (1965).
58. Halpern, B., and J. W. Westley. Chem. Commun. **12**, 246 (1965).
59. Halpern, B., and J. W. Westley. Biochem. Biophys. Res. Commun. **19**, 361 (1965).
60. Halpern, B., and J. W. Westley. Tetrahedron Letters **21**, 2283 (1966).
61. Halpern, B., and J. W. Westley. Australian J. Chem. **19**, 1533 (1966).

62. Hann, C. S.   Biochim. Biophys. Acta **181**, 342 (1969).
63. Horning, M. G., A. M. Moss, and E. C. Horning.   Biochim. Biophys. Acta **148**, 597 (1967).
64. Hunter, I. R., K. P. Dimick, and J. W. Corse.   Chem. Ind. (London) **1956**, 294.
65. Hunter, I. R., and E. F. Potter.   Anal. Chem. **30**, 293 (1958).
66. Ikekawa, N.   J. Biochem. (Tokyo) **54**, 279 (1963).
67. Ikekawa, N., O. Hoshino, R. Watanuki, H. Orimo, T. Fujita, and M. Yoshikawa.   Anal. Biochem. **17**, 16 (1966).
68. Irreverre, F., K. Morita, S. Ishii, and B. Witkop.   Biochem. Biophys. Res. Commun. **9**, 69 (1962).
69. Ishii, S., and B. Witkop.   J. Am. Chem. Soc. **85**, 1832 (1963).
70. Jaakonmäki, P. I., and J. E. Stouffer.   J. Gas Chromatog. **5**, 303 (1967).
71. James, A. T., and A. J. P. Martin.   Analyst **77**, 915 (1952).
72. Johnson, D. E., S. J. Scott, and A. Meister.   Anal. Chem. **33**, 669 (1961).
73. Karagounis, G., and E. Lemperle.   Z. Anal. Chem. **189**, 131 (1962).
74. Karagounis, G., and G. Lippold.   Naturwiss. **46**, 145 (1959).
75. Karmen, A., and L. Giuffrida.   Nature **201**, 1204 (1964).
76. Karrer, P., in *Organic Chemistry* (Amsterdam: Elsevier, 1950, 794.
77. Klebe, J. F., H. Finkbeiner, and D. M. White.   J. Am. Chem. Soc. **88**, 3390 (1966).
78. Lamkin, W. M., and C. W. Gehrke.   Anal. Chem. **37**, 383 (1965).
79. Landowne, R. A., and S. R. Lipsky.   Federation Proc. **22**, 235 (1963).
80. Landowne, R. A., and S. R. Lipsky.   Nature **199**, 141 (1963).
81. Levesque, C. L., and A. M. Craig.   Ind. Engin. Chem. **40**, 96 (1948).
82. Losse, G., A. Losse, and J. Stöck.   Z. Naturforsch. **17b**, 785 (1962).
83. Makasumi, S., C. H. Nicholls, and H. A. Saroff.   J. Chromatog. **12**, 106 (1963).
84. Makasumi, S., and H. A. Saroff.   J. Gas Chromatog. **3**, 21 (1965).
85. Marcucci, F., E. Mussini, F. Poy, and P. Gagliardi.   J. Chromatog. **18**, 487 (1965).
86. Martin, A. J. P., and R. L. M. Synge.   Biochem. J. **35**, 1358 (1941).
87. Mason, P. S., and E. D. Smith.   J. Gas Chromatog. **4**, 398 (1966).
88. Melamed, N., and M. Renard.   J. Chromatog. **4**, 339 (1960).
89. Mill, P. J., and W. R. C. Crimmin.   Biochim. Biophys. Acta **23**, 432 (1957).
90. Moore, S., and W. H. Stein.   J. Biol. Chem. **211**, 893 (1954).
91. Moore, S., D. H. Spackman, and W. H. Stein.   Anal. Chem. **30**, 1185 (1958).
92. Mussini, E., and F. Marcucci.   J. Chromatog. **20**, 266 (1965).
93. Nicholls, C. H., S. Makasumi, and H. A. Saroff.   J. Chromatog. **11**, 327 (1963).
94. Pierce, A. E.   *Silylation of Organic Compounds* (Rockford, Illinois: Pierce Chemical Co., 1968).

95. Pirkle, W. H.   J. Am. Chem. Soc. **88**, 1837 (1966).
96. Pisano, J. J., W. J. A. Vanden Heuval, and E. C. Horning.   Biochem. Biophys. Res. Commun. **7**, 82 (1962).
97. Pitt-Rivers, R., and J. E. Rall.   Endocrinology **68**, 309 (1961).
98. Pollock, G. E.   Anal. Chem. **39**, 1194 (1967).
99. Pollock, G. E., and A. H. Kawauchi.   Anal. Chem. **40**, 1356 (1968).
100. Pollock, G. E., and V. I. Oyama.   J. Gas Chromatog. **4**, 126 (1966).
101. Pollock, G. E., V. I. Oyama, and R. D. Johnson.   J. Gas Chromatog. **3**, 174 (1965).
102. Richards, A. H., and W. B. Mason.   Anal. Chem. **38**, 1751 (1966).
103. Rühlmann, K.   J. Prakt. Chem. **9**, 86 (1959).
104. Rühlmann, K.   J. Prakt. Chem. **9**, 315 (1959).
105. Rühlmann, K.   Chem. Ber. **94**, 1876 (1961).
106. Rühlmann, K., and W. Giesecke.   Angew. Chem. **73**, 113 (1961).
107. Rühlmann, K., and G. Michael.   Bull. Soc. Chim. Biol. **47**, 1467 (1965).
108. Saroff, H. A., and A. Karmen.   Anal. Biochem. **1**, 344 (1960).
109. Saroff, H. A., A. Karmen, and J. W. Healy.   J. Chromatog. **9**, 122 (1962).
110. Sen, N. P., and P. L. McGeer.   Biochem. Biophys. Res. Commun. **13**, 390 (1963).
111. Shahrokhi, F., and C. W. Gehrke.   J. Chromatog. **36**, 31 (1968).
112. Shahrokhi, F., and C. W. Gehrke.   Anal. Biochem. **24**, 281 (1968).
113. Shlyankov, S. V., M. Y. Karpeiskii, and E. F. Litvin.   Biokhimiya **28**, 664 (1963); English translation **28**, 544 (1963).
114. Smith, E. D., and H. Sheppard, Jr.   Nature **208**, 878 (1965).
115. Stalling, D. L., and C. W. Gehrke.   Biochem. Biophys. Res. Commun. **22**, 329 (1966).
116. Stalling, D. L., C. W. Gehrke, and R. W. Zumwalt.   Biochem. Biophys. Res. Commun. **31**, 616 (1968).
117. Stalling, D. L., G. Gille, and C. W. Gehrke.   Anal. Biochem. **18**, 118 (1967).
118. Stefanovic, M., and B. L. Walker.   Anal. Chem. **39**, 710 (1967).
119. Stein, W. H., and S. Moore.   J. Biol. Chem. **176**, 337 (1968).
120. Stern, R. L., B. L. Karger, W. J. Keane, and H. C. Rose.   J. Chromatog. **39**, 17 (1969).
121. Stevenson, G. W., and J. M. Luck.   J. Biol. Chem. **236**, 715 (1961).
122. Stouffer, J. E., P. I. Jaakonmäki, and T. J. Wenger.   Biochim. Biophys. Acta **127**, 261 (1966).
123. Sussman, S.   Ind. Engin. Chem. **38**, 1228 (1946).
124. Sweeley, C. C., R. Bentley, M. Makita, and W. W. Wells.   J. Am. Chem. Soc. **85**, 2497 (1963).
125. Tannenbaum, S. R., W. G. Thilly, and P. Issenberg.   Anal. Chem. **40**, 1723 (1968).
126. Vanden Heuvel, W. J. A., J. L. Patterson, and K. L. K. Braly.   Biochim. Biophys. Acta **144**, 691 (1967).

127. Vitt, S. V., M. B. Saporovskaya, I. P. Gudkova, and V. M. Belikov. Tetrahedron Letters 30, 2575 (1965).
128. Virtanen, A. I., and N. Rautanen. Biochem. J. 41, 101 (1947).
129. Wagner, J., and G. Rausch. Z. Anal. Chem. 194, 350 (1963).
130. Wagner, J., and G. Winkler. Z. Anal. Chem. 183, 1 (1961).
131. Walle, T. Acta Pharm. Suecica 5, 353 (1968).
132. Weinstein, B., in *Methods of Biochemical Analysis*, Vol. 14. Ed. by D. Glick (New York: Interscience, 1966), p 203.
133. Wellby, M. L., and B. S. Hetzel. Nature 193, 752 (1963).
134. Werner, S. C., and I. Radichevich. Nature 197, 877 (1963).
135. Westley, J. W., B. Halpern, and B. L. Karger. Anal. Chem. 40, 2046 (1968).
136. Weygand, F., and R. Geiger. Chem. Ber. 89, 647 (1956).
137. Weygand, F., and R. Geiger. Chem. Ber. 92, 2099 (1959).
138. Weygand, F., H. Geiger, and W. Swodenk. Angew. Chem. 68, 307 (1956).
139. Weygand, F., P. Klinke, and I. Eigen. Chem. Ber. 90, 1896 (1957).
140. Weygand, F., G. Klipping, and D. Palm. Chem. Ber. 93, 2619 (1960).
141. Weygand, F., B. Kolb, and P. Kirchner. Z. Anal. Chem. 181, 396 (1961).
142. Weygand, F., and E. Leising. Chem. Ber. 87, 248 (1954).
143. Weygand, F., A. Prox, L. Schmidhammer, and W. König. Angew. Chem. Intern. Ed. Engl. 2, 183 (1963).
144. Weygand, F., and H. Rinno. Chem. Ber. 92, 517 (1959).
145. Weygand, F., and A. Röpsch. Chem. Ber. 92, 2095 (1959).
146. Wieland, I., and E. Bende. Chem. Ber. 98, 504 (1965).
147. Youngs, C. G. Anal. Chem. 31, 1019 (1959).
148. Zlatkis, A., and J. F. Oró. Anal. Chem. 30, 1156 (1958).
149. Zlatkis, A., J. F. Oró, and A. P. Kimball. Anal. Chem. 32, 162 (1960).
150. Zomzely, C., G. Marco, and E. Emery. Anal. Chem. 34, 1414 (1962).
151. Del Favero, A., A. Darbre, and M. Waterfield. J. Chromatog. 40, 213 (1969).
152. Funakoshi, K., and H. J. Cahnmann. Anal. Biochem. 27, 150 (1969).
153. Gehrke, C. W., H. Nakamoto, and R. W. Zumwalt. J. Chromatog. 45, 24 (1969).
154. Gelpi, E., W. A. Koenig, J. Gibert, and J. Oró. J. Chromatog. Sci. 7, 604 (1969).
155. Halász, I., and C. Horváth. Anal. Chem. 36, 2226 (1964).
156. Islam, A., and A. Darbre. J. Chromatog. 43, 11 (1969).
157. Kruppa, R. F., R. S. Henly, and D. L. Smead. Anal. Chem. 39, 851 (1967).
158. Neri, P., and P. Tarli. Quad. Sclavo Diagn. 5, No. 1 (1969).
159. Roach, D., and C. W. Gehrke. J. Chromatog. 43, 303 (1969).
160. Roach, D., C. W. Gehrke, and R. W. Zumwalt. J. Chromatog. 43, 311 (1969).
161. Roach, D., and C. W. Gehrke. J. Chromatog. 44, 269 (1969).
162. Stouffer, J. E. J. Chromatog. Sci. 7, 124 (1969).
163. Waterfield, M.D., and A. Del Favero. J. Chromatog. 40, 294 (1969).

Chapter 3

Gas Chromatography of Peptides
by B. Kolb

3.1   Introduction

The introduction of gas-liquid chromatography (GLC) as a method for the analysis of amino acids, peptides, and related compounds has greatly increased the possibilities for new advances in peptide chemistry. Considerable effort has been made on behalf of amino acid analysis by GLC, and various types of derivatives have been investigated for this purpose. For quantitative analysis, however, all methods based on GLC have had to compete with the refined methods of ion-exchange chromatography with its high performance in automation, precision, and even speed of analysis (e.g., the technique of ligand analysis). For this reason GLC of amino acids has found practical application during recent years more for some special problems, where it might even be superior to other chromatographic techniques, than for the quantitative determination of component amino acids in a complex mixture. But GLC may now be used also for this application as a complementary technique, due to the analytical approach developed mainly by Gehrke and co-workers.[1]

Moreover, as compared with other chromatographic techniques, GLC offers a number of advantages for peptide analysis. These are related to the high separation efficiency and the versatility of this technique, both of which are necessary and useful for the wide variety of compounds in this field. One of the basic advantages of GLC is the fact that separation is achieved both by differences in the activity coefficients and in the vapor pressures of the respective compounds, thus permitting separations according to polarities and boiling points. However, if a complex mixture of numerous compounds with a wide range of molecular weights is analyzed by GLC—preferably by temperature programming—the order of elution is mainly by increasing molecular weights. The chemical properties of the compounds can alter retention

129

only to a certain extent through their intermolecular interaction with the liquid phase. Such a complex mixture, for example, is represented by a partial hydrolysate of a protein, including the respective amino acids and the entire range of possible peptides resulting from statistical cleavage of the peptide bonds. Using temperature-programmed GLC it is possible to scan the entire range from the low volatile amino acids up to the dipeptides and even higher peptides; the different classes of compounds are eluted in different temperature ranges with little over-lapping of these ranges. In this way, such a complex mixture may be roughly divided into the different classes of compounds, starting with the amino acids group and followed by the dipeptides which are mixed with tripeptides and higher peptides only at the limits of their respective ranges. It is evident that such a classification of compounds, achieved with a temperature-programmed GLC column, might hold an advantage over other chromatographic procedures by which the compounds are separated mainly according to their chemical properties, resulting in a chromatogram in which amino acids and peptides are all mixed together.

Figure 1 shows the separation of a synthetic mixture of amino acids and dipeptides as N-TFA*-methyl ester derivatives using a temperature-programmed glass capillary column. On this particular column, prepared for a special separation problem, the two different ranges of amino acids (from alanine up to phenylalanine) and dipeptides (from alanyl-alanine up to phenylalanyl-phenylalanine) can be seen. This particular mixture obviously did not contain all possible amino acids and dipeptides.

Temperature-programmed separation of a wide-range mixture is not the only possibility for use of GLC, however; difficult separations of closely related isomers also may be carried out, preferably under isothermal conditions, using highly efficient capillary columns. Figure 2 shows such a separation in a synthetic mixture of some closely related diastereoisomeric N-TFA-dipeptide methyl esters on a glass capillary column coated with polypropylene glycol liquid phase. This particular mixture represents all possible diastereoisomeric dipeptide combinations between two DL-amino acids only.

One of the most important features of the application of GLC to peptide analysis, however, is its combination with mass spectrometry for further analysis of the peptides it has separated. This GLC-mass spectrometry combination is a powerful tool in up-to-date investigations on this subject, and it is a necessity if a mixture of unknown peptides has to be analyzed as represented—for example, by a partial

---

*N-TFA- = N-trifluoroacetyl-.

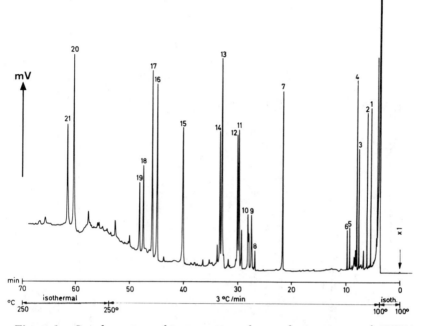

Figure 1. Gas-chromatographic separation of a synthetic mixture of N-TFA-methyl ester derivatives of amino acids and dipeptides.

Instrument: Perkin-Elmer Fraktometer F20; Column: 50-m x 0.25-mm glass capillary pretreated according to Grob[21] and coated with dimethyl silicone OV-101 (95%) and Carbowax 20M (5%); Carrier gas: 1 ml/min nitrogen, split ratio 1:30; Temperature: programmed as given; Detector: FID.

1 = Ala, 2 = Val, 3 = Leu, 4 = Ile, 5 = Ser (TMS)*, 6 = Thr (TMS)*, 7 = Phe, 8 = Ala-Ala(LL), 9 = Val-Val(LL), 10 = Val-Val(DL), 11 = Ile-Val(LL), 12 = Ile-Val(DL), 13 = Leu-Leu(LL), 14 = Leu-Leu(DL), 15 = Met-Ala(LL), 16 = Ala-Phe(LL), 17 = Ala-Phe(DL), 18 = Phe-Pro(LL), 19 = Phe-Pro(DL), 20 = Phe-Phe(LL), 21 = Phe-Phe(DL).

hydrolysate of a protein. While it should be possible to identify amino acids and dipeptides by measuring relative retentions or retention indices or by adding reference compounds, this procedure appears to be impossible for investigations of higher peptides because of the tremendous number of possible compounds. If, for example, only 10 amino acids are taken into account, 100 dipeptides and 1,000 tripeptides are to be expected by their combination.

Mass spectrometry of peptides was investigated by Stenhagen[2] and most extensively later on by Weygand, Prox, and co-workers[3, 4, 5, 6, 7] on the example of N-TFA-peptide methyl esters separated by GLC after their compounds had been trapped from the column effluent. Mass

---

*TMS = trimethylsilyl derivatives.

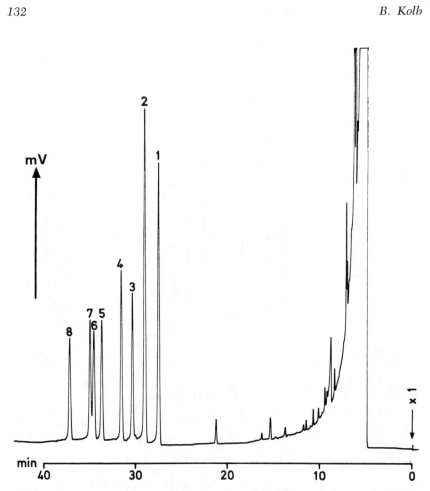

Figure 2. Gas-chromatographic separation of *N*-TFA-methyl ester derivatives of some closely related dipeptides.

Instrument: Perkin-Elmer Fraktometer F20; Column: 50-m x 0.25-mm glass capillary pretreated according to Grob[21] and coated with polypropylene glycol liquid phase; Carrier gas: 0.5 ml/min nitrogen, split ratio 1:50; Temperature: 170°C isothermal; Detector: FID.

1 = Ala-Ala(DL), 2 = Ala-Ala(LL), 3 = Val-Ala(LL), 4 = Val-Ala(DL), 5 = Ala-Val(LL), 6 = Val-Val(LL), 7 = Ala-Val(DL), 8 = Val-Val(DL).

spectrometry of peptides has also been reported by Lederer[8] and Shemyakin,[9] but since the aspects of mass spectrometry are covered in Chapter 4 of this book, they will not be discussed further in this paper.

Gas-chromatographic analysis of peptide derivatives was first reported by Weygand, Kolb, Prox, Tilak, and Tomida[10] in 1960. In this early paper the main topics—separation of diastereoisomers, important for studying racemization, and analysis of partial hydrolysates for sequence analysis, both still valid—had been demonstrated in several experiments. Emphasis was already on the application of mass spectrometry for further identification of trapped compounds.

After these early results, a number of studies extended the usefulness of GLC in this field. They particularly concerned GLC in elucidation of the amino acid sequence in some peptides and in investigations involving racemization during peptide synthesis, besides other applications for some special problems. These aspects and results are now discussed more in detail.

3.2 Formation of Volatile Derivatives

Since free amino acids and peptides are not sufficiently volatile, they must be converted to volatile derivatives before GLC begins. Preparation of derivatives is a major problem which so far has not been solved for all the possible peptides. Part of the trouble arises from the fact that many important amino acids in the peptide chain contain a number of other functional groups in addition to the characteristic α-amino and carboxyl groups. Derivatives of widely differing volatilities are formed and in addition troublesome side reactions are often encountered. Since, however, there is in principle no difference in approach to the preparation of volatile derivatives either of amino acids or peptides, it is to be expected that the results and experiences with amino acid derivatives will stimulate development of similar procedures for the corresponding peptides, too. Peptides which are not yet accessible to analysis by GLC are those containing histidine, arginine, or amino acids like asparagine or glutamine which still contain a functional amide group. The special effects encountered with peptides as compared with amino acids arise from a peptide's higher molecular weight and its thus-reduced volatility. To compensate for this lower volatility, only those protecting groups very stable at high temperatures, which cause a high volatility and are at least readily prepared, should be used. These conditions reduce the wide variety of protecting groups used with amino acids for peptide application.

Biemann and Vetter[11] separated by GLC the amino alcohols resulting from the reduction of N-acyl peptide ethyl esters with lithium–aluminum hydride. They made use of mass spectrometry for identification but no further application of this procedure has been reported.

3.2.1 Amino Protecting Groups

With few exceptions, the trifluoroacetyl group has been used as the amino protecting group for GLC for several reasons. Trifluoroacetyl derivatives of amino acid- and peptide esters are very stable at high temperatures, and the trifluoroacetyl group causes a higher volatility than any other acetyl group.[12] Only the heptafluorobutyryl derivatives have been found to give a higher volatility[13] but they have not yet found wide application, though they are attractive for tandem use with mass spectrometry,[77] because the respective peptide ester derivatives can be separated at a lower column temperature with reduced column bleeding.

A further important advantage of the trifluoroacetyl group is its use as a common protecting group in peptide chemistry, since a large supply of reference compounds must be available for analysis by GLC, and most of them are not commercially available and must be synthesized. If the trifluoroacetyl protecting group is also employed for synthesis, the resulting N-TFA-peptide methyl esters can be immediately used for gas-chromatographic analysis as reference compounds. If, however, the well-known carbobenzyloxy protecting group is used for synthesis, it might be replaced by the trifluoroacetyl group with trifluoroacetic acid[14] although even carbobenzyloxy dipeptide methyl esters are suitable for GLC.[15] A number of other derivatives might be of some interest, but since only the N-TFA-derivatives of peptides have been investigated to such an extent so far, the discussion in this paper will be limited to these derivatives. Trifluoroacetylation is performed either with trifluoroacetic acid methyl ester together with triethylamine in weakly basic solution[10] or with trifluoroacetic anhydride.[16] The second amino group in the side chain with the amino acids lysine and ornithine is trifluoroacetylated together with the α-amino group under the same conditions.

3.2.2 Carboxyl Protecting Groups

So far, for peptides the simple methyl esters have been used most exclusively. Esterification is performed either before acylation with anhydrous methanol using dry HCl gas as a catalyst, thus giving the respective peptide methyl ester hydrochlorides, or after acylation with diazomethane. If the compounds contain an additional carboxyl such as glutamic or aspartic acid, it is also best methylated under the same conditions at the same time. The methyl ester hydrochlorides are immediately formed by partial hydrolysis of a peptide if it is carried out in methanolic hydrochloride solution.

3.2.3 Groups Protecting the Third Function

The simple N-TFA-peptide methyl ester derivatives are not sufficient if amino acids with a third functional group are incorporated in the peptide under investigation.

N-TFA-Peptide methyl esters with hydroxyl-containing amino acids, such as serine, threonine or hydroxyproline, are best converted to the trimethylsilyl ether derivatives (TMS) by refluxing with hexamethyl-disilazane for about 20 minutes.[17] Without silylation multiple peaks due to decomposition have been observed. N-TFA-Peptide methyl esters containing tyrosine have been reported to result in asymmetrical peaks,[17] probably due to the strong hydrogen bonds of the phenolic hydroxyl group. While it is possible to overcome this problem by methylation with diazomethane, silylation again is preferable. Serine and threonine-containing peptides must be silylated in any case, here, and tyrosine in the respective peptides is silylated under the same conditions.

Some problems have been encountered with peptides containing cysteine, because they have been found to be decomposable under the conditions prevailing in GLC analysis. For this reason the free mercapto group should be protected to avoid β-elimination as well as oxidation to the disulfide link. Bayer and Koenig[18] made use of the S-benzyl group for this purpose and reported GLC analysis of some cysteine-containing N-TFA-dipeptide and tripeptide esters. The peptide under investigation was first benzylated before partial hydrolysis, which was carried out with concentrated hydrochloric acid at 37°C for 72 hours. After evaporation *in vacuo*, esterification with diazomethane, and trifluoroacetylation, analysis was performed by GLC and mass spectrometry.

Weygand *et al.*[17] recommend desulfurization by heating the peptides with Raney nickel in 90% methanol. Thus the respective alanine derivatives are formed for each cysteine unit. All other sulfur-containing amino acids also undergo the sulfur elimination; methionine, for example, is converted to α-aminobutyric acid. Another example is in a report by Weygand *et al.*[17] on the sequence analysis of glutathione. Bis-N-TFA-glutathione disulfide tetraethyl ester was first desulfurized, and after partial hydrolysis, methylation and trifluoroacetylation, the resulting N-TFA-dipeptide methyl esters of L-Glu-γ–L-Ala and L-Ala-Gly were identified by GLC. They had resulted from the sequence L-Glu-γ–L-Ala-Gly, thus proving the original sequence L-Glu-γ–L-Cys-Gly.

Weygand *et al.*[17] also turned attention to the use of deuterated methanol for the desulfurization reaction, since mass spectrometry after gas-chromatographic analysis should make it possible to decide whether an alanine-containing peptide does originate from the respective cysteine-containing peptide or not. Later on,[7] this procedure was also employed for the elucidation of the amino acid sequence of a methionine-containing cyclic nonapeptide. The peptide fragments resulting from partial hydrolysis which now contained the corresponding α-aminobutyric acid (Abu) instead of the methionine could be successfully identified after derivatization by combined GLC and mass spectrometry (see Figure 7).

It should be mentioned that the same problems have been encountered with GLC of the respective hydroxyl- and sulfhydryl-containing amino acids, and that, according to Gehrke,[1] they all can be successfully analyzed by GLC as the bis-N-TFA ester derivatives. But since the trifluoroacetylated hydroxyl and sulfhydryl groups are easily decomposed by moisture or even traces of alcohol in the solvents, they should be injected as a solution in trifluoroacetic anhydride. This procedure is less attractive for routine work because the powerful reagent will attack the liquid phase of the column at elevated temperature.

3.3   Instrumentation

3.3.1   *Columns*

In the beginning of investigations on this subject the usual high-loaded packed columns were found satisfactory since only N-TFA-dipeptide methyl esters were concerned and they were sufficiently volatile to result in reasonable analysis times on such columns. However, very long retention times had to be taken into account, too, for the high-molecular-weight dipeptides. All retention data found in the literature have been calculated from such high-loaded columns (see Table I below), with isothermal operation.

Gas-chromatographic analysis of the whole range, however, including low-molecular-weight dipeptides or even amino acids up to high-molecular-weight oligopeptides, is carried out preferably by using low-loaded columns with about 1–5% liquid phase on silanized support material and operated with temperature program, as shown in some of the figures. Due to the high temperature range, only liquid phases with good thermal stability can be used. Examples of these include the various dimethyl silicon polymers SE-30, OV-1, and OV-101 or the methyl phenyl silicones SE-52 and OV-17, which are more polar in nature. A wide range of silicon polymers, however, with exceptional thermal stability and an incremental increase in polarity have been developed recently and are available now for high-temperature application; all of these materials are of potential interest for peptide application, too. Besides the various silicon liquid phases, packed columns with apiezon grease also have been employed.[17] A packed column with 0.5% FFAP on Chromosorb-W was used for a special separation of two diastereoisomeric N-TFA-tripeptide methyl esters.[19] A polyester column with 15% liquid phase was used for separation of the TMS-derivatives of N-TFA-L-Ser–L-Ala-methyl ester and N-TFA-L-Thr–L-Ala-methyl ester. This separation could not be achieved on a nonpolar silicon column but a separation factor of 1.9* did result with the polyester column, the serine compound having the longer retention time. Only such a high-loaded polyester column has been used so far for peptide analysis, and probably due to the long retention times, no further application has been reported. Low-loaded polyester columns, though, should be of potential interest for peptide analysis, too, since they have been found very useful in amino acid analysis.

Capillary columns up to now have been used isothermally only for some special applications—mainly for the difficult separation of isomers.[6, 18, 20, 22] With the usual stainless steel capillaries, tailing and asymmetrical peaks have been observed,[20] thus reducing the high efficiency of those columns. These troubles can be avoided by the use of glass capillaries, which result in sharp and symmetrical peaks, as can be seen from Figures 1, 2 and 4. The glass capillaries used for these chromatograms have been prepared by covering the glass surface with a layer of carbon black before coating with the liquid phase, according to the procedure developed by Grob.[21] This procedure has proved eminently successful in deriving a coherent film of the liquid phase, particularly if polar liquid phases are employed.

Besides one application with the silicon SF-96 liquid phase,[18] capillaries with either polypropylene glycol or polyphenyl ether OS-138 as liquid phase have been employed, particularly for the separation of diastereoisomers,[20, 22] where these liquid phases of moderate polarity have been quite useful. For more general applications, however, embracing the whole range of amino acids and dipeptides, the above-mentioned high-temperature silicones might be very useful for capillary work also, especially when operated with temperature program as it is shown in Figure 1. This particular capillary with the mixed liquid phase of the dimethyl silicon OV-101 and Carbowax-20M was prepared in order to achieve a better separation among the different classes of amino acids, dipeptides, and tripeptides. It has been observed that a small amount of a strong polar material such as a polyglycol, mixed with a nonpolar liquid phase (for example, OV-101) results in an increasing shift between the class of dipeptides and tripeptides to higher retention times. This effect is particularly useful in that region of a peptide chromatogram where dipeptides are mixed with tripeptides because the class of tripeptides as a whole can be set apart from the dipeptide class so that identification of the dipeptides is simplified. This effect can be seen also in the Figure 1 example of separation between

---

*Quotient of the retention times.

the amino acid phenylalanine and the dipeptide Ala-Ala; with the pure dimethyl silicon OV-101 this particular dipeptide would be eluted before this particular amino acid. It is that region with a nonpolar liquid phase where the amino acids overlap with the dipeptides. The same effect, however, might possibly be achieved with one of the above-mentioned high-temperature silicones of moderate polarity, but this application needs further investigation. The chromatogram in Figure 1 demonstrates the possible advantages of such a temperature-programmed glass capillary with a high-temperature silicon liquid phase.

3.3.2 Detectors

The standard detectors in GLC, flame ionization (FID) and hot-wire (HWD) detector, are sufficient for these applications. The FID is essential both for low-loaded packed columns and capillary columns due to the low capacity of such columns, whereas the HWD can be used for high-loaded packed columns only. No application has been reported for gas chromatographic analysis of peptide derivatives using the electron-capture detector (ECD). Contrary to its use for amino acid derivatives, such an application appears to be of limited value for peptide derivatives due to the necessary high column temperature with the inherent danger of detector contamination by column bleeding, which influences both response and sensitivity. However, the use of heptafluorobutyryl derivatives could be advantageous because of their higher volatility and their high electron affinity.

The most universal detector appears to be a mass spectrometer, which may indeed be considered a detector for GLC if the column is directly connected with the mass spectrometer by means of a molecular separator (see also Section 3.5.1). With such an instrument, the gas chromatogram is recorded by the total ion monitor, as shown in Figure 3, while a mass spectrum may be scanned from each peak of interest. However, due to the excessive column bleeding at elevated temperatures, which results in a high background from the liquid phase in the mass spectrum, this direct combination is limited to investigations of lower peptides and is less suitable for investigations of higher peptides at the high temperatures necessary for gas-chromatographic separation. For this reason the method of trapping the compounds after the column for further investigations by mass spectrometry appears to be preferable for higher peptides, for example, tri- and tetrapeptides. Recently Bayer and Koenig[18] reported also the use of a combined gas chromatograph-mass spectrometer instrument for separation of peptide derivatives, but probably due to the difficulties discussed above the given

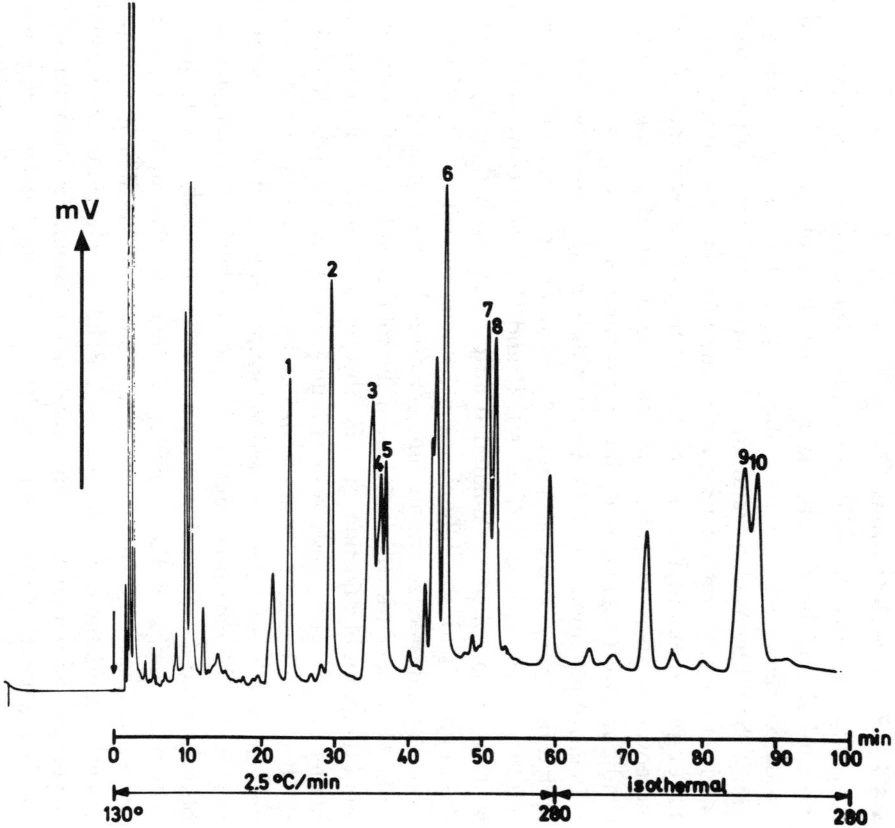

Figure 3. Gas-chromatographic separation of N-TFA-methyl ester derivatives of amino acids and dipeptides.

Instrument: Perkin-Elmer Model 270 GC-DF mass spectrometer; Column: 12' x ⅛" stainless steel, packed with 5% SE-30 on Chromosorb-G, AW-DMCS 80/100 mesh; Carrier gas: 30 ml/min helium, split ratio after the column 1:3; the minor part going to the molecular separator; Temperature: programmed as given; Detector: total ion monitor.

1 = Met, 2 = Phe, 3 = Leu-Ala(LL + DL), 4 = Val-Val(LL), 5 = Val-Val(DL), 6 = Pro-Val(LL), 7 = Ala-Phe(LL), 8 = Ala-Phe(DL), 9 = Phe-Phe(LL), 10 = Phe-Phe(DL).

mass spectra were scanned using the direct inlet system of this instrument.

3.4 Data Presentation

Retention data should be given in such a manner that they can be converted for use in experiments with other apparatuses and under different conditions. This can be done on an absolute basis by measurement of the partition coefficients or specific retention volume, but for practical application retention data are given mostly on a relative basis by measurements either of relative retentions relative to a standard solute or of retention indices relative to the homologous series of *n*-alkanes (see Supplement).

3.4.1 *Relative Retentions*

For determining the relative retentions of a series of substances, a standard should be chosen, the retention time of which is near the middle of the series. For interlaboratory comparisons the same substance should be used as standard solute to avoid recalculations. Fortunately, all relative retentions given in the literature and concerning *N*-TFA-dipeptide methyl esters are related to myristic acid methyl ester; moreover, all these measurements have been carried out under similar conditions with high-loaded silicon oil columns.

Relative retentions of *N*-TFA-dipeptide methyl esters have been reported from Weygand's laboratory[10, 17, 23] and from Tomida *et al.*[24] The results are summarized in Tables I, II and III. The relative retentions in Table I have been calculated from columns with dimethyl silicon liquid phase, with the difference that some of the values are calculated from the adjusted retention time, measured from the air-peak time, while the others are calculated from the total retention time, measured from the point of injection. It is for this reason that these numbers are higher at low values and lower at high values than those calculated on the basis of adjusted retention times. Relative retentions should always be calculated from the adjusted retention times, and if the air-peak time cannot be measured, due to the use of a flame detector, it should be calculated (see Supplement). These deviations, however, may be within the limits of the accuracy of these measurements, since for several reasons relative retentions are known to be difficult to reproduce in interlaboratory comparisons. Relative retentions are strongly dependent on temperature, but much of the commercial equipment of the last-generation instruments is lacking in good oven design with respect to gradients along the column and accuracy of temperature read-out if thermocouples are used. In addition to these temperature

Table I
Relative Retention of N-TFA-Dipeptide Methyl Esters
on Dimethyl Silicon Liquid Phase
(Myristic Acid Methyl Ester = 1.00)

Literature reference	200°C 17, 23*	200°C 24†	210°C 24†	220°C 24†	225°C 17, 23*
L-Ala–L-Ala	0.30	0.32	0.34	0.35	0.32
L-Ala–Gly	0.31	0.34	0.37	0.38	0.35
L-Ala–L-Val	0.44	0.40	0.48	0.49	0.48
L-Ala–L-Leu	0.58	0.58	0.59	0.61	0.59
L-Ala–L-Glu	–	1.29	1.24	1.23	–
L-Ala–L-Phe	2.14	2.08	–	1.93	1.95
TMS-L-Ala–L-Tyr‡	–	–	–	–	5.4
Gly–L-Ala	0.35	0.37	0.38	0.41	0.38
Gly–Gly	0.39	0.39	0.41	0.43	0.41
Gly–L-Val	0.53	0.53	0.55	0.56	0.56
Gly–L-Leu	0.70	0.69	0.70	0.69	0.70
Gly–L-Ile	0.71	–	–	–	0.71
Gly–L-Glu	1.68	1.57	1.51	1.47	1.51
Gly–L-Pro	–	–	–	–	0.99
Gly–L-Met	–	1.75	1.67	1.64	–
Gly–L-Phe	2.66	2.51	2.42	2.28	2.37
LVal–L-Ala	0.42	0.43	0.45	0.47	0.46
L-Val–Gly	0.46	0.46	0.49	0.50	0.49
L-Val–L-Val	0.64	0.62	0.63	0.66	0.66
L-Val–L-Leu	0.83	0.77	0.79	0.80	0.82
L-Val–L-Ile	0.84	0.82	0.82	0.84	0.83
L-Val–L-Glu	–	1.75	1.63	1.62	–
L-Val–L-Met	2.03	–	–	–	1.90
L-Val–L-Orn**	–	2.68	2.48	2.38	–
L-Val–L-Phe	2.95	2.84	2.70	2.58	2.60
L-Leu–L-Ala	0.54	0.53	0.55	0.55	0.55
L-Leu–Gly	0.58	0.58	0.60	0.61	0.61
L-Leu–L-Val	0.77	0.76	0.78	0.78	0.78
L-Leu–L-Leu	1.01	0.96	0.95	0.96	0.98
L-Leu–L-Ile	1.04	–	–	–	1.01
L-Leu–L-Pro	1.23††	–	–	–	1.19
L-Leu–L-Glu	–	2.15	2.06	1.97	–
L-Leu–L-Met	2.60	–	–	–	2.35
L-Leu–L-Phe	3.73	3.48	3.23	3.11	3.14
L-Ile–L-Ala	0.56	0.56	0.58	0.61	0.59
L-Ile–Gly	–	0.62	0.64	0.66	–
L-Ile–L-Val	0.83	0.81	0.80	0.84	0.83
L-Ile–L-Leu	1.07	1.01	1.01	1.01	1.04
L-Ile–L-Ile	1.15	1.07	1.05	1.05	1.06
L-Ile–L-Met	2.79	–	–	–	2.45
L-Ile–L-Phe	–	3.65	3.45	3.23	–
L-Pro–Gly	–	0.74	0.76	0.77	0.88
L-Pro–L-Val	–	1.13	1.13	1.15	–
L-Pro–L-Leu	–	1.37	1.34	1.33	1.35

Table I, continued

Literature reference	200°C 17, 23*	200°C 24†	210°C 24†	220°C 24†	225°C 17, 23*
L-Pro-L-Phe	—	—	4.94	4.60	—
L-Orn–L-Leu**	—	2.88	2.70	2.56	—
L-Orn–L-Ala**	—	1.68	1.60	1.55	—
L-Glu-α–L-Ala	1.14	1.14	1.12	1.11	1.10
L-Glu-α–Gly	—	1.26	1.23	1.23	1.20
L-Glu-α–L-Val	—	1.61	1.55	1.51	1.50
L-Glu-α–L-Leu	2.13	2.01	1.91	1.86	1.88
L-Glu-α–L-Ile	2.25	—	—	—	1.97
L-Glu-α–L-Pro	—	2.65	2.42	2.36	—
L-Glu-α–L-Glu	—	4.68	4.15	3.96	—
L-Glu-α–L-Phe	—	7.43	6.52	6.20	6.27
L-Glu-γ–L-Ala	—	—	1.51	1.50	1.48
L-Glu-γ–Gly	—	—	—	—	1.58
L-Glu-γ–L-Val	—	—	2.21	2.11	—
L-Glu-γ–L-Leu	3.20	—	2.76	2.62	2.68
L-Glu-γ–L-Ile	3.23	—	—	—	2.75
L-Glu-γ–L-Glu	—	—	6.05	5.46	—
L-Glu-γ–L-Phe	—	—	9.95	8.70	8.78
L-Phe–L-Ala	1.97	—	—	—	1.83
L-Phe–Gly	2.14	2.06	2.00	1.92	1.96
L-Phe–L-Val	—	2.73	2.61	2.48	—
L-Phe–L-Leu	3.66	3.39	3.22	3.01	3.11
L-Phe–L-Pro	—	3.86	3.66	3.53	—
L-Phe–L-Glu	—	7.80	6.70	6.45	—
L-Phe–L-Met	—	—	—	—	7.25
L-Phe–L-Phe	—	12.3	10.9	9.8	9.9

Literature reference	190°C 17*	201°C 17*	210°C 17*	222°C 17*	
TMS-L-Ser–L-Ala‡	0.64	0.65	0.65	0.66	—
TMS-L-Ser–L-Val	0.97	0.97	0.95	0.95	—
TMS-L-Ser–L-Leu	1.22	1.19	1.17	1.14	—
TMS-L-Ser–L-Ile	1.33	1.28	1.26	1.23	—
TMS-L-Ser–L-Pro	1.58	1.53	1.49	1.46	—
TMS-L-Ser–L-Met	—	3.10	2.94	2.72	—
TMS-L-Ser–L-Phe	4.47	4.08	3.82	—	—
TMS-L-Thr–L-Ala	0.60	0.61††	0.63	0.65	—
TMS-L-Thr–L-Val	0.93	0.92	0.92	0.92	—
TMS-L-Thr–L-Leu	1.15	1.11	1.10	1.09	—
TMS-L-Thr–L-Ile	1.25	1.23	1.20††	1.17	—
TMS-L-Thr–L-Pro	1.61	1.55	1.53	1.50	—
TMS-L-Thr–L-Met	—	2.88	2.74	2.49	—
TMS-L-Thr–L-Phe	—	4.10	3.86	3.56	—

*Calculated from adjusted retention time
†calculated from the total retention time
‡TMS = trimethylsilyl derivatives
**bis-N-TFA-derivatives
††Value from 204°C (Reference 10).

problems, the column material may influence relative retentions by differences in the activity of support materials and in chemical properties of the liquid phase if it is not all from the same batch. It is known that even the history of a column may influence the relative retention of a polar compound to a certain degree. With these limitations in mind, a screening of the numbers of Table I results in a reasonable agreement between the numbers of both laboratories.

The compounds in Table I are listed in such a manner that with fixed N-terminal amino acid *aa*, the carboxyl-terminated amino acid *x* is varied, resulting in dipeptide series *aa-x*. From these series it can be seen immediately that the sequence of the carboxyl-terminated amino acid *x* is the same in each series. Moreover, the compounds are shifted in each series by a rather constant factor. On the basis of this result it is possible to extend the system of relative retentions to cover other dipeptides the relative retentions of which have not yet been measured, provided that at least one compound from each of the series is available as a reference compound for this particular series. Weygand, Kolb and Kirchner[23] used a graph for these calculations and later on Tomida, Tokuda, Ohashi, and Nakajima,[24] also. This system, however, is more simply explained on the basis of retention indices and is therefore described later. In Table II, some trimethylsilyl ether derivatives of N-TFA-dipeptide methyl esters on Apiezon grease as liquid phase are compiled.

The combination of two amino acids however may result not only in dipeptides, but also in diketopiperazines, which have also been found

Table II
Relative Retention of Trimethysilyl Ether Derivatives of
N-TFA-Dipeptide Methyl Esters on Apiezon Grease
(Myristic Acid Methyl Ester = 1.00[17])

	190°C	*201°C*	*210°C*	*222°C*
TMS-L-Ser–L-Ala	0.26	0.27	0.28	0.29
TMS-L-Ser–L-Val	0.41	0.42	0.43	0.43
TMS-L-Ser–L-Leu	0.54	0.54	0.54	0.54
TMS-L-Ser–L-Ile	0.59_5	0.59	0.59	0.58
TMS-L-Ser–L-Pro	0.76	0.76	0.76	0.75
TMS-L-Ser–L-Met	–	1.80	1.72	1.62
TMS-L-Ser–L-Phe	–	2.73	2.63	2.44
TMS-L-Thr–L-Ala	0.23	0.24	0.25	0.26
TMS-L-Thr–L-Val	0.36	0.37	0.37	0.39
TMS-L-Thr–L-Leu	0.46	0.47	0.47	0.48
TMS-L-Thr–L-Ile	0.52	0.52	0.52	0.53
TMS-L-Thr–L-Pro	0.73	0.73	0.73	0.73
TMS-L-Thr–L-Met	–	1.52	1.47	1.40
TMS-L-Thr–L-Phe	–	2.31	2.22	2.09

in partial hydrolysates.[6] The result is confusion in the evaluation of such a chromatogram, since the diketopiperazines are running together with the N-TFA-dipeptide methyl esters. Relative retentions of some of these diketopiperazines, which have been measured under conditions comparable to the relative retentions of the N-TFA-dipeptide methyl esters in Table I, are listed in Table III.

Table III
Relative Retention of Diketopiperazines on Dimethyl Silicon Liquid Phase (Myristic Acid Methyl Ester = 1.00 [17])

	225°C
DL-valine anhydride	0.44
L-Pro–L-Leu anhydride	1.96
L-Ala–L-Phe anhydride	3.92
sarcosine anhydride	0.48

Westley, Close, Nitecki, and Halpern[25] reported the gas-chromatographic separation of 13 diketopiperazines which, however, are not included in Table III. Their retention data were given under conditions which do not enable recalculations for the purpose of getting values comparable to the N-TFA-peptide methyl esters. Diketopiperazines have also been found as physiologically active compounds from various kinds of microorganisms and have been analyzed also by GLC.[26]

3.4.2   The Retention Index System

Relative retentions are quite useful for qualitative identifications. However, the use of one reference compound only is a serious limitation, because relative retentions can be calculated from isothermal chromatograms only and not from temperature-programmed analyses. This drawback is avoided by the use of the retention index system by Kováts,[27, 28] which is also a type of relative retention but related to the two neighboring n-alkanes of the compound of interest. With the series of the n-alkanes as standard, the whole temperature range of GLC is covered. By definition, the retention index of the n-alkanes will be 100 times the carbon number, i.e., 600 for hexane, 700 for heptane, etc. The retention index system offers two distinct advantages. First, the indices can be calculated from linear temperature-programmed analyses, and further, the index values are descriptive. For example, if the retention index of N-TFA-L-Ile–L-Val-OMe* is 1652 with a dimethyl silicon liquid phase, this value immediately shows that it will emerge between hexadecane (I = 1600) and heptadecane (I = 1700) on such a column,

---

*-OMe = -methyl ester.

regardless of whether it is a packed or a capillary one. For further information however, the interested reader may refer to the cited literature, particularly to a review written by Ettre.[28] In the supplement of this paper, a simple graphic method is given for practical calculation of the index values.

Retention indices are known to be only slightly dependent on temperature; this is shown in Table IV for a temperature interval of 30°C in the example of some N-TFA-dipeptide methyl esters, calculated from a 50-m glass capillary with DC-200 silicon liquid phase. The variation of relative retentions for a 25°C change in temperature can be taken from the numbers in Table I.

Table IV
Temperature Dependence of Retention Indices
of N-TFA-Dipeptide Methyl Esters[°]

		170°C	*200°C*
N-TFA-Met-Ala-OMe	L,L	1778	1774
	D,L	1783	1777
N-TFA-Met-Val-OMe	L,L	1872	1870
	D,L	1884	1881
N-TFA-Met-Leu-OMe	L,L	1938	1935
	D,L	1949	1942
N-TFA-Ala-Phe-OMe	L,L	1907	1906
	D,L	1932	1928
N-TFA-Val-Phe-OMe	L,L	1994	1990
	D,L	2022	2015

[°]Conditions are the same as given in Table V.

As a mean, the index values decreased by only a few units with increasing temperature, although these deviations were of about the same order of magnitude as the precision of measurement, which was about ± two index units.

Table V summarizes the retention indices of some of the N-TFA-dipeptide methyl ester series with the same DC-200 glass capillary. A chromatogram of a synthetic mixture of N-TFA-dipeptide methyl esters with this particular column is given in Figure 4.

Assignment of the LL- and DL-compounds was made in several cases with the pure compounds. In general however assignment was made only by comparison of peak areas because synthesis of these compounds, starting with the pure L-amino acids, was carried out under conditions where a considerable racemization during synthesis could be expected in order to get the DL-compound for calculation, also. If both diastereoisomers had been resolved by the column, the minor peak was taken as the respective DL-compound. This procedure appears

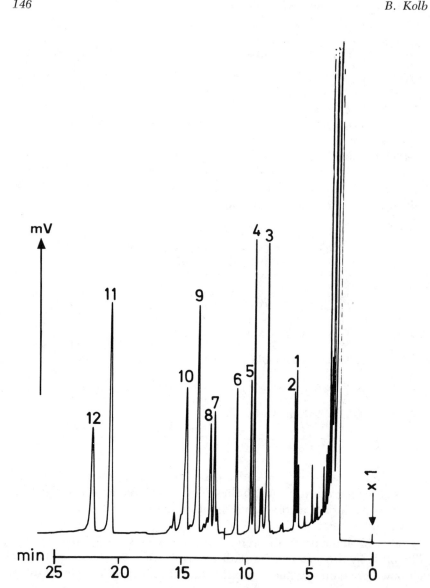

Figure 4. Gas-chromatographic separation of a synthetic mixture of N-TFA-methyl ester derivatives of some dipeptides.

Instrument: Perkin-Elmer Fraktometer F20; Column: 50-m x 0.25-mm glass capillary, pretreated according to Grob[21] and coated with dimethyl silicone DC-200 liquid phase; Carrier gas: 2 ml/min nitrogen, split ratio 1:25; Temperature: 200°C isothermal; Detector: FID.

1 = Val-Val(LL), 2 = Val-Val(DL), 3 = Ile-Leu(LL), 4 = Ile-Pro(LL), 5 = Ile-Pro(DL), 6 = Pro-Leu(LL), 7 = Met-Val(LL), 8 = Met-Val(DL), 9 = Ala-Phe(LL), 10 = Ala-Phe(DL), 11 = Leu-Phe(LL), 12 = Leu-Phe(DL).

Table V

Retention Indices of N-TFA-Dipeptide Methyl Esters (OMe) with Dimethyl Silicon Liquid Phase*

		-Ala-OMe	-Gly-OMe	-Val-OMe	-Leu-OMe	-Ile-OMe	-Pro-OMe	-Phe-OMe
170°C								
N-TFA-Ala-	L,L	1383	1396	1485	1558	1572	1612	1907
	D,L	1379		1492	1566	1580	1624	1932
N-TFA-Gly-	L,L	1414	1448	1540	1611	1624	1690	1970
N-TFA-Val-	L,L	1475	1489	1573	1645	1657	1680	1994
	D,L	1478		1587	1657	1670	1696	2022
N-TFA-Leu-	L,L	—	1560	1638	1705	1714	1754	2061
	D,L	—		1650	1716	1726	—	2083
N-TFA-Ile-	L,L	1599	1585	1652	1718	1729	1760	2064
	D,L	1564		1661	1724	1736	1770	2082
N-TFA-Pro-	L,L	1623	1619	1736	—	1812	1898	2177
N-TFA-Met-	L,L	1778	1804	1872	1938	1948	1988	—
	D,L	1783		1884	1949	1955	1994	—
200°C								
N-TFA-Met-	L,L	1774	1799	1870	1935	1949	1994	2290
	D,L	1777		1881	1942	1958	2000	2308
N-TFA-Phe-	L,L	1886	1909	1985	2046	2062	2087	2406
	D,L	1896		1997	2054	2073	2099	2424

*Conditions: 50 m × 0.25 mm glass capillary with DC-200 silicon liquid phase.

reasonable, since the reaction mixture was used for measurement without further purification to avoid fractionation between both diastereoisomers. This also explains the impurities in the chromatograms of the synthetic mixtures, which have not been further identified. It has been found that in general the LL-compound of N-TFA-dipeptide methyl esters is eluted before the corresponding DL-compound with the exception of N-TFA-Ala–Ala-OMe and dipeptides with proline as the N-terminal amino acid. In these cases the order of elution is reversed. This result agrees with retention data given in the literature,[20, 25] which, however, were measured with different columns.

As shown by Kovats, the index system is helpful for extrapolations, since within certain rules the prediction of the retention indices is fairly accurate. On the basis of the above-mentioned peptide series aa-x, such an extrapolation seems to be possible also in this class of compounds. Screening the numbers in Table V makes it obvious that the compounds in each series are shifted by a rather constant amount of index units. This increment can be taken from one of these peptide series, which is then used as a reference series and can be applied to another one to allow calculation of the indices of the respective compounds in this latter series. At least one compound in this particular series must be available, of course, as a basis for such calculations.

3.4.2.1 Example

The index of N-TFA-L-Ala–L-Phe-OMe should be calculated, first of all. One compound from the series L-Ala–L-x must be available—for example, N-TFA-L-Ala–L-Val-OMe with the index 1485. Then the index increment between the two corresponding compounds is taken from a reference series, for example, from the series L-Leu–L-x. The index difference between the two corresponding compounds in this reference series N-TFA-L-Leu–L-Phe-OMe ($I = 2061$) and N-TFA-L-Leu–L-Val-OMe ($I = 1638$) is 423 index units. Adding this value to the index 1485 of N-TFA-L-Ala–L-Val-OMe results in a calculated retention index of 1908 of N-TFA-L-Ala–L-Phe-OMe, which is in excellent agreement with the experimental value of 1907.

This relationship may be explained by the supposition that most of the N-TFA-dipeptide methyl esters have the same structure. Only the two different side chains give respective differences and each side chain results in a constant increment which obviously depends also on the position in the peptide chain. It is evident that this explanation presupposes the absence of intramolecular interaction of the side chain and steric effects; dipeptides which are found by experiment to be markedly different from the calculated value should be examined with this possibility in mind. Such deviations, for example, are found with dipeptides containing proline or glycine which may be related to the cyclic structure of proline and to the absence of a bulky side group with glycine, resulting in a stronger interaction of the polar groups of the molecule with the liquid phase. However, considering the wide range both of chemical properties of the different types of side chains and of the wide temperature range necessary for GLC of these compounds, such extrapolations should not be extended to an excessive amount. It is for these reasons that the above example may not be at all representative of the accuracy of those calculations. Due to the high number of possible dipeptides, however, such a method for calculating the retention data on the basis of a rather limited amount of reference compounds is of potential interest, and there is some probability that a similar relationship exists in the class of tripeptides, too, where it may be even more interesting.

3.5 Application to Sequence Analysis

From the beginning of application of GLC to peptide analysis this new technique was used for the purpose of sequence analysis of peptides. In the first report of Weygand, Kolb, Prox, Tilak, and Tomida[10] this approach was already being demonstrated in the example of six synthetic peptides of known composition.

Sequence analysis by GLC of the peptide under investigation is based on partial hydrolysis—preferably acid hydrolysis—either in concentrated hydrochloric acid[10, 18] or in a solution of 8.5N hydrogen chloride–anhydrous methanol in a sealed tube for 5–17 hours at 70°C.[6, 29] During this degradation all the amino acids as well as all the possible peptide fragments are produced, starting with dipeptides up to oligopeptides. Obviously they are not produced in equal amounts, since the various peptide bonds vary greatly in their susceptibility to acid hydrolysis. If partial hydrolysis is carried out in methanolic hydrochloride solution, the degradation products are formed as the respective ester hydrochlorides and need only trifluoroacetylation for further gas-chromatographic analysis.

Partial hydrolysis of a linear peptide chain—for example a tetrapeptide A–B–C–D—results in the following fragmentation products: two tripeptides A–B–C and B–C–D, three dipeptides A–B, B–C, and C–D, and the four amino acids A, B, C, and D, besides the original tetrapeptide. From the relation of one tetrapeptide, two tripeptides and three dipeptides in such a hypothetical example, it can be concluded that the peptide under investigation should be a linear tetrapeptide. Its amino acid sequence can be established on the basis of the identified dipeptides because the terminal amino acids of the original peptide are found only once in the fragmentation dipeptides, either in the respective amino- or carboxyl-terminated position. All other amino acids—namely, those amino-terminated in one dipeptide and carboxyl-

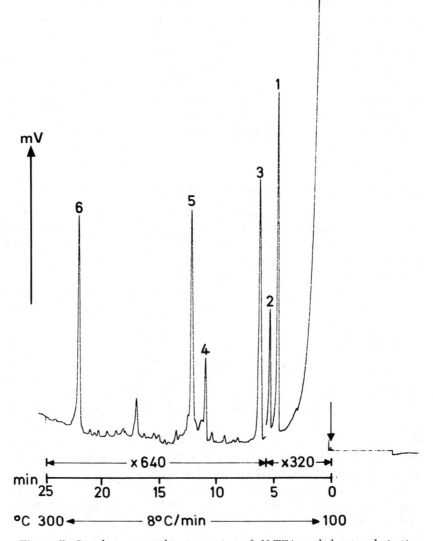

Figure 5. Gas-chromatographic separation of N-TFA-methyl ester derivatives resulting from a partial hydrolysate of the tetrapeptide L-Leu–L-Val–L-Ala–L-Ala. Instrument: Perkin-Elmer Model 900; Column: 6′ x ⅛″ stainless steel, packed with 2.5% dimethyl silicone OV-1 on Chromosorb-G, AW-DMCS 80/100 mesh; Carrier gas: 30 ml/min helium; Temperature: programmed as given; Detector: FID.
1 = L-Ala–L-Ala, 2 = L-Val–L-Ala, 3 = L-Leu–L-Val, 4 = L-Val–L-Ala–L-Ala, 5 = L-Leu–L-Val–L-Ala, 6 = L-Leu–L-Val–L-Ala–L-Ala.

terminated in another—are found twice. Summing up all these identified dipeptides with overlapping immediately results in the original amino acid sequence. This relation is demonstrated in Figure 5 in the example of the partial hydrolysate of such a tetrapeptide, which has been analyzed by GLC using a temperature-programmed packed column with OV-1 dimethyl silicon liquid phase. Identification of the three dipeptides in the chromatogram permits verification of the original amino acid sequence:

L-Leu–L-Val
  L-Val–L-Ala
    L-Ala–L-Ala

L-Leu–L-Val–L-Ala–L-Ala

In addition to the three dipeptides, the two possible tripeptides and even the original tetrapeptide can be analyzed successfully. The corresponding amino acid derivatives, however, have been lost in the solvent peak under the prevailing gas-chromatographic conditions.

The next example, in Figure 6, shows the chromatogram resulting from partial hydrolysis of the rather simple tripeptide L-Pro–Gly–L-Phe. The *N*-TFA-methyl ester derivatives of the three amino acids and the two possible dipeptides have been analyzed. Identification of both dipeptides results again in the amino acid sequence of the original tripeptide. This simplest example of all will now be used to discuss the identification problems in analysis by GLC.

On the basis of an amino acid analysis only, all of the following possibilities for the amino acid sequence must be taken into account: Gly–L-Phe–L-Pro, Gly–L-Pro–L-Phe, L–Phe–Gly–L-Pro, L-Phe–L-Pro–Gly, L-Pro–Gly–L-Phe, and L-Pro–L-Phe–Gly. This rather simple example demonstrates best the great variety of possibilities and the problems in identification by GLC. This is because the retention data of all the dipeptides which would result from these combinations must be known to confirm that under the prevailing conditions of GLC they all would have been resolved successfully. Only if all other combinations can be thus excluded is the identification confirmed. The necessary retention data may be taken from tabulated values by running reference compounds under the same conditions or, with some precautions, from calculated values. Obviously, any further information—for example, the knowledge of the terminal amino acid—would reduce considerably the number of possibilities.

In practice, however, the analysis of the dipeptides only in such a partial hydrolysate is not always sufficient, particularly if one or more of the amino acids are incorporated several times in the peptide chain at different positions, since sufficiently overlapping fragments may not

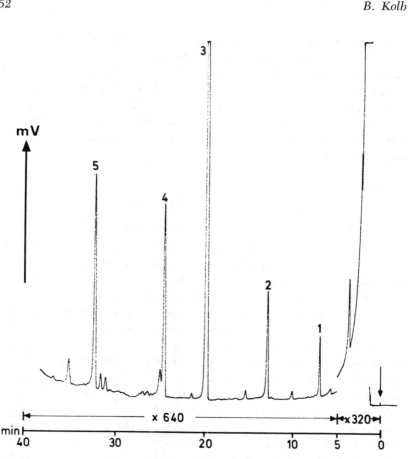

Figure 6. Gas-chromatographic separation of N-TFA-methyl ester derivatives resulting from a partial hydrolysate of the tripeptide L-Pro–Gly–L-Phe.
Instrument: Perkin-Elmer Model 900; Column: 12′ x ⅛″ stainless steel, packed with 5% SE-30 on Chromosorb-G, AW-DMCS 80/100 mesh; Carrier gas: 30 ml/min helium; Temperature: programmed as given; Detector: FID.
1 = Gly, 2 = Pro, 3 = Phe, 4 = L-Pro–Gly, 5 = Gly–L-Phe.

be available in this case. The tripeptides or even the tetrapeptides must then also be used for identification to cover the whole peptide chain with overlapping fragments. Due to the high amount of compounds in this class of peptides, however, retention data are no longer sufficient and a need arises for further methods of identification. This can be done by trapping the compounds from the column effluent for further investigations if a packed column is used. The trapped compound may again be submitted to partial hydrolysis and its amino acid sequence may be verified by means of GLC again, or by other chromatographic techniques.

3.5.1 Sequence Analysis by Gas Chromatography-Mass Spectrometry Combination

The most promising approach for further investigation of peptides separated by GLC is by means of mass spectrometry (see Chapter 4 and *cf.* Section 3.3.2), as was pointed out by Weygand, Prox, and co-workers in several contributions.[3-7,29] It is mainly the advances in mass spectrometry of peptides that renders the combination of gas chromatography with mass spectrometry (GLC/MS) of utmost importance for sequence analysis now. The GLC/MS technique has been used with exceptional success for elucidation of the structure of several peptides from natural origin. Some of the results will be discussed in the following pages, thus complementing Chapter 4.

Prox and Weygand[6] reported the sequence analysis of a homodetic cyclic nonapeptide from linseed by such a combined technique. The peptide fragments from a partial methanolysate were separated as trifluoroacetyl methyl esters on a temperature-programmed, packed silicon column, and 20 peptides, including even a pentapeptide, could be separated. After the compounds were trapped in a glass tube, they were all further identified by mass spectrometry. From these results the amino acid sequence could be derived, as shown by overlapping the identified fragments:

```
L-Val–L-Pro
L-Val–L-Pro–L-Pro
L-Val–L-Pro–L-Pro–L-Phe
      L-Pro–L-Pro
      L-Pro–L-Pro–L-Phe
            L-Pro–L-Phe
            L-Pro–L-Phe–L-Phe
                  L-Phe–L-Phe
                  L-Phe–L-Phe–L-Leu
                        L-Phe–L-Leu
                        L-Phe–L-Leu–L-Ile
                              L-Leu–L-Ile
                              L-Leu–L-Ile–L-Ile
                              L-Leu–L-Ile–L-Ile–L-Leu
                              L-Leu–L-Ile–L-Ile–L-Leu–L-Val
                                    L-Ile–L-Ile
                                    L-Ile–L-Ile–L-Leu
                                    L-Ile–L-Ile–L-Leu–L-Val
                                          L-Ile–L-Leu
                                          L-Ile–L-Leu–L-Val
```

L-Val–L-Pro–L-Pro–L-Phe–L-Phe–L-Leu–L-Ile–L-Ile–L-Leu

Some problems were encountered in this particular investigation because it was difficult to decide between leucine and isoleucine. The

GLC did not separate the respective leucine and isoleucine peptides and even mass spectrometry could not make it possible in every case to decide between isomers. The final decision was possible by use of a capillary column for the further separation of trapped compounds. These difficulties, however, also appear to be overcome now by mass spectrometry.[30]

Later on Weygand[7] reported the sequence analysis of a methionine-containing cyclic nonapeptide which had been found as a contaminant in the above-mentioned cyclopeptide. Since methionine is decomposed under the conditions of partial methanolysis with methanol–hydrochloric acid and for this reason methionine-containing peptides were missing in the gas chromatogram, the peptide was first desulfurized with Raney nickel before partial hydrolysis. After derivatization 16 peptides could be separated by GLC, as shown in Figure 7, and identified by mass spectrometry after the compounds from the column effluent had been trapped. Methionine in this case was represented by α-aminobutyric acid (Abu) in the respective fragment peptides.

Arrangement of the identified peptide fragments, again with overlapping, results in the original amino acid sequence of this peptide, all of the amino acids probably having L-configuration:

Pro-Pro
Pro-Pro-Phe
Pro-Phe
Pro-Phe-Phe
Phe-Phe
Phe-Phe-Val
Phe-Val
Phe-Val-Ile
Phe-Val-Ile-Met
Val-Ile
Val-Ile-Met
Val-Ile-Met-Leu
Ile-Met
Met-Leu
Met-Leu-Ile
Leu-Ile

⌐Pro-Pro-Phe-Phe-Val-Ile-Met-Leu-Ile⌐

In the same contribution[7] Weygand mentioned the sequence analysis of two further cyclopeptides from *Amanita phalloides*, performed by the same technique. One of these peptides, antamanid, is a cyclodekapeptide and the other, phallin A, a cyclohexapeptide.

Verification of the amino acid sequence of antamanid was reported later by Wieland *et al.*[29] and more extensively by Prox *et al.*[79] From these investigations some problems with side reactions during partial hydrolysis have been reported.

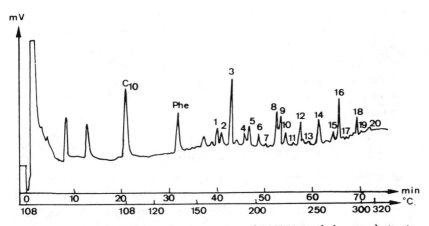

Figure 7. Gas-chromatographic separation of N-TFA-methyl ester derivatives resulting from a partial hydrolysate of a desulfurated cyclic nonapeptide [Reprinted, by permission, from Zeitschrift für Analytische Chemie **243**, 2 (1968). © Springer-Verlag.]

1 = Abu-Leu + Ile-Abu*, 2 = Val-Ile, 3 = Leu-Ile, 4 = Pro-Pro (diketopiperazine), 5 = Pro-Pro, 6 = Phe-Val, 7 = Phe-Leu, 8 = Pro-Phe, 9 = Val-Ile-Abu*, 10 = Abu-Leu-Ile*, 11 = Leu-Leu-Ile, 12 = Phe-Phe, 13 = Phe-Val-Abu*, 14 = Phe-Val-Ile, 15 = Val-Ile-Abu-Leu*, 16 = Pro-Pro-Phe, 17 = Phe-Phe-Val + Phe-Val-Phe, 18 = Phe-Val-Ile-Abu*, 19 = Pro-Phe-Phe, 20 = ?.

After partial hydrolysis and derivatization, this cyclodekapeptide results in more than 30 components in the gas chromatogram. Arranging again some of the characteristic peptide fragments with overlapping results in the amino acid sequence, all amino acids having L-configuration:

$$\text{Val-Pro-Pro-Ala}$$
$$\text{Ala-Phe-Phe}$$
$$\text{Phe-Phe-Pro}$$
$$\text{Phe-Pro-Pro}$$
$$\text{Pro-Pro-Phe}$$
$$\text{Phe-Phe-Val}$$
$$\text{Phe-Val-Pro}$$

$$\boxed{\text{Val-Pro-Pro-Ala-Phe-Phe-Pro-Pro-Phe-Phe}}$$

However, several peptide fragments which also had been identified—Phe-Val-Phe, Phe-Phe-Ala, Phe-Val-Ala, and Phe-Phe-Phe—did not fit in this structure. The explanation for occurrence of these misleading fragments is given by the supposition that during acid hydrolysis of such a cyclic peptide in methanol a smaller cyclic peptide is formed by restriction of the cycle; the rest of the molecule is subsequently eliminated. This reaction is induced by a transannular interaction and re-

---

*Abu = α-aminobutyric acid instead of methionine.

sults in a new peptide linkage which does not exist in the original peptide. Elimination of the smaller part of the cycle may also result in a diketopiperazine if only two amino acids are affected. The combined technique of GLC-mass spectrometry together with liquid chromatography was also used by Bayer and Koenig[18] for investigations concerning failure sequences in peptide synthesis under the Merrifield method.[31]

3.6 Separation of Diastereoisomers*

One of the most interesting results at the beginning of the application of GLC to N-TFA-peptide methyl esters was the successful separation of diastereoisomeric N-TFA-dipeptide methyl esters, first reported by Weygand, Kolb, Prox, Tilak, and Tomida,[10] who achieved the separation of the N-TFA-methyl ester derivatives of L-Ala–L-Phe and L-Ala–D-Phe using merely a 2-m packed silicon column at 200°C.

Following this early result, a series of diastereoisomeric pairs of N-TFA-dipeptide methyl esters was separated by Weygand and co-workers,[20, 22] mainly with capillary columns. These encouraging results were immediately applied to determine the configuration of amino acids in naturally occurring as well as synthesized peptides, particularly for investigations concerning racemization during peptide syntheses.

Due to the separation of diastereomeric N-TFA-dipeptide methyl esters into two peaks, each containing two of the four possible configurations—either LL and DD or LD and DL if all of them are present—the determination of the configuration of one amino acid in a dipeptide is possible only if the configuration of the other is known. The respective pairs of enantiomers have not been resolvable by GLC up to now, although this should not be impossible at all, since Gil-Av[32] was successful in separating enantiostereoisomeric N-TFA-amino acid methyl esters.

So far, apart from a few diastereoisomeric tripeptides,[19] only the dipeptide class has been investigated to such an extent with regard to diastereoisomer separation. Higher peptides must first be degraded by partial hydrolysis to dipeptides for determination of the configuration of the respective amino acids. For each dipeptide the configuration of one amino acid must be known exactly.

Under this method Tomida et al.[24] have verified the D-configuration of phenylalanine in gramicidin J_1 by identification of the N-TFA-methyl ester derivatives of L-Pro–L-Val, L-Val–L-Orn, L-Orn–L-Leu, L-Leu–D-

---

*See also Section 2.2.11 in this volume.

Phe and D-Phe–L-Pro. Performed in the chromatogram of a partial hydrolysate, this identification also has proved the same sequence as in gramicidin S.

Weygand and co-workers[20, 22, 33, 34] applied this method to synthetic tripeptides for investigations concerning the degree of racemization during synthesis of these tripeptides, but these investigations are discussed in more detail in Section 3.6.2. During partial hydrolysis in the methanolic hydrogen chloride system, some racemization occurs by hydrolytic cleavage of a peptide bond in the carboxyl-terminated amino acid of the resulting dipeptide. This slight racemization may be compensated for by similar treatment of the appropriate amino acid under the same conditions, taking into account the degree of racemization as a blank.[36]

It is evident, however, that the N-TFA-dipeptide methyl esters resulting from partial hydrolysis of a higher peptide are not the only possible derivatives with diastereoisomeric properties suitable for determining the configuration of amino acids by GLC. The peptide of interest may also be submitted to total hydrolysis, and the resulting amino acids may then be converted to suitable derivatives by introduction of a second asymmetric center to form diastereoisomers for separation by GLC. This can be done extremely well with the simple N-TFA-amino acid methyl esters if a glass capillary coated with an optically active liquid phase, as reported by Gil-Av,[32] is used; under these conditions the separation of the enantiomeric L- and D-amino acid derivatives is achieved by forming a hydrogen-bonded "diastereoisomeric" association complex with the liquid phase. Such a derivative may also be an ester with an optically active alcohol as reported by Charles[37] and Gil-Av,[38, 39] using the 2-butyl or 2-n-octyl esters. Pollock and co-workers[40–42, 78] used a similar procedure.

Finally, the resulting amino acids from a total hydrolysate may be coupled with another amino acid of known configuration, forming N-TFA-dipeptide esters again. In this case only one of both possible diastereoisomers should be present if the respective amino acid from the total hydrolysate is optically pure and provided that any additional racemization during the step of peptide synthesis can be excluded. For this reason, Halpern and Westley[43] used N-TFA-L-prolyl chloride as an optically active N-acylating agent: proline does not racemize during acylation or peptide synthesis since the formation of oxazolone intermediates is not possible. The amino acid mixture resulting from total hydrolysis was first esterified and then acylated, and the resulting N-TFA-L-prolyl peptide methyl esters were separated by GLC.

Prox and Weygand[6] used a similar procedure to confirm the L-configuration of proline in a cyclic nonapeptide (see Section 3.5.1). After total hydrolysis the resulting amino acids were first trifluoroacetylated and then converted to the respective N-TFA-amino acid–L-valine methyl esters with L-valine methyl ester under conditions where no racemization was to be expected. The L-configuration of proline was confirmed through GLC but a considerable amount of about 10% D-proline was also detected, and this was attributed to the excessively long hydrolysis time necessary for total hydrolysis. This result now is of general interest because it shows that all methods used for investigation of the optical purity of compounds based on total hydrolysis of the respective peptide, suffer the disadvantage of occurrence of some degree of racemization during acid hydrolysis. This must be expected much more under the conditions of total hydrolysis than under the gentler conditions of partial hydrolysis. Thus the latter method should be given preference, although if necessary this racemization may be offset to a certain extent by a similar treatment of the pertinent amino acids, as mentioned above.

As shown above, the configuration of one amino acid in an N-TFA-dipeptide can be determined if the configuration of the other is known, provided that both diastereoisomers are resolved by GLC. Since during peptide synthesis only the N-terminal amino acid in its activated state is affected by racemization[22, 44] while no racemization is to be expected with the other amino acid, GLC opens the possibility for measuring the degree of racemization. This was shown by Weygand and co-workers in a series of extensive studies.

If half of the dipeptide is optically pure, then any optical impurity in the other half will lead to the two resolved components. Thus, when starting a dipeptide synthesis with the pure L-amino acids, and after some racemization during the synthesis, only the diastereoisomeric LL and DL combinations are to be expected and not the corresponding enantiomers DD and LD.

Racemization is a big problem in peptide chemistry, and measuring the degree of racemization during peptide synthesis is another one. The drawback of all methods used up to now has been the need to purify the compounds under investigation in order to check their optical purity. On the other hand, during the clean-up procedure after completion of the synthesis any fractionation must be strictly avoided to get a reliable result. This is the problem if the optical rotation is measured, since pure compounds must be at hand and weighed for this purpose. The same necessity exists for enzymatic methods, because blocking of the enzyme by impurities must be avoided. Enzymatic methods are quite useful for investigations of purified peptides, but for the above-mentioned reasons not as good for investigations on the crude reaction mixture itself.

These are the types of problems which can be solved most expertly by GLC, and suitable techniques are available for such studies. The reaction mixture of a peptide synthesis may be used immediately for GLC, even if it includes solvents, unreacted starting materials, additives such as salts and bases, and other impurities. All these compounds are either separated by the column or, if they are not volatile, they remain in the injection system of the gas chromatograph. Moreover, all measurements can be done with high precision on a submicrogram scale. Thus GLC may also be used to monitor synthetic reactions with respect to the optical purity of the final product.

For all these reasons, analysis by GLC is more sensitive, simpler, and more reliable, needs less time and material, and is more flexibly adopted to the various problems than all other tests used so far for these applications, which all suffer the above-mentioned limitations. These other tests are based on the synthesis of some representative peptides where the degree of racemization is measured by polarimetry (Young test),[45] by weighing the amount of any racemate after quantitative isolation by fractional crystallization (Anderson test),[46] or by separation of the respective diastereoisomers by multiplicative distribution together with polarimetry of the DL-fraction if complete separation is not achieved (Kenner test).[47]

However, the exclusive use of the N-TFA-methyl ester derivatives for the gas-chromatographic method does not mean that only these derivatives can be employed for the foregoing peptide synthesis. After completion of the peptide synthesis, another protecting group may be replaced by the trifluoroacetyl group without fractionation or racemization. This is possible at least with the widely used carbobenzyloxy group, which can be split off and replaced by the trifluoroacetyl group through treatment of the peptide with trifluoroacetic acid.[14] It should also be possible, however, to use the N-carbobenzyloxy dipeptide methyl esters directly for gas-chromatographic analysis.[15]

For investigations concerning racemization during peptide synthesis it is sufficient to restrict oneself to the synthesis of a representative di- or tripeptide, the diastereoisomers of which are well-separated on a suitable column. The influence on racemization of all parameters involved may be studied on the example of such a model, and calculation of the percentage of racemization is based on the sum of both peak areas considered as twice the area percentage of the DL-isomer.[20, 22] It has been shown[22] that the peak area percentage of both diastereoisomers corresponds to the weight percentage of both compounds, as can be expected with such closely related isomers. Therefore no response factors with their inherent problems are necessary.

Some simple test procedures have been developed by Weygand and co-workers and have been used on a routine basis for a series of extensive studies.[19, 20, 22, 33, 34, 35, 36, 44] Tests reported have covered all aspects of racemization, such as type of applied synthesis method, influence of the type of solvents, influence of reaction time, temperature, additives such as salts and tertiary bases, and the influence of inductive and steric effects. It is beyond the scope of this paper to go into details of these investigations, and the results are quoted only to an extent that will enable the interested reader to find the original literature.

3.6.1 Dipeptide Test

The synthesis of L-valyl–L-valine, either as N-carbobenzyloxy methyl ester or N-trifluoroacetyl methyl ester derivative, was found to be a suitable model for such a test procedure because, due to the steric hindrance of the bulky side groups, the rate of reaction should be lowered as compared to other compounds, and thus the influence of the various factors on racemization should be well-recognized and measured. Both diastereoisomers are well-separated on a capillary column with polyphenyl ether OS-138 or polypropylene glycol liquid phase (see Figure 2). The carbobenzyloxy protecting group was replaced by the trifluoroacetyl group before gas-chromatographic analysis, as mentioned above.[14]

The following methods of peptide synthesis have been studied with this test procedure in early papers on this subject:[20, 22] methods of activated esters (p-nitrophenyl ester,[48] cyanomethyl ester,[49, 50] thiophenyl ester in glacial acetic acid,[51] and vinyl ester[52]); dicyclohexylcarbodiimide method;[53] azide method;[54] carbonyldiimidazole method;[55] phosphorazo method;[56] mixed anhydride method;[57] Woodward's method;[58] and Patchornik's method.[59] The influence of temperature, triethylamine, and various salts as additives and the effect of various solvents were studied on the example of the dicyclohexylcarbodiimide method.

As a general result of these investigations it may be mentioned that the methods show a negligible amount of racemization, or none at all, provided a urethane-type protecting group such as the carbobenzyloxy group is used. An exception is the method of cyanomethyl ester with 2% racemization and the thiophenyl ester in glacial acetic acid with 29%. Contrary to these results, all methods using N-acyl protecting groups like the N-trifluoroacetyl group show more or less racemization with the exception of the azide method and the vinyl ester method. The degree of racemization depends on solvents, reaction temperature, and the method used to introduce the methyl ester of L-valine. There is less racemization if the free ester is added rather than generated from its hydrogen chloride with triethylamine, since a portion of the triethylamine hydrogen chloride, depending on the solvent, remains always in solution. According to the results found by the high degree of racemization when imidazole or N-benzylimidazole is added using the dicyclohexylcarbodiimide method, it was concluded that synthesis of peptides containing histidine should be prepared only by way of acyl amino acids with urethane-type structure or from N-acyl peptides by the azide method.

3.6.2 Tripeptide Test

For investigations concerning racemization if higher peptides are built up by linking smaller peptides, it is sufficient to confine the study to a representative tripeptide. This tripeptide is synthesized from an amino-protected dipeptide Z-L-aa$_3$-L-aa$_2$-OH* with an amino acid ester H-L-aa$_1$-OR, resulting in the tripeptide derivative Z-L-aa$_3$-L-aa$_2$-L-aa$_1$-OR. No racemization is to be expected in that peptide, which is used as the amino component, and thus it might be substituted by a simple amino acid ester. Suitable techniques for analyzing possible content of the racemized amino acid aa^2 in that tripeptide have been developed at Weygand's laboratory.

In some early studies on this subject, the tripeptide derivative Z-L-Leu–L-Phe–L-Val-OtBu was found to be a suitable model. It was prepared by coupling Z-L-Leu–L-Phe-OH (Z = either benzyloxycarbonyl or p-methoxybenzyloxycarbonyl) with H-L-Val-OtBu (L-valine-t-butyl ester). The resulting tripeptide derivative was submitted to partial hydrolysis and after conversion into the respective N-TFA-methyl ester derivatives, the resulting dipeptide derivative N-TFA-L-Phe–L-Val-OMe was analyzed by GLC with regard to the possible content of the respective DL-isomer. A capillary column with polyphenyl ether OS-138 liquid phase was used.

With this procedure, which may be obsolete now due to a technique described later, 16 different methods for peptide synthesis were investigated extensively by Weygand, Prox, and Koenig.[44] No racemization was found with the methods using N-hydroxypiperidine ester, according to Young,[60] or using the vinyl ester[52] but the methods using the azides[54] and the p-nitrophenyl esters[48] were found free of racemization only if excess of bases was avoided. Only a trace amount of racemization was found using the method of Jakubke[61] with 8-hydroxychinoline ester, and no racemization was found with the thiophenyl ester method using tetrahydrofurane as solvent. More or less racemization was observed using ethoxyacetylene,[62] carbonyldiimidazole,[55] carbonylditriazole,[63] and dicyclohexylcarbodiimide[53] as condensating

---

*Z = amino protecting group.

agents, while a high amount of racemization occurred when the phosphorazo method[56] and inamines were used. Examples of inamine methods include (t-butyl ethinyl)-dimethyl amine[64] and the water-soluble diimide, 1-ethyl-3-(3-dimethylamino-propyl)-carbodiimide-hydrogen chloride.[65] A slight racemization was also encountered with the methods of Woodward[58] and Patchornik,[59] and the methods of mixed anhydrides.[57]

The dicyclohexylcarbodiimide method has been found free of racemization if the reaction is carried out at $-20°C$ in tetrahydrofurane or dimethylformamide with the addition of two equivalents of N-hydroxysuccinimide.[35] Inamines have been found free of racemization only if used for synthesis of benzyloxycarbonyl dipeptide esters, whereas a high amount of racemization has been found when this method has been used for synthesis of either N-TFA-dipeptide esters or for the linkage of an N-benzyloxycarbonyl peptide with an amino acid ester or a peptide ester.[33]

Beyerman and Weygand et al.[36] used the same technique to investigate the influence of some representative bifunctional catalysts on racemization in peptide synthesis via aminolysis of cyanomethyl, p-nitrophenyl, thiophenyl, and vinyl esters. 1,2,4-Triazole, which was investigated most extensively, did not cause racemization, but imidazole, which had also been recommended for acceleration of peptide synthesis, led to racemization in several cases.

Later on, the tripeptide test was redesigned and further simplified by Weygand, Hoffmann and Prox,[19] because it had been found that diastereoisomers of N-TFA-tripeptide methyl esters can also be separated by GLC. The N-TFA-methyl ester derivatives of the tripeptide prolyl–valyl–proline has proved quite useful for this purpose, due to the excellent separation of the respective diastereoisomers N-TFA-L-Pro–L-Val–L-Pro-OMe and N-TFA-L-Pro–D-Val–L-Pro-OMe using a 2-m packed column with 0.5% FFAP on Chromosorb-G. If the N-TFA-derivatives are used for synthesis, both partial hydrolysis and trifluoroacetylation are avoided; if the carbobenzyloxy protecting group is used, it may be replaced by the trifluoroacetyl group as outlined above. In the same paper, results are given for a range of 0.5–9.9% D-valine in that tripeptide which was synthesized from L-proline methyl ester and N-TFA-L-Pro–L-Val-N-hydroxysuccinimide ester or from L-proline methyl ester and N-TFA-L-Pro–L-Val-OH with either dicyclohexylcarbodiimide or 1-cyclohexyl-3-(-2-morpholine ethyl)-carbodiimide-metho-p-toluenesulfonate. Dicyclohexylcarbodiimide together with N-hydroxysuccinimide was used as condensating agent under various temperatures and tetrahydrofurane or methylene chloride acted as solvents.

3.7 Application to Special Problems

Racemization during peptide synthesis with the dicyclohexylcarbodiimide method was related to the formation of oxazolone intermediates which racemize quickly. 2-Trifluoromethyl-4-isopropylpseudooxazole-5-one (I) was in fact detected by GLC as an intermediate in the reaction of N-TFA-L-valine with the methyl ester of L-valine (II) and dicyclohexylcarbodiimide in absolute tetrahydrofurane at 20°C.[20]

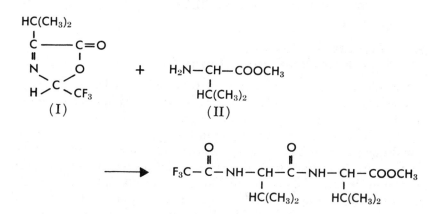

Since the pseudooxazolone, which is rapidly formed, reacts with the methyl ester of valine to yield the methyl ester of N-TFA-DL-valyl-L-valine, it disappears with increasing time. As a result 74% LL- and 26% DL-derivatives have been found by GLC.[20] This type of asymmetric synthesis may be studied with respect to the influence of the reaction parameters and type of side chain.

Weygand[12] turned attention to the possible use of this type of asymmetric synthesis for the introduction of tritium in the α-position of the amino terminal amino acid. This proceeds on condition that the respective DL-amino acid is reacted with trifluoroacetic anhydride with addition of some tritium-labeled water. By means of proton-exchange the tritium-labeled pseudooxazolone is formed. After reaction with an optically active amino acid methyl ester—for example, of the L-configuration—this pseudooxazolone results in the respective tritium-labeled dipeptide derivative which may be further separated by recrystallization or by using preparative GLC.

GLC has also been used to study the ratio of α:γ-peptides as a result of splitting N-TFA-glutamic anhydride by amino acid methyl esters. It has been found that the α-dipeptide derivatives have shorter retention times (see Table I) than the γ-derivatives and that this behavior is related to the spherical geometry of the α-derivatives.[17]

Similarly, the ratio of $\alpha:\beta$-peptide esters resulting from N-TFA-aspartic acid anhydride can be determined, and any racemization during these reactions can be measured.[12]

Asparagyl peptide esters are known for $\alpha:\beta$ transformation during saponification in alkaline solution, and the results found by GLC (Weygand)[12] agreed fairly well with investigations of Bernhard et al.[66] These latter studies had been based on measuring reaction kinetics; the results illustrated the sequence of reactions including diastereoisomeric imides, which could be detected by GLC.

3.8   Peptide Analysis by Pyrolysis Gas-Chromatography

Nonvolatile organic compounds may be analyzed by pyrolysis gas chromatography, thus avoiding chemical derivatization. The samples are thermally fragmented in a stream of inert gas, and the reaction products are passed immediately onto the column for further separation. Fragmentation by pyrolysis is carried out either by flash pyrolysis, with a wire in contact with the sample and heated by an electrical current, or by use of a heated tube as a furnace through which carrier gas is flowing. Pyrolysis gas chromatography is quite useful provided the breakdown products are closely related to the structure of the original compound.

Pyrolysis gas chromatography of amino acids has been applied by several authors, first by Janák[67] and Ulehla,[68] but because pyrolysis conditions are difficult to reproduce, such an approach to the analysis of amino acids would not be expected to be suitable for quantitative analysis. However, it is interesting to note that the fragments found in the pyrolysis of amino acids are also formed in fragmentation of the corresponding peptides.[67, 68, 69, 70] For this reason, pyrolysis gas chromatography could be very useful if the pyrolysis chromatogram of a peptide could be obtained by simply adding the characteristic fragments of the corresponding amino acids. This, however, meets with some difficulties because the relative intensities of the resulting peaks, which are characteristic for each amino acid and important for assignment, seem to be affected by the peptide linkage. Some of the simple amino acids yielded relatively simple chromatograms when pyrolyzed as pure compounds whereas some other amino acids—for example, arginine—have been reported[71] to give no less complex chromatograms than those obtained with proteins.

To overcome the problems posed by the poor reproducibility of pyrolysis conditions, Simon and Giacobbo[72] used a device formed by depositing a thin, roughly monomolecular layer of film on a ferromagnetic conductor. The film can be heated in 20–40 msec to an accurately defined temperature, given by the respective Curie point. With this procedure, the fragmentation patterns of 28 amino acids have been published.[72] However, when applied to peptide analysis the same problems were encountered. This is shown in Figure 8 where the fragmentation pattern of the dipeptide prolylphenylalanine and of a 1:1

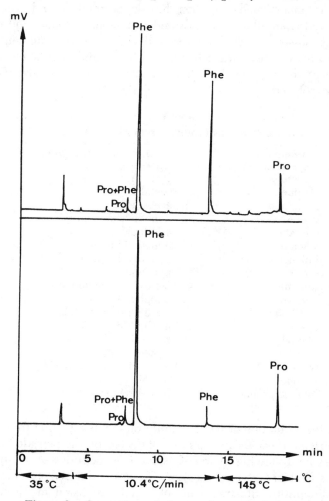

Figure 8. Comparison of the pyrolysis gas chromatograms of prolyl-phenylalanine (A) and a 1:1 mixture of proline and and phenylalanine (B).

GLC: Perkin-Elmer 226; Golay K 1540 (150′, 0.01″); 1.2 ml of N$_2$ per minute; 4 minutes at 35°C; programmed at 10.4°C per minute; 145°C [Reprinted, by permission, from Angewandte Chemie. International Edition **4**, 938 (1965). © Verlag-Chemie.]

mixture of proline and phenylalanine as sodium salts are compared. As can be seen, the relative intensities of the characteristic fragments are markedly changed. These investigations seem to have done much more for the approach of pyrolysis technique than for peptide analysis.

Merrit and Robertson[73] have reported the results found by pyrolysis of 17 amino acids and 10 peptides. Among other common, frequently occurring fragments such as carbon dioxide, acetonitrile and propenylnitrile derived from all amino acids and ammonia, methane, and nitrogen, it was possible to find unique and characteristic fragments for each amino acid. These are listed in Table VI and should be useful for identification purposes provided they are reproducible also if peptides are pyrolyzed.

Table VI

Unique Pyrolysis Products from Amino Acids[73]

Amino acid	Unique pyrolysis product
glycine	acetone
alanine	acetaldehyde
β-alanine	acetic acid
valine	2-methylpropanal
norvaline	n-butanal
leucine	3-methyl butanal
isoleucine	2-methyl butanal
serine	pyrazine
threonine	2-ethylethyleneimine
taurine	thiophene
methionine	methyl propyl sulfide
cystine	methyl thiophene
phenylalanine	benzene
tyrosine	toluene
tryptophan	ammonia, carbon dioxide
proline	pyrrole
hydroxyproline	N-methyl pyrrole

However, the formation of these unique fragments resulting from pyrolysis of pure amino acids appears to be affected by the peptide linkage, too, since the following pairs of isomeric dipeptides—Ala–Gly and Gly–Ala as well as Gly–Leu and Leu–Gly—gave different fragments.

Other dipeptides, the fragments of which are listed in Table VII, agreed with the results found with pure amino acids (Table VI).

These deviations of peptide pyrolysis as compared to amino acid pyrolysis appear to depend on the position of the amino acid in the peptide. This result would tend more to confuse the evaluation of the fragment pattern found with peptide pyrolysis than to provide a valuable means of studying end group positions of certain amino acids in the peptide

Table VII
Unique Pyrolysis Products from Peptides[73]

Peptide	Unique pyrolysis product
glycyl-glycine	acetone
glycyl-valine	acetone, 2-methylpropanal
glycyl-proline	acetone, pyrrole
glycyl-methionine	acetone, methyl propyl sulfide
glycyl-serine	acetone, pyrazine
glycyl-tryptophan	acetone, ammonia
glycyl-alanine	acetone, 2-methyl pyrrole
alanyl-glycine	acetone, acetaldehyde, ammonia
glycyl-leucine	acetone, cyclopentane
leucyl-glycine	acetone, acetic acid

sequence, as suggested by the authors. Extending these investigations also to natural peptides, for example, pyrolysis of crystalline bovine insulin, resulted in acetone from glycine, benzene from phenylalanine, toluene from tyrosine, pyrrole from proline, propene from glutamic acid, and carbonyl sulfide and carbon disulfide from sulfur-containing amino acids.

3.9 Supplement

3.9.1 *Determination of Retention Data*

All retention data, either for relative retentions or for retention indices determined under isothermal conditions, are derived from the adjusted retention time measured from the "air peak" time and not the time of injection. When using a flame detector, however, the air peak is not recorded in the chromatogram but can be calculated on the basis of a homologous series. If retention indices are used for data presentation, it is convenient to use the homologous series of n-alkanes for the calculation of both the air peak time and the retention indices.

The present paper does not pretend to go into the problems of sophisticated theories. Only the simplest methods are discussed here for utilizing the advantages of retention indices for practical work, because retention indices might be quite helpful in gas-chromatographic analysis of peptides, too. It is the experience of the author that only the simplest methods do have a chance to be used in practical application; for this reason the methods that follow are restricted mainly to graphic techniques.

3.9.2 *Calculation of the Air Peak Time*

The calculation of air peak time utilizes the known fact that a linear relationship exists between the logarithm of the adjusted retention time and the number of carbon atoms in a homologous series. From the methods described in the literature[74, 75, 76] using this relationship, only the simplified method of Peterson and Hirsch[74, 75] is now discussed (see Figure 9).

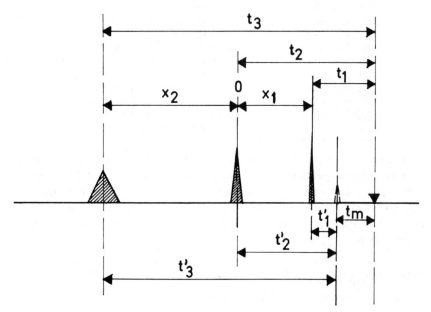

Figure 9. Calculation of the air-peak time using the simplified method of Peterson and Hirsch.[74, 75]

t_1, t_2, t_3 = retention time of the three homologs, measured from the point of injection

t_1', t_2', t_3' = adjusted retention time of the three homologs, measured from the air peak

t_m = retention time of the air peak.

The three peaks in Figure 9 should represent three homologs—for example, three n-alkanes—and the peak maximum of the second homolog is designated as an arbitrary reference point O, from which the distance of the other homologs x_1 and x_2 is measured. The adjusted retention time of the second homolog t_2' is calculated from the formula:

$$t_2' = \frac{x_1 \cdot x_2}{x_2 - x_1}$$

The "air peak" time t_m results from the difference of the retention time t_2 and the adjusted retention time t_2' of this second homolog.

$$t_m = t_2 - t_2'$$

From this result, the adjusted retention times of all other peaks in the chromatogram can be calculated.

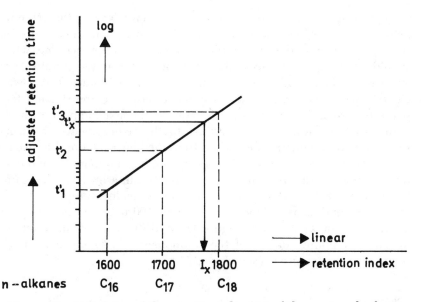

Figure 10. Calculation of the retention index (I_x) of the compound x from an isothermal chromatogram.

    t_1', t_2', t_3' = adjusted retention time of the three homologous n-alkanes

               C_{16}, C_{17}, C_{18}

    t_x' = adjusted retention time of the compound x.

3.9.3 Determination of the Retention Indices

3.9.3.1 Calculation from isothermal chromatograms

The basic relationship is the linear relationship that exists between the logarithm of the adjusted retention times of the members of a homologous series and their carbon number. For the purpose of determining the retention indices it is only necessary to make a semilogarithmic plot as shown in Figure 10. Here the carbon number is multiplied by the factor 100 and the distance between the n-alkanes is divided into 100 units. On the logarithmic scale of the ordinate the adjusted retention times of the respective n-alkanes are plotted first and the linear reference line of the n-alkanes is drawn. Further on, the adjusted retention times t_x' of each of the interesting components x in the chromatogram is calculated as outlined above, and its retention index I_x is found by the help of this reference line as shown in Figure 10.

3.9.3.2 Calculation from temperature-programmed chromatograms

This method is indeed simpler than the calculation from isothermal chromatograms since in this case even the adjusted retention times are not necessary and the retention times themselves or the retention temperatures can also be used. However, more than three n-alkanes should be available then to establish a reasonable reference line, which should be linear, too, provided that an ideal linear temperature program is employed. This means that the program is started at a sufficiently low temperature. Since in practice, however, the starting temperature is set according to the separation of the first components in the chromatogram rather than according to the requirements of theory, some positive or negative deviations from the linear relationship usually occur. From such a temperature-programmed chromatogram the retention times of the n-alkanes are now plotted in a linear scale against the carbon number. The carbon number is again multiplied by the factor 100, and each distance divided into 100 units. A reference line is drawn again, as shown in Figure 11. With the help of this reference line the retention index I_x of the component x is found again if the retention time t_x is used (see Figure 11).

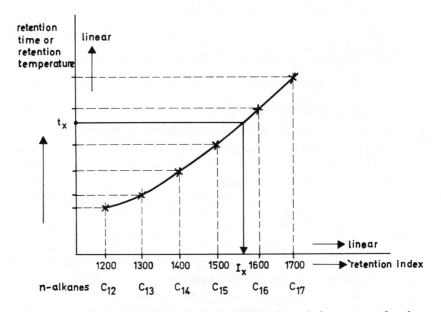

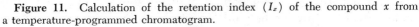

Figure 11. Calculation of the retention index (I_x) of the compound x from a temperature-programmed chromatogram.

Since it is inconvenient in practice to mix the sample under investigation with the test mixture of the n-alkanes, the chromatograms of the test mixture and that of the sample must be recorded separately. The accuracy of the retention indices thus determined depends mainly on the reproducibility of retention times with the respective instrument under temperature-programmed conditions.

REFERENCES

1. Gehrke, C. W., and D. L. Stalling. Separ. Sci. **2**(1), 101 (1967). Gehrke, C. W., D. Roach, R. W. Zumwalt, D. L. Stalling, and L. L. Wall. *Quantitative Gas Liquid Chromatography of Amino Acids In Proteins and Biological Substances* (Columbia, Miss.: Analytical Bio-Chemistry Laboratories, 1968).

2. Stenhagen, E. Z. Anal. Chem. **181**, 462 (1961).

3. Weygand, F., A. Prox, W. König, and H. H. Fessel. Angew. Chem. **75**, 724 (1963).

4. Weygand, F., A. Prox, H. H. Fessel, K. Kun Sun. Z. Naturforsch. **20b**, 1169 (1965).

5. Prox, A., and K. Kun Sun. Z. Naturforsch. **21b**, 1028 (1966).

6. Prox, A., and F. Weygand, in *Peptides,* Ed. by H. C. Beyerman, A. van de Linde, and W. Maassen van den Brink (Amsterdam: North-Holland Publishing Co., 1967), p 158.

7. Weygand, F. Z. Anal. Chem. **243**, 2 (1968).

8. Lederer, E., and B. C. Das, in *Peptides,* Ed. by H. C. Beyerman, A. van de Linde, and W. Maassen van den Brink (Amsterdam: North-Holland Publishing Co., 1967), p 131.

9. Shemyakin, M. M., Y. Ovchinnikov, and A. A. Kiryushkin, in *Peptides,* Ed. by H. C. Beyerman, A. van de Linde, and W. Maassen van den Brink (Amsterdam: North-Holland Publishing Co., 1967), p 155.

10. Weygand, F., B. Kolb, A. Prox, M. A. Tilak, and I. Tomida. Hoppe-Seylers Z. Physiol. Chem. **322**, 38 (1960).

11. Biemann, K., and W. Vetter. Biochem. Biophys. Res. Commun. **3**, 587 (1961).

12. Weygand, F., in *Peptides,* Ed. by H. C. Beyerman, A. van de Linde, and W. Maassen van den Brink (Oxford: Pergamon Press, 1965), p 359.

13. Weygand, F. Z. Anal. Chem. **205**, 406 (1964).

14. Weygand, F., and W. Steglich. Z. Naturforsch. **14b**, 472 (1959).

15. Fales, H. M., and J. J. Pisano, in *Biomedical Applications of Gas Chromatography,* Ed. by H. A. Szymanski (New York: Plenum Publishing Corp., 1964), p 82.

16. Cruickshank, P. A., and J. C. Sheehan. Anal. Chem. **36**, 1191 (1964).

17. Weygand, F., A. Prox, E. C. Jorgensen, R. Axén, and P. Kirchner. Z. Naturforsch. **18b**, 93 (1963).

18. Bayer, E., and W. A. Koenig. J. Chromatog. Sci. **7**, 95 (1969).

19. Weygand, F., D. Hoffmann, and A. Prox. Z. Naturforsch. **23b**, 279 (1968).

20. Weygand, F., A. Prox, L. Schmidhammer, and W. König. Angew. Chem. **75**, 282 (1963). Angew. Chem. Intern. Ed. Engl. **2**, 183 (1963).

21. Grob, K. Helv. Chim. Acta **48** (6), 1362 (1965); **51**(4), 718 (1968).

22. Weygand, F., A. Prox, L. Schmidhammer, and W. König, in *Peptides,* Ed. by H. C. Beyerman, A. van de Linde, and W. Maassen van den Brink (Oxford: Pergamon Press, 1963), p 97.

23. Weygand, F., B. Kolb, and P. Kirchner. Z. Anal. Chem. **181**, 396 (1961).
24. Tomida, I., T. Tokuda, J. Ohashi, and M. Nakajima. J. Agricult. Chem. Soc. Japan **39**, 391 (1965).
25. Westley, J. W., V. A. Close, D. N. Nitecki, and B. Halpern. Anal. Chem. **40**, 1888 (1968).
26. Tamura, S., A. Suzuki, Y. Aoki, and N. Otaka. Agr. Biol. Chem. **28(9)**, 650 (1964).
27. Wehrli, A., and E. Kováts. Helv. Chim. Acta **42**, 2709 (1959).
28. Ettre, L. S. Anal. Chem. **36(8)**, 31A (1964).
29. Wieland, T., G. Lueben, H. Ottenheym, J. Faesel, J. X. de Vries, W. Konz, A. Prox, and J. Schmid. Angew. Chem. **80**, 209 (1968).
30. Prox, A. Private communication.
31. Merrifield, R. B. J. Am. Chem. Soc. **85**, 2149 (1963); and **86**, 304 (1964).
32. Gil-Av, E., B. Feibush, and R. Charles-Sigler, in *Gas Chromatography*. Ed. by A. B. Littlewood (London: Butterworths and Co., Ltd., 1966).
33. Weygand, F., W. König, R. Buyle, and H. G. Viehe. Chem. Ber. **98**, 3632 (1965).
34. Weygand, F., and W. König. Z. Naturforsch. **20b**, 710 (1965).
35. Weygand, F., D. Hoffmann, and E. Wünsch. Z. Naturforsch. **21b**, 426 (1966).
36. Beyerman, H. C., W. Maassen van den Brink, F. Weygand, A. Prox, W. König, L. Schmidhammer, and E. Nintz. Rec. Trav. Chim. **84**, 213 (1965).
37. Charles, R., G. Fischer, and E. Gil-Av. Israel J. Chem. **1**, 234 (1963).
38. Gil-Av, E., R. Charles-Sigler, G. Fischer, and D. Nurok. J. Gas Chromatog. **4**, 51 (1966).
39. Gil-Av, E., R. Charles, and G. Fischer. J. Chromatog. **17**, 408 (1965).
40. Pollock, G. E., and L. H. Frommhagen. Anal. Biochem. **24**, 18 (1968).
41. Pollock, G. E., V. I. Oyama, and R. D. Johnson. J. Gas Chromatog. **3**, 174 (1965).
42. Pollock, G. E., and V. I. Oyama. J. Gas Chromatog. **4**, 126 (1966).
43. Halpern, B., and J. W. Westley. Chem. Anal. **47**, 589 (1965); Biochem. Biophys. Res. Commun. **19**, 361 (1965); Tetrahedron Letters **21**, 2283 (1966); and Chem. Commun. **2**, 34 (1966).
44. Weygand, F., A. Prox, and W. König. Chem. Ber. **99**, 1451 (1966).
45. Williams, M. W., and G. T. Young. J. Chem. Soc. **1963**, 881.
46. Anderson, G. W., and F. M. Callahan. J. Am. Chem. Soc. **80**, 2902 (1958).
47. Clayton, D. W., J. A. Farrington, G. W. Kenner, and J. M. Turner. J. Chem. Soc. **1957**, 1398.
48. Bodanszky, M. Nature **175**, 685 (1955); Bodanszky, M., M. Szelke, E. Tömörkeny, and E. Weisz. Chem. Ind. (London) **1955**, 1517.
49. Schwyzer, R., et al. Helv. Chim. Acta **38**, 69, 80, 83 (1955).
50. Weygand, F., and W. Swodenk. Chem. Ber. **90**, 639 (1960).
51. Weygand, F., and W. Steglich. Chem. Ber. **93**, 2983 (1960).

52. Weygand, F., and W. Steglich. Angew. Chem. **73**, 99, 757 (1961).
53. Sheehan, J. P., and G. P. Hess. J. Am. Chem. Soc. **77**, 1067 (1955).
54. Curtius, T. Ber. Deut. Chem. Ges. **35**, 3226 (1902); Harris, J. I., and T. S. Work. Biochem. J. **46**, 582 (1950); and Nyman, N. A., and R. M. Herbst. J. Org. Chem. **15**, 108 (1950).
55. Anderson, G. W., and R. Paul. J. Am. Chem. Soc. **80**, 4423 (1958); Paul, R., and G. W. Anderson. *ibid.* **82**, 4596 (1960).
56. Goldschmidt, S., and H. Lautenschlager. Liebigs Ann. Chem. **580**, 68 (1953); Goldschmidt, S., and K. K. Gupta. Chem. Ber. **98**, 2831 (1965).
57. Wieland, T., and R. Sehring. Liebigs Ann. Chem. **569**, 122 (1950); Boissonnas, R. A. Helv. Chim. Acta **34**, 874 (1951); Wieland, T., and H. Bernhard. Liebigs Ann. Chem. **572**, 190 (1951); Vaughan, J. R., Jr. J. Am. Chem. Soc. **73**, 1389, 3547, 5553 (1951).
58. Woodward, R. B., R. A. Olofson, and H. Meyer. J. Am. Chem. Soc. **83**, 1010 (1961).
59. Wolma, Y., P. M. Gallop, and A. Patchornik. J. Am. Chem. Soc. **83**, 1263 (1961); and **84**, 1889 (1962).
60. Beaumont, S. M., B. O. Hanford, and G. T. Young. Acta Chim. Acad. Sci. Hung. **44**, 37 (1965); Beaumont, S. M., B. O. Hanford, J. H. Jones, and G. T. Young. Chem. Commun. **1965**, 53. Weygand, F., W. König, E. Nintz, D. Hoffmann, P. Huber, N. M. Khan, and W. Prinz. Z. Naturforsch. **21b**, 325 (1966).
61. Jakubke, H. D. Z. Naturforsch. **20b**, 273 (1965).
62. Arens, J. F. Rec. Trav. Chim. **74**, 769 (1955).
63. Beyerman, H. C., and W. Maassen van den Brink. Rec. Trav. Chim. **80**, 1372 (1961).
64. Buyle, R., and H. G. Viehe. Angew. Chem. **76**, 572 (1964); and Angew. Chem. Intern. Ed. Engl. **3**, 582 (1964); Weygand, F., W. König, R. Buyle, and H. G. Viehe. Chem. Ber. **98**, 3632 (1965).
65. Sheehan, J. C., P. A. Cruickshank, and G. L. Boshart. J. Org. Chem. **26**, 2525 (1961).
66. Bernhard, S. A., *et al.* J. Am. Chem. Soc. **84**, 2421 (1962).
67. Janák, J. Nature **185**, 684 (1960).
    Janák, J., in *Gas Chromatography.* Ed. by R. P. W. Scott (London: Butterworths and Co., Ltd., 1960), p 387.
    Janák, J. Collection Czech. Chem. Commun. **25**, 1780 (1960).
68. Ulehla, J. Sb. Cesk. Akad. Zemedel. Ved. **7**, 567 (1960).
69. Andrew, T. D., C. S. G. Phillips, and J. A. Semlyen. J. Gas Chromatog. **1**, 27 (1963).
70. Winter, L. N., and P. W. Albro. J. Gas. Chromatog. **2**, 1 (1964).
71. Stack, M. V. Biochem. J. **96**(3), 56P (1965).
72. Simon, W., and H. Giacobbo. Angew. Chem. Intern. Ed. Engl. **4**, 938 (1965).
73. Merritt, C. Jr., and D. H. Robertson. J. Gas Chromatog. **5**, 96 (1967).
74. Ettre, L. S. (ed.). *Open Tubular Columns* (New York: Plenum Publishing Corp., 1965).

75. Peterson, M. L., and J. Hirsch.   J. Lipid Res. **1**, 132 (1959).
76. Gold, H. J.   Anal. Chem. **34**, 174 (1962).
77. Andersson, B. Å.   Acta Chem. Scand. **21**, 2906 (1967).
78. Pollock, G. E., and A. H. Kawauchi.   Anal. Chem. **40**, 1356 (1968).
79. Prox, A., J. Schmid, and H. Ottenheym.   Liebigs Ann. Chem. **722**, 179 (1969).

Chapter 4

Mass Spectrometry in Peptide Chemistry

by B. C. Das and E. Lederer

4.1　Introduction

Determination of the primary structure of a protein molecule involves the sequential analysis of amino acids in a very large number of oligopeptides that result from a series of specific or random cleavages performed in the molecule. Conventional methods employed for the determination of amino acid sequences in oligopeptides are time-consuming and tedious. Mass spectrometric techniques have attracted considerable attention over the past few years and have found promising applications in the determination of the sequence of amino acid residues in N-acyl oligopeptide derivatives. The minute quantities needed for obtaining a mass spectrum and the rapidity of the measurements, as well as possibilities of computer-aided interpretation enable this method to compete favorably with other well-established methods.

The principles of instrumentation in mass spectroscopy and the behavior of ionized organic molecules under electron impact have been elaborately discussed in many standard texts.[1] Mass spectral fragmentation patterns of individual free amino acids,[2,3] amino acid alkyl esters,[4] and N-acetyl amino acids[5] and their alkyl esters[6] have been described and reviewed (Jones, Reference 7). In view of the general interest associated with the problem of amino acid sequence determination in peptides and proteins, the following pages principally review the application of mass spectrometry in the field of peptide derivatives. The attention of the reader is also directed to some recent reviews[7] of this rapidly expanding field. See also Section 4.8.

4.2　Chemical Modifications of Peptides for Mass Spectrometry

Because of their poor volatility and susceptibility to thermal decomposition, free peptides are not suitable for mass spectrometric investigation. Chemical modification of the polar groups in peptides is necessary before subjecting them to mass spectrometric analysis.

4.2.1 Polyamino Alcohols

Preliminary studies on the use of mass spectrometry for the determination of amino acid sequences were published by Biemann et al.[8] They reduced the peptide bonds and the terminal carboxyl by lithium aluminum hydride to obtain polyamino alcohols containing diaminoethane moieties in place of the amide groups. These polyamino alcohols obtained from small peptides (the largest molecules used in this method have been pentapeptides) are fairly volatile and give interpretable mass spectra.

$$
\begin{array}{c}
\text{R} \qquad\qquad \text{R}' \qquad\qquad \text{R}'' \\
| \qquad\qquad\quad | \qquad\qquad\quad | \\
\text{H}_2\text{N}-\text{CH}-\text{CO}-\text{NH}-\text{CH}-\text{CO}-\text{NH}-\text{CH}-\text{COOH} \xrightarrow{\text{LiAlH}_4}
\end{array}
\qquad (I)
$$

$$
\begin{array}{c}
\qquad\qquad \text{R} \quad a \qquad\qquad \text{R}' \quad b \qquad\qquad \text{R}'' \quad c \\
d \text{-----}|\text{-}\blacktriangleright \qquad\quad e \text{-----}|\text{-}\blacktriangleright \qquad\quad f \text{-----}|\text{-}\blacktriangleright \\
\text{H}_2\text{N}-\text{CH}\!-\!\!|\text{CH}_2-\text{NH}-\text{CH}\!-\!\!|\text{CH}_2-\text{NH}-\text{CH}\!-\!\!|\text{CH}_2\text{OH}
\end{array}
$$

Cleavages at the C—C bonds of the diaminoethane moieties occur on electron impact and a series of ions are formed. Thus, fragmentation of I at a, b or c leads to ions with retention of positive charge on either side. However, fragments retaining the positive charge on the left side are usually more abundant because in that case the charge is on a more highly substituted carbon atom. Fission at d, e, or f leads to an ion the positive charge of which is also on a carbon atom where it is readily stabilized by the adjacent nitrogen atom. The polyamino alcohols exhibit hardly any peak at the molecular weight, but a significant M + 1 peak is observed instead in the mass spectrum. From a consideration of all these peaks, the most significant ones in the spectrum, sizes and arrangements of the different R-groups and therefore the structure of the original peptide can be deduced.

The correctness of assignment of the peaks in the mass spectra of polyamino alcohols may be corroborated by mass spectra of the corresponding deuteriated products obtained by reducing the peptide with lithium aluminum deuteride. Comparison of the two spectra shows that the peaks are shifted in agreement with the fact that each of the original carbonyl groups has now become CD_2 instead of CH_2.

When the polyamino alcohol contains side-chain hydroxyl groups (originating from the reduction of the polyfunctional amino acids such as glutamic acid, aspartic acid, serine, threonine, or hydroxyproline residue which may be present in the peptide) further chemical modification becomes necessary. This has been done by replacement of the hydroxyl group by chlorine (treatment with thionyl chloride) followed by reduction with $LiAlH_4$ or $LiAlD_4$. The relative complexity of the chemical treatments involved and the presence of a fairly large number of peaks in the mass spectra are probably the reasons why this method did not gain much popularity.

4.2.2 N-Acyl Oligopeptide Esters

With improvement of techniques for direct insertion of samples into the ion source of the mass spectrometer, it has been possible to examine peptide derivatives with intact amide bonds. Terminal amino and carboxyl groups of a peptide are protected by acylation and esterification, respectively, leading to an N-acyl peptide ester which gives increased volatility. Such acylation and esterification of peptides can be performed on a very small scale.[9, 10] Methyl,[11] ethyl[12] or t-butyl[13] esters have been examined, and the different N-protected peptides investigated include acetyl,[14] trifluoroacetyl,[15, 16] heptafluorobutyryl,[17] benzyloxy-carbonyl,[18, 19] phthaloyl,[20] ethoxy carbonyl,[21] hexanoyl,[22] 2,4-dinitrophenyl,[23] decanoyl,[18] or stearoyl[18] derivatives. Of these, the N-acetyl (or trifluoroacetyl) peptide methyl esters are simplest to prepare and quite convenient for mass spectral investigation.

Peptides containing functional groups in the side chains may present difficulties owing to decreased volatility. Such side-chain functional groups need to be modified before mass spectral investigation. Side-chain amino groups (lysine, ornithine) and carboxyl groups (aspartic acid, glutamic acid) are modified simultaneously with the terminal groups during acylation-esterification procedure. Alcoholic groups (serine, threonine, etc.) may be converted into their O-acetyl derivatives[24] or may better be methylated (*vide infra*). Tyrosine-containing peptides give satisfactory mass spectra only after methylation of the phenolic hydroxyl.[25] Difficulties encountered with arginine-containing peptides may be solved (at least partially) by conversion of arginine residues into ornithine or δ-(2-pyrimidinyl) ornithine residues[26, 27] which give derivatives with sufficient volatility for mass spectra.

4.3 Major Fragmentation Patterns of Peptides and Their Derivatives

A peptide derivative containing the structure

$$X\text{-CO-} \dots \text{-NH-CH} \underset{R'}{\overset{}{|}} C \underset{O}{\overset{\|}{}} \text{NH} \text{CH} \underset{R''}{\overset{}{|}} C \underset{O}{\overset{\|}{}} \text{-} \dots \text{COOCH}_3$$

$$\quad\quad\quad\quad 1 \quad 2 \quad 3$$

may undergo fragmentation in the mass spectrometer by cleavage of any one of the bonds 1, 2, or occasionally 3 with charge retention on the left or on the right side. For sequence determinations only those ions containing the N-terminal part of the chain are important; fortunately, they are the most frequent ones.

4.3.1   *Fragmentation of the Peptide Bond*

In the mass spectra of N-acylated peptide derivatives peaks are present corresponding to the rupture of the peptide bond:

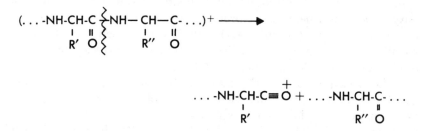

In higher oligopeptides these peaks due to the acylium ions are often the most intense ones and seem to occur by a sequential splitting of one amino acid residue after the other, starting from the molecular ion; metastable peaks[28] show that this is at least partially the case. For instance, in fortuitine II (see Section 4.4.1.1) the successive loss of the [4]Val and [3]Val residue is indicated by metastable peaks at m/e 437.8 and m/e 465.1.[29]

The splitting of the peptide bond, which is the most "useful" for sequence determination is more pronounced in higher peptides than in di- and tripeptides, etc.

4.3.2   *Fragmentation of the C—CO Bond*

The above-mentioned peaks due to the fission of the peptide bond are often accompanied by more-or-less intense peaks 28 mass units (m.u.) lower, derived from C—CO cleavage of the following type:

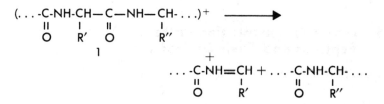

In some cases, metastable peaks show that these "minus-28 peaks" are formed from the corresponding acyl fragment by elimination of CO:

The "minus-28 peaks" can undergo a further fission to give "normal" peptide bond fission peaks in the following way:

The "minus-28 peaks" are frequently accompanied by "minus-27 peaks," due to a fragmentation accompanied by a hydrogen rearrangement:

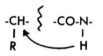

Weygand *et al.*[16] have studied this fission by deuterium-labeling and propose the following mechanism:

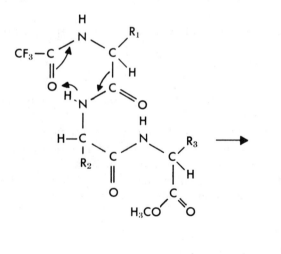

$$CF_3-C\overset{+}{=}NH-CHR_1 + O=C=N-CHR_2-\ldots$$
$$\underset{OH}{|}$$

The "minus-28 peaks" and "minus-27 peaks" are found in α-peptides; their absence may be used to recognize β-peptide linkages in aspartyl peptides and γ-linkages in glutamyl peptides, as shown by Van Heijenoort *et al.*[25] for several isomeric pairs of peptide derivatives containing dicarboxylic amino acids (or also β-alanine derivatives).

4.3.3 *Fragmentation of Side Chains*

The mass spectra of peptide derivatives contain more-or-less intense peaks due to the loss of parts of the various side chains of the amino acids. The most frequent are the following.

4.3.3.1 **Aliphatic monoamino-monocarboxylic acids**

The branched-chain amino acids valine, leucine and isoleucine lose part of their side chain with elimination of an olefin and transfer of one H atom to the peptide chain.

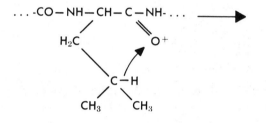

Loss of 56 mass units thus can indicate the presence of leucine and loss of 42 m.u. the presence of valine.

Weygand *et al.*[16] have observed that the side-chain fragmentation of the leucyl and isoleucyl units in a peptide chain differ significantly; it is thus possible to distinguish these isomers in a sequence.

Side chains containing functional substituents in β-position (serine, threonine, cysteine) may suffer elimination of the function:[13]

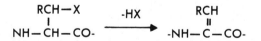

4.3.3.2 **Monoamino–dicarboxylic acids**

According to the Russian authors[13] peptides containing the residues of asparagine or the β-ester of aspartic acid undergo elimination of the β-substituent. It has been observed that the elements of ammonia or alcohol are eliminated followed by loss of CO.

They have noted that the same elimination process is shown by the γ-esters of glutamic acid and by glutamine. For the γ-substituted amino acids (derivatives of glutamic acid and methionine), elimination of the entire side chain also partially occurs.[13]

4.3.3.3 Aromatic and heterocyclic amino acids

Peptides containing aromatic and/or heterocyclic amino acids (phenylalanine, tyrosine, tryptophan, histidine) are characterized by: (a) partial elimination of the side chain as $ArCH_2$ during the amino acid type of fragmentation; (b) initial rupture of the $N-C_\alpha$ bond of the aromatic/heterocyclic amino acid residue followed by the amino acid type of fragmentation; (c) elimination of the side chain in the form of $ArCH_2^+$ (for example, tropylium in the case of phenylalanine). All three processes are clearly manifested in tryptophan-containing peptides.[13]

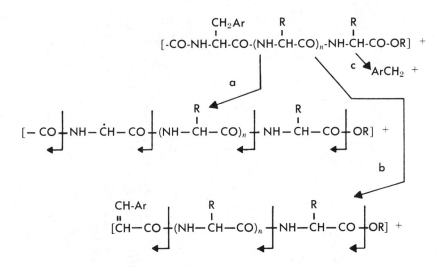

This is exemplified by the mass spectrum (Figure 1) of decanoyl $(d_0 + d_3)$-Gly-Pro-Trp-Leu-OMe (M = 639; 642).[25] The appearance of a pair of peaks at m/e 510 and 513 in the mass spectrum of this peptide is noteworthy. They arise by loss of 129 mass units (C_9H_7N by mass measurement) from the tryptophan part of the peptide chain. This together with the appearance of an intense peak at m/e 130 indicates the presence of a tryptophan unit in the peptide molecule. Subsequent loss of terminal Leu-OCH$_3$ from m/e 510 and 513 can then

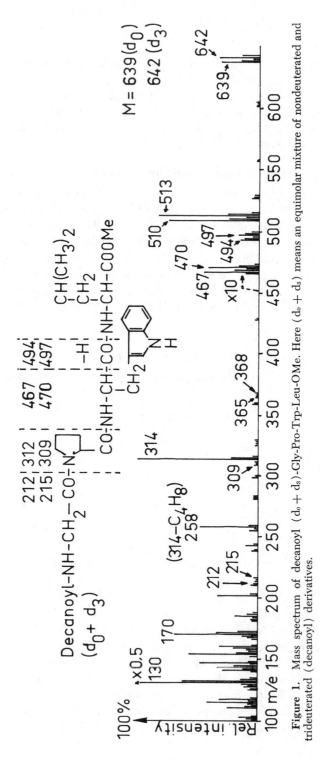

Figure 1. Mass spectrum of decanoyl $(d_0 + d_3)$-Gly-Pro-Trp-Leu-OMe. Here $(d_0 + d_3)$ means an equimolar mixture of nondeuterated and trideuterated (decanoyl) derivatives.

182

give rise to ions at m/e 365 and 368. Two other peaks observed at m/e 314 and 170 can be explained as due to the ions shown below.

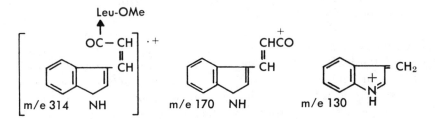

Peptides containing C-terminal phenylalanine or tryptophan methyl ester have been found to give peaks corresponding to the mass ions of methyl cinnaminate and methyl β-indolyl acrylate.[16, 30]

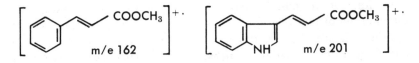

4.3.3.4 Basic amino acids

Histidine peptides can give rise to peaks 14 m.u. higher; this is due to the basicity of the imidazole imino group which undergoes inter-molecular methylation. Shemyakin *et al.*[13] have observed that peptides containing two histidine residues exhibit two satellite peaks with 14 and 28 m.u. in excess. Senn *et al.*[31] have shown by deuterium labeling that the additional methyl group comes either from the methylating reagent or from the methyl ester group for the peptide.

For peptides with N-terminal lysine or ornithine,[13] one of the strongest peaks is due to ions of the type

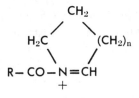

n = 1 or 2.

4.4 Sequence Determination of N-Acyl Peptide Esters

From the above discussion it becomes evident that the amino acid sequence of the peptide can be deduced from the mass spectrum if the peaks due to peptide bond cleavages can be identified. The recognition of these sequence-determining peaks can be facilitated by proper choice of the protecting groups. The acylation of the N-terminus with an equimolecular mixture of acetic and trideuteroacetic acids (or mixed CD_3- and CH_3-decanoic acids)[25] causes the ions with an intact N-terminus (*i.e.*, the sequence ions) to appear as pairs of peaks of equal intensity separated by three mass units. Other mixed acyl groups such as an equimolecular mixture of hepta- and octadecanoic acids[18] can be used, in which case all acyl-containing ions in the spectra give pairs of peaks 14 m.u. apart, thus facilitating interpretation. Since in some natural oligopeptides doublets with a mass difference of 14 could also be due to the presence of different amino acid homologs, *e.g.*, valine- or isoleucine, of gramicidins A, B, and C,[32] it is preferable to use mixed CD_3- and CH_3-bearing acyl chains.

The effect of the length of the N-acyl group on the mass spectral fragmentation of peptides has been studied[18] by comparing the spectra of synthetic N-acetyl-, -decanoyl, and -stearoyl Ala-Val-Gly-Leu methyl esters. It was observed that there was no difference in the basic fragmentation of the peptides due to variation of the length of the N-acyl groups. Despite this, the spectrum of the stearoyl derivative is clearly simpler in that the peaks corresponding to cleavage of the peptide bonds are more easily recognized since they are shifted to a high mass region clear of the majority of other fragments. Thus, for example, in the case of N-stearoyl peptides all peaks below m/e 267 [CH_3 $(CH_2)_{16}$–CO^+] can be disregarded for the purpose of sequence determination.

Interpretation of the spectra can also be simplified by comparing the fragmentation of the methyl and trideuteriomethyl ester of the acyl peptide, which identifies ions retaining the C-terminus.[22]

To facilitate interpretation of the complex mass spectra of peptide derivatives, Weygand *et al.*[16] have developed a data processing system known as "Differenzschema." This consists of a systematic examination of the difference in mass among the peaks in the spectrum. Starting from the molecular ion, a search is made for meaningful mass differences which correspond to loss of amino acid residues. The relationship among sequence-determining peaks may be checked by appropriate metastable ions observed in the spectra.

Recent advances in technique have made possible the automatic determination of amino acid sequences from an analysis of the complete high-resolution mass spectra of peptides with the aid of computers.[31,86,87]

The computer program devised by Senn *et al.*[31] first searches for the N-terminal amino acid by checking exact mass combinations of the

N-terminal protecting group (CH_3CO: 43.01839, CF_3CO: 96.99012, etc.) and the exact mass of each of the possible amino acid residues against the observed accurate masses of ions in the spectrum.

The search for each additional amino acid residue follows the same pattern, and the process continues until addition of the exact mass of the C-terminal protecting group locates the molecular ion. In the program used by Biemann *et al.*[86] the N-terminal amino acid is identified by checking combinations of the exact mass of the protecting group plus -NHCH- and the exact masses of each of the possible amino acid side chains against the accurate masses found. By addition of the mass of CO (27.9949), the ion corresponding to the peptide bond cleavage is looked for. The next amino acid is identified by adding the mass of -CONHCH- plus the accurate masses of the side chains of all possible amino acids, testing the data for a fit in amino acid residue each time. The process is repeated until the C-terminal has been reached.

For elucidating the sequence, the above two groups have utilized computer programs starting from the N-terminal substituent. Barber *et al.*,[87] on the other hand, have made use of a program which elaborates the amino acid sequence starting from the C-terminus.

Computer-aided interpretation of complete high-resolution mass spectra obviates the necessity of using any special protecting groups for the identification of the sequence ions. A simple *N*-acetyl group is sufficient since the computer identifies all peaks by their elemental compositions.

4.4.1 *Natural Peptidolipids*

4.4.1.1 **Fortuitine**

Fortuitine, m.p. 199–202°C, $[\alpha]_D = -72°$ ($CHCl_3$), isolated from *Mycobacterium fortuitum* is a peptidolipid containing approximately equimolecular quantities of eicosanoic and docosanoic acids linked to an oligopeptide chain reported to contain $Val_3Thr_2Ala_1Pro_1$. The carboxyl terminal of the oligopeptide was shown to be esterified by methanol.[33] The preliminary structure

Ac

$CH_3(CH_2)_nCO$-Val-Val-Val-Thr-Thr-Ala-Pro-OMe

(n = 18, 20)

for fortuitine, had been proposed.[34]

A closer examination of the results of quantitative determinations of the amino acids of a total hydrolysate of fortuitine showed, however, that a nitrogenous portion of the molecule was not yet accounted for. Mass spectrometry has led to the identification of this "unknown con-

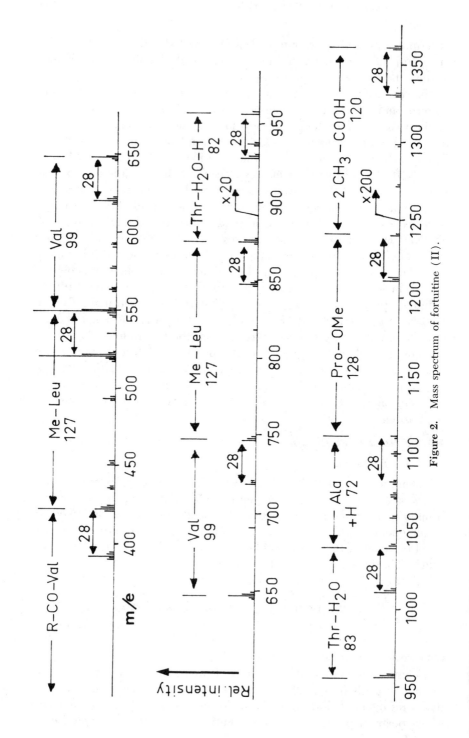

Figure 2. Mass spectrum of fortuitine (II).

186

stituent" as two molecules of N-methyl leucine and has allowed their place in the nonapeptide chain to be determined without ambiguity.

The mass spectrum (Figure 2) of fortuitine shows two parent peaks at 1331 and 1359, due to its being a mixture of two homologs containing a C_{20} and C_{22} fatty acid respectively. All ions containing the lipid moiety therefore give sets of peaks 28 mass units apart, thus facilitating interpretation.[29]

The parent peaks, at 1331 and 1359, each lose $2 \times CH_3COOH = 120$ to give peaks at 1211 and 1239. This shows the presence of two acetyl groups; both threonine molecules are O-acetylated. Two small peaks at 1083 and 1111 are apparently due to the loss of -Pro-OMe (128 m.u.). A prominent set of peaks occurs at 1011 and 1039. This is 72 m.u. less than 1083 and 1111, due to the loss of Ala + H (migration of one H atom).

The next set of peaks occurs at 928 and 956. This loss of 83 m.u. from 1011 and 1039 corresponds to the loss of one molecule of anhydrothreonine, again after migration of one H atom. This gives the sequence R-Thr-Ala-Pro-OMe.

The further interpretation of Figure 2 gives clearly the whole structure of fortuitine as *II*, corresponding to $C_{70}H_{125}N_9O_{15} = 1331$ and $C_{72}H_{129}N_9O_{15} = 1359$, respectively.

$$CH_3(CH_2)_n\text{-CO-Val-MeLeu-Val-Val-MeLeu-Thr-Thr-Ala-Pro-OMe} \qquad (II)$$

n = 18, 20; M⁺ 1359 (for n = 20)   Ac Ac

The above interpretation was confirmed by mass measurements of the peaks at m/e 521, 648, 719, 846, and 928.

This work showed for the first time that mass spectrometry could be very useful for sequence determination of oligopeptide derivatives and that the possible mass range for such work was as high as 1400, and that even nonapeptide derivatives could be tackled.

4.4.1.2   Peptidolipid of *M. johnei*

A peptidolipid analogous to fortuitine had been isolated from *M. johnei* by Lanéelle and Asselineau;[35] a preliminary structure, R-CO-(Phe)₂,Ala,Ala-OMe, had been considered. Mass spectrometry showed quite unequivocally, however, that two molecules of isoleucine had escaped the Moore and Stein analysis (isoleucyl–isoleucine being very difficult to hydrolyze) and that the correct structure of this compound is III.[36]

$$CH_3(CH_2)_n\text{-CO-Phe-Ile-Ile-Phe-Ala-OCH}_3 \qquad (III)$$

n = 14, 16, 18, 20

Here again the most prominent peaks corresponded to the splitting of the peptide bond.

4.4.1.3 The peptidolipin NA group

Peptidolipin NA, m.p. 232–233°C, $[\alpha]_D = +42°$, isolated from *No-cardia asteroides,* is a peptidolipid for which Guinand and Michel[37,38] have proposed structure *IV,* corresponding to the molecular formula $C_{50}H_{89}N_7O_{11}$ (mol wt 963).

In this case mass spectrometry fully confirmed this structure; one of the more prominent fragmentations observed was the elimination of CO_2 from the lactone ring, thus leading to an open-chain acyl oligo-peptide ion quite analogous to the cases discussed above; from then on, the sequential splitting of peptide bonds could be recognized.[39]

The principal compound, peptidolipin NA *IV* is accompanied by two or three analogs, in which one amino acid of the macrocyclic lac-tone is replaced by another. The structure determination of these com-pounds by mass spectrometry was particularly easy, based on the experience obtained with the parent compound. It was thus established that one of the "companions" is a Val⁶-peptidolipin NA (V)[40] (Figure 3), another is an α-aminobutyryl¹–peptidolipin NA (VI).[41]

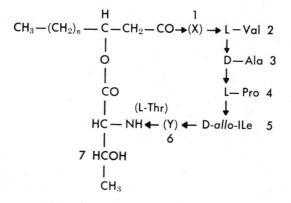

Peptidolipin   NA (IV)   X = L-Thr
                                    Y = L-Ala
                                    n = 16

Val⁶-peptidolipin   NA (V)   X = L-Thr
                                    Y = L-Val
                                    n = 16, 17, 18

Abu¹-peptidolipin   NA (VI)   X = L-Abu
                                    Y = L-Ala
                                    n = 16, 17, 18

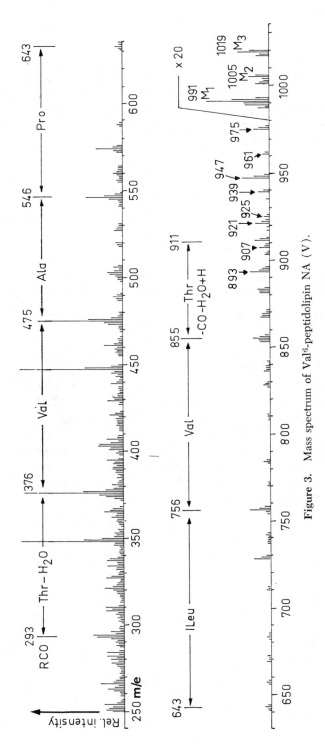

Figure 3. Mass spectrum of Val⁵-peptidolipin NA (V).

189

4.4.1.4 Staphylomycin

This cyclic antibiotic (VII) contains a lactone ring and resembles the peptidolipins NA in this respect. Kiryushkin et al.[42] have observed that here, too, the fragmentation starts with the splitting off of CO_2 and a hydrogen atom leading to the acyl peptide acylium ion (VIII) which then is fragmented in a sequential manner.

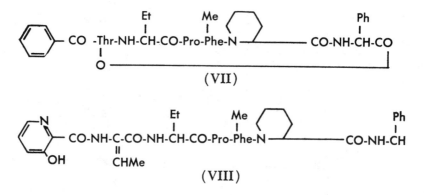

(VII)

(VIII)

4.4.1.5 Ostreogrycin A

The structure of ostreogrycin A, an antibiotic produced by the soil organism *Streptomyces ostreogriseus*, has been confirmed by the detailed analysis of the mass spectrum of its hydrogenation product "perhydro A," $C_{28}H_{49}N_3O_7$ (IX). The compositions of all the abundant ions in the mass spectrum[43] of perhydro A which have been determined by exact mass measurements correspond to species which have lost two or more oxygen atoms. On the basis of this evidence it has been concluded that the behavior of perhydro A is similar to that of peptidolipin

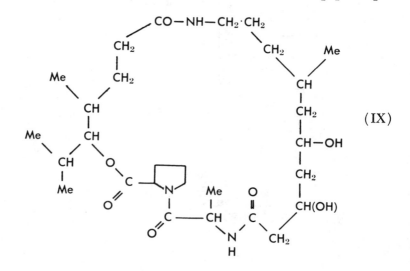

(IX)

NA upon electron impact and eliminates carbon dioxide (small M-44 peak at m/e 495) from the molecular ion with the associated ring-opening. Important fragments at m/e 425 (loss of proline residue) and 354 (loss of alanine residue) in the high mass region of the spectrum can then be formed by cleavage of peptide N-C bonds, exactly as observed for other *N*-acyl oligopeptides.

4.4.1.6 Isariin

Isariin, a metabolite of the fungus *Isaria cretacea van Beyma,* has been shown by Vining and Taber[44] to contain 1 mole each of glycine, L-alanine, D-leucine and D-β-hydroxydodecanoic acid, and 2 moles of L-valine, linked in a cyclic structure. Partial acid hydrolysis indicated that glycine was attached via an amide bond to the carboxyl function of the hydroxy acid. Mild alkaline hydrolysis afforded isariic acid, an open-chain *N*-acyl peptide in which L-valine was identified as the C-terminal amino acid. It was concluded that isariin contained the sequence L-valine → D-β-hydroxydodecanoic acid → glycine, but the order of the remaining amino acids was not determined.

The mass spectrum[45, 46] of isariin gave a molecular ion peak at m/e 637 ($C_{33}H_{59}N_5O_7$ confirmed by mass measurement); although the fragmentation pattern as a whole was somewhat complex, the sequence Leu-Ala-Val could be deduced and structure (X) established; methyl isariate, the open-chain ester obtained from isariin gave a clearer mass spectrum which allowed confirmation of structure (X).

Mass spectrometry has also been useful in the structure elucidation of isarolide A, B, C—a mixture of cyclic peptidolipids isolated from a new species of *Isaria*.[47]

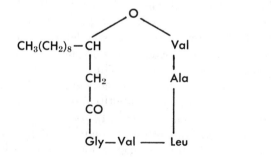

(X)

4.4.1.7 The mycoside C group

The mycosides, discovered by Smith and Randall,[49–50] have been defined as "type-specific glycolipids of mycobacterial origin."[51] They all contain typical 6-deoxyhexoses.[50–52] Here we shall discuss only the mycoside C compounds which also contain amino acids; they are thus peptidoglycolipids. Initially they were thought to be restricted to strains of *M. avium*,[48-52] but they have since been found also in *M. marianum*,[53] *M. butyricum*,[54] and *M. scrofulaceum*.[55]

Total acid hydrolysis of all mycoside C compounds gives three categories of components:

(1) a mixture of long-chain fatty acids and several molecules of acetic acid

(2) at least two 6-deoxyhexoses

(3) three different amino acids, all of the D-series: D-phenylalanine, D-allothreonine, and D-alanine[56] as well as one amino alcohol, such as alaninol.[57,58]

In a first stage[59,60] a partial formula of an acyl tripeptide, RCO-Phe-*allo*-Thr-Ala, emerged where the detailed structure of an RCO-long-chain, unsaturated, and oxygenated acyl group could not be determined because it usually decomposes during the strong acid hydrolysis necessary for splitting the RCO-Phe-amide linkage. The amide of phenylalanine, however, is relatively easy to obtain by partial acid hydrolysis of all C-mycosides. The methyl ester (RCO-Phe-OMe) gives a good mass spectrum,[54] confirming the structure of the acyl radical deduced by mass spectrometry of the acyl methyl ester. The sequence RCO-Phe-*allo*-Thr-Ala, first elaborated chemically by standard methods, was then confirmed by mass spectrometry.[54]

The presence of a fourth nitrogenous constituent had also been noticed and from a small fraction of the mycoside C of *M. butyricum* a "new amino acid" could be isolated and identified as N-methyl-O-methyl-L-serine.[61] This, however, is not a major constituent and up to now no pure molecular species containing the hypothetical sequence RCO-Phe-*allo*-Thr-Ala-N-MeSer(OMe) has been isolated.

During a study of the constituents of an atypical photochromogenic strain no. 1217, a mycoside C was isolated.[62] Lanéelle[58] identified a fourth nitrogenous constituent of it as alaninol, which had partly escaped notice due to its volatility. This leads to the sequence RCO-Phe-*allo*-Thr-Ala-alaninol, which is in agreement with the mass spectrum.[63]

All mycoside C compounds described until now contain one molecule of 6-deoxytalose, which is usually di- or triacetylated, and one molar proportion of partly or entirely O-methylated rhamnoses.[52]

The ease with which the 6-deoxytalose is split off in alkaline medium was first interpreted as proof of an ester linkage of a C-terminal carboxyl group with the C-1 hydroxyl group of this sugar.[60] Then came the discovery[64,65] that glycosides of serine and threonine, such as exist in glycoproteins, easily suffer a β-elimination reaction with liberation of the sugar molecule and formation of dehydroalanine from serine and dehydro-α-aminobutyric acid from threonine. This led to a revision of the location of the 6-deoxytalose in the mycoside C compounds, because it was found that here, too, β-elimination takes place with liberation of the sugar and destruction of *allo*-threonine.[54]

It now seems that in all mycoside C compounds the talose derivative is in glycosidic linkage on the OH group of one molecule of *allo*-threonine as also shown by mass spectrometry. This leaves only the rhamnose derivative to be located; chemical degradation and mass spectrometry show unambiguously that it is linked to the OH group of the terminal alaninol molecule.[54, 63]

The structure and the major mass spectral fragmentations of a mycoside C$_b$ from *M. butyricum* are represented by formula (XI) on facing page. (Lederer, Reference 7.)

The structure of the long-chain fatty acids which form the RCO group of these mycosides seems to vary from strain to strain; their chemistry has been largely defined by mass spectrometry of their amide with phenylalanine, as already mentioned. They all represent new types in fatty acid structures.

4.4.1.8 The mycoside C compounds of M. avium and M. marianum

The structure R-CO-Phe-*allo*-Thr-Ala-alaninol is found in most of the mycoside C compounds. A reexamination of fractions of mycoside C described previously as containing a pentapeptide -Phe-(*allo*-Thr-Ala)$_2$-,[60] or even a heptapeptide -Phe-(*allo*-Thr-Ala)$_3$,[53] has shown that a pentapeptide as quoted above does exist and that it is linked to a terminal amino alcohol molecule. Mass spectrometry of these more complex compounds was not successful but after partial acid hydrolysis and acetylation, a good mass spectrum of a 2-*O*-acetyl-3, 4-di-*O*-methyl-rhamnoside of *N*-acetyl alaninol was obtained. This suggested a structure analogous to (XI), but having the pentapeptide sequence Phe-(*allo*-Thr-Ala)$_2$ instead of the tripeptide sequence Phe-*allo*-Thr-Ala.[66]

4.4.2 Synthetic Glycopeptides

Mester *et al.*[66a] have investigated some synthetic *N*-acylamino-acyl-2-deoxy-2-acetamido-3,4,6-tri-*O*-acetyl-β-D-glucosylamines and have shown that the presence of an amino acid–hexosamine link in a glycopeptide can be recognized by mass spectrometry of a suitable derivative.

4.4.3 Cyclic Peptides

Since the time of some preliminary experiments of Heyns and Grützmacher[14, 67] on cyclo (Gly-L-Ala-D-Phe-Gly-D-Ala-L-Phe), Millard[68] has studied in more detail, by high-resolution mass spectrometry, the be-

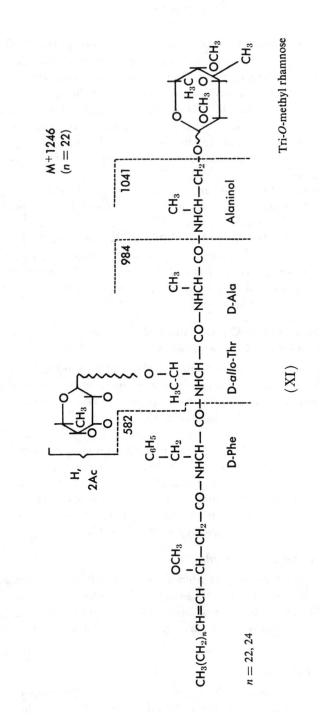

$$M^+ 1246$$
$$(n = 22)$$

Tri-O-methyl rhamnose

Alaninol

1041

D-Ala

984

D-*allo*-Thr

D-Phe

582

H,
2Ac

$CH_3(CH_2)_nCH=CH-CH-CH_2-CO-NHCH-CO-NHCH-CO-NHCH-CO-NHCH-CH_2-O$

$n = 22, 24$

(XI)

havior of the above cyclic peptide and of cyclo(Gly-L-Leu-Gly-L-Leu-Gly), cyclo(Gly-L-Leu-D-Leu-Gly-Gly-), cyclo(Gly-L-Leu-Gly-Gly-L-Leu-Gly), and cyclo(Gly-L-Phe-L-Leu-Gly-L-Phe-L-Leu).

In some instances useful information can be obtained concerning the sequence of amino acid residues in these cyclic peptides. Ring opening followed by the usual peptide fragmentations can occur, for example, at a phenylalanine residue with transfer of hydrogen, leading to an open-chain ion:

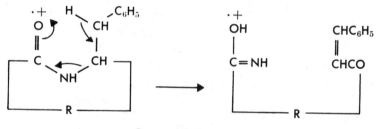

R = remainder of the chain

Another process of ring opening can occur at a position adjacent to a bulky amino acid side chain, ascribed by Millard[68] to "the release of excess vibrational energy as rotational energy" and leading to a radical ion.

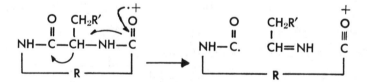

The ions obtained after ring opening can then undergo further fragmentation, as described for linear peptide derivatives. The amino acid sequence can be deduced from the mass spectrum, but the possibility of ring opening at more than one site complicates the interpretation.

The structure of a cyclononapeptide (XII) isolated from linseed could be determined[69] by the application of gas chromatography and mass spectrometry, as described in Section 3.5.1.

$$\text{Leu-Ile-Ile-Leu-Val-Pro-Pro-Phe-Phe} \qquad (\text{XII})$$

Similar procedure was employed to establish the structure of antamanid (M^+1146), a cyclodecapeptide isolated from the fungus *Amanita phalloides*.[69a] (See Section 3.5.1)

A consideration of mass spectrometry and chemical degradation has been very useful in the assignment of structures to the cyclic peptide

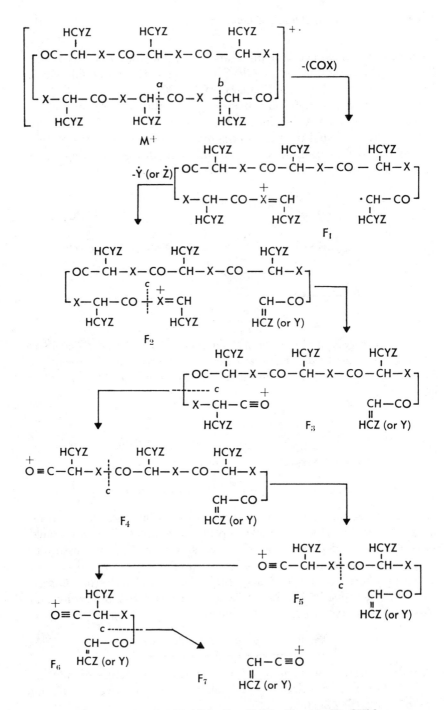

Scheme 1. $X = O$, NH, NMe; $Y = H$, Me; $Z = $ Me, Et, $CHMe_2$.

alkaloids scutanin,[70] pandamine,[71] zizyphin,[72] ceanthonine B,[73] and several others.[74]

4.4.4 Depsipeptides

The Institute of Chemistry of Natural Products of Moscow has made extensive studies of the mass spectrometry of a number of cyclic di-, tetra-, hexa-, octa- and dodecadepsipeptides, analogs of the enniatin and valinomycin antibiotics.[75, 76, 77]

These compounds undergo three main types of fragmentation:[76]

(1) The "COX type" of fragmentation is associated with elimination of elements of the ester (-COO-) or amide (-CONH- or -CONMe-) group from the molecular ion due to sucessive or simultaneous bond rupture according to types *a* and *b* (see Scheme 1). The resultant ion-radical F_1 may become further stabilized by eliminating in the form of a radical (Y or Z) part of the side chain from the carbon atom formerly attached to the heteroatom X. The ion F_2 thus produced undergoes a series of type *c* degradations with consecutive loss of amino or hydroxy acid residues and formation of fragments F_3–F_7;

(2) The morpholine type of fragmentation, the first stage of which is associated with fragmentation of the molecular cyclohexadepsipeptide ion (M^+) to form either the 2,5-dioxomorpholine ion (F_8) and a neutral fragment, or the neutral 2,5-dioxomorpholine molecule (MO) and the depsipeptide ion-radical F_9 resulting from simultaneous or consecutive two-fold *c*-type bond rupture (see Scheme 2). The 2,5-dioxomorpholine ions (F_8) may derive not only directly from M^+, but from the neutral 2,5-dioxomorpholine molecule that results from thermal degradation of the initial cyclohexadepsipeptide under the experimental conditions.

(3) The acylaminoketene type of fragmentation is characterized by bond rupture in the two ester or amide groups of the molecular cyclohexadepsipeptide ion according to routes *b'* and *c'* with the formation of the acylaminoketene ion, which then undergoes further degradation by splitting off a methyl group, CO, the side chain, etc. (Scheme 3).

McDonald and Shannon *et al.*[78–81] have also studied the fragmentation mechanisms of cyclic, natural depsipeptides—the sporidesmolides I, II, and III, and angolide.[81]

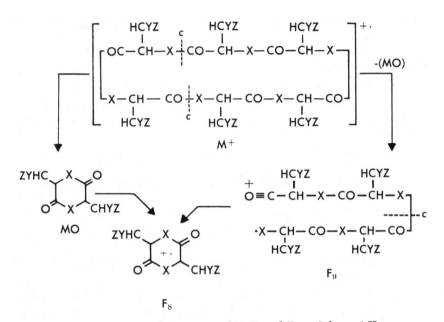

Scheme 2.   For the meaning of X, Y, and Z see Scheme 1.[77]

4.5   O,N-Permethylation of Peptide Derivatives

Poor volatility of higher N-acyl oligopeptide methyl esters has been a major problem in the mass spectrometric investigation of these compounds. Whereas fortuitine (II) with its nine amino acids and heavy acyl groups had given an excellent spectrum (M^+1359),[29] several de-

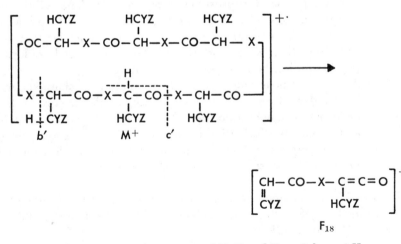

Scheme 3.   For the meaning of X, Y, and Z see Scheme 1.[77]

rivatives of synthetic hepta- or octapeptides were apparently not volatile enough to give useful mass spectra. It was suggested by Van Heijenoort *et al.*[25] that an important factor concerning the low volatility of peptide derivatives might be hydrogen bonding due to the presence of -CO-NH groups. An argument in favor of this hypothesis was that the tetrapeptide methyl ester H-Ile-Pro-Sar-MeVal-OMe containing no peptide hydrogen (*i.e.*, absence of -CO-NH-) gave a mass spectrum exhibiting a molecular ion peak even without acylation of the terminal amino group. Also, the satisfactory volatility of the nonapeptide derivative fortuitine (II) could be explained by the presence of three tertiary amide bonds (two -MeLeu- and one -Pro-). These observations suggested that if a procedure leading to permethylation of the -CO-NH- groupings of oligopeptide derivatives could be found, the resulting modified peptide might be more volatile and particularly suitable for the determination of the amino acid sequence by mass spectrometry.

The methylation of the peptide bond had not yet been described and was considered to be a very difficult enterprise. It was thus very satisfying to find that the well-known methylation procedure of Kuhn *et al.*[83] using silver oxide and methyl iodide in dimethylformamide gave complete *N*-methylation of peptide derivatives in a few hours.[84, 85] In the same operation the C-terminal carboxy group and OH groups of threonine, serine and tyrosine are also methylated.

In the first paper on "*N*-methylation of *N*-acyl oligopeptides"[84] the usefulness of the procedure for obtaining more volatile peptide derivatives was shown by the fact that the heptapeptide derivative (XIII) was not volatile enough to give a mass spectrum, but after complete *O,N*-permethylation (XIV) gave a satisfactory mass spectrum which exhibited a molecular ion peak at m/e 1082.

Heptapeptide : Ac-Gly-Phe-Phe-Tyr-Thr-Pro-Lys-OMe (mol wt 970)

$$\text{Me} \downarrow \quad \text{Ac}$$
$$\downarrow \text{MeI/Ag}_2\text{O} \qquad \text{(XIII)}$$

Ac|MeGly|MePhe|MePhe|MeTyr|MeThr|Pro-MeLys-OMe

peaks m/e   43   114   275   436   627   756   ├─ Me   (M$^+$1082)

Me Me (627, 756)   Ac

$$\downarrow - 32 \text{ (MeOH)}$$
$$724 \qquad \text{(XIV)}$$

This molecular weight indicates that the seven -CONH- groups as well as the threonine hydroxyl have been methylated by the methyl iodide–silver oxide treatment. The mass spectral fragmentation pattern (Figure 4) is also consistent with the amino acid sequence of this

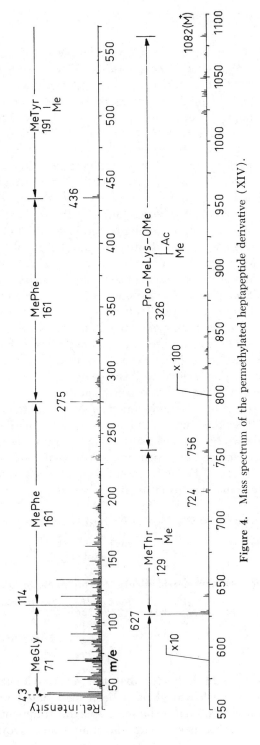

Figure 4. Mass spectrum of the permethylated heptapeptide derivative (XIV).

synthetic heptapeptide derivative (XIV) and shows that the reaction -CONH- → -CO-NMe- proceeds smoothly to give a completely methylated product.

Another advantage of the use of permethylated peptide derivatives for sequence determination was noticed by Thomas *et al.*[85] The fragmentation pattern of the permethylated derivatives is much simpler than that of the corresponding nonmethylated derivatives because the spectra consist almost exclusively of the "sequence determining" peaks resulting from -CO-NCH₃ cleavage; this can be seen by comparing the spectra of Figure 5 and Figure 6.

The enhancement of volatility obtained by permethylation may be exemplified by peptide (XV) (mol wt 483), which before methylation could be volatilized in the mass spectrometer at 210°C, whereas the permethylated product (XVI) (mol wt 539) gave a spectrum at a source-sample temperature of 150°C. The simplification of the spectra obtained by permethylation is certainly due in part at least to the decrease of pyrolytic reactions as a result of the lower temperature used.

$$\text{Ac-Orn(Ac)-Orn(Ac)-Pro-OMe} \qquad \text{(XV)}$$

$$\text{Ac-Me}_2\text{Orn(Ac)-Me}_2\text{Orn(Ac)-Pro-OMe} \qquad \text{(XVI)}$$

The permethylation technique will prove especially valuable in combination with the computer techniques described by Biemann *et al.*,[86] Senn *et al.*,[31] and Barber *et al.*[87] Moreover, the computer could easily give useful information with impure peptide preparations, where the usual technique is of little help.

4.5.1 Comments on Methylation Technique

O,N-Permethylation of *N*-acyl peptides was first accomplished by employing methyl iodide–silver oxide (dried over P_2O_5 under vacuum) in dimethylformamide.[84] It is interesting to note that if moist silver oxide is used, only *O*-methylation is observed; thus Guinand and Michel[38] have obtained the *O*-permethylated derivative of peptidolipic acid (XVII) by using moist silver oxide,

$$\text{CH}_3\text{-(CH}_2)_{16}\text{-CH-CH}_2\text{-CO-Thr-Val-Ala-Pro-alle-Ala-Thr-OH}$$
$$\overset{|}{\text{OH}} \qquad \text{(XVII)}$$

whereas the same compound gives the *O,N*-permethylated derivative when methylated with dry silver oxide.[85]

Several amino acid residues give undesirable results[88, 89] by the silver oxide–methyl iodide method. Thus:

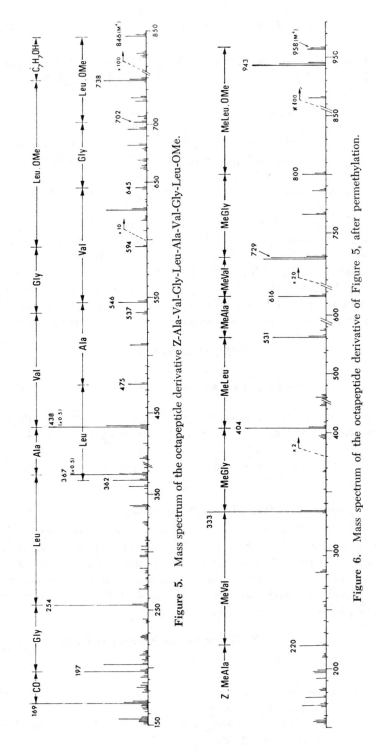

Figure 5. Mass spectrum of the octapeptide derivative Z-Ala-Val-Gly-Leu-Ala-Val-Gly-Leu-OMe.

Figure 6. Mass spectrum of the octapeptide derivative of Figure 5, after permethylation.

(a) Methylation of peptides containing aspartic acid gives complex mixtures of products.[88]

(b) Glutamic acid residues are frequently "well-behaved," but some peptide derivatives undergo partial chain cleavage with formation of a pyroglutamyl residue.[88, 89]

(c) Tryptophan residues normally form a dimethyl derivative after methylation; but one exception was noted with the tryptophan at position 9 of the gramicidins A and B (Table I, see below) which gave an artifact 30 mass units heavier.

Vilkas and Lederer[90] have demonstrated that the methylation method described by Hakomori[91] for glycolipids (NaH in dimethylsulfoxide + CH_3I) is also applicable for the *O,N*-permethylation of peptide derivatives. This method was applied to a wide variety of peptides by Thomas[89] who found that in addition to having all the advantages of the silver oxide method, it may be successfully used for peptide derivatives containing aspartic acid, glutamic acid, or tryptophan without formation of undesired by-products.

More recently, it has been shown that *O,N*-permethylation can also be effected[92] by employing methyl iodide and sodium hydride in dimethylformamide as described by Coggins and Benoiton.[93] However, this method also suffers from the inconvenience in that partial C-methylation occurs in the case of certain amino acids, particularly aspartic acid and glycine.[111]

4.5.2 Some "Troublesome" Amino Acids

Before the development of the permethylation technique it had been shown that peptides containing several trifunctional amino acids could be analyzed by mass spectrometry, provided that the dicarboxylic amino acids (Asp, Glu) were esterified on their free carboxy group, tyrosine was present as its *O*-methyl ether, lysine was ε-acylated; derivatives of cystine and histidine had given mass spectra.[25] As a result only arginine was left as the most "troublesome" one. Then Shemyakin *et al.*[26] showed that arginine peptides can be condensed with a β-diketone (*e.g.*, acetylacetone) to yield pyrimidine derivatives which give good mass spectra (see also Vetter-Diechtl *et al.*[27]). Shemyakin *et al.*[26] further reported that a treatment of arginine peptides with hydrazine produces the corresponding ornithine derivatives.

After introduction of the permethylation technique new problems arose with some of the trifunctional amino acids; let us consider these briefly and define the solutions that can be adopted.

Free amino groups present either at the N-terminal end of the peptide chain, or as the ω-function of lysine or ornithine are usually acylated to ensure volatility of oligopeptide derivatives to be analyzed by mass spectrometry;[25] methylation of free amino groups would, of course, lead to quaternary methyl iodide salts with low volatility. Therefore acylation must precede methylation.

Thomas et al.[94] propose acetylation of the peptide with methanolic acetic anhydride which leads only to N-acetyl derivatives leaving OH groups free; these are then methylated in the subsequent permethylation procedure.

When arginine peptides are acylated and methylated by the usual procedure, chloroform-soluble products of sufficient volatility can be obtained, but the mass spectra show sequence peaks only up to but not including the arginine residue. After treatment with hydrazine, following the procedure of Shemyakin et al.,[26] the corresponding ornithine peptides are obtained and can then be permethylated.

Thomas et al.[94] have recently described experimental details for arginine peptides consisting of (1) treatment with hydrazine, (2) acetylation, and (3) permethylation of the corresponding ornithine peptides. The tripeptide derivative (XVIII) containing two arginine

H-Arg-Arg-Pro-OBu^t

(XVIII)

residues thus gave the tetramethyltriacetyldiornithine methyl ester (XVI). When arginine methyl ester is at the C-terminal (peptides obtained from proteins by tryptic hydrolysis frequently contain C-terminal arginine), the lactam of ornithine is obtained; thus, the pentapeptide methyl ester (XIX) gave the permethylated lactam (XX) (Fig. 7).

H-Phe-Ser-Pro-Phe Arg-OMe

(XIX)

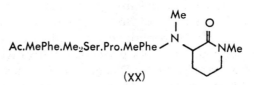

(XX)

Histidine derivatives give a quaternary base after permethylation, but in the mass spectrometer CH_3I is lost thermally so that mass spectra can be obtained (D. W. Thomas, unpublished results) at least in smaller peptides.

Methionine peptides show a peculiar behavior after permethylation. Agarwal et al.[88] have found that methylation of acetyl methionine gives

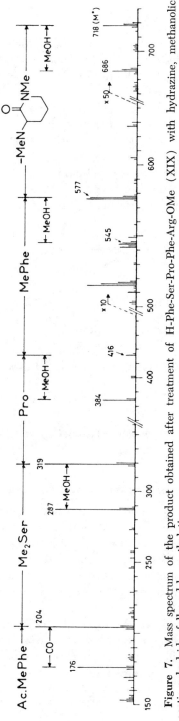

Figure 7. Mass spectrum of the product obtained after treatment of H-Phe-Ser-Pro-Phe-Arg-OMe (XIX) with hydrazine, methanolic acetic anhydride, followed by methylation.

a volatile crystalline product, $C_8H_{13}NO_3$, which they have shown to be the cyclopropyl derivative (XXI). Thomas *et al.*[94] have not observed this reaction. Instead, methionine derivatives seem on methylation to give sulfonium derivatives.

The spectra obtained with such permethylated methionine peptides show peptide cleavage peaks only up to but not including the methionine residue (an analogy to the situation observed with methylated arginine peptides; see above).

A solution to the problem of methionine can be found[94] by treating methionine peptides—for instance, (XXII)—with Raney nickel, thus producing an α-aminobutyric acid (Abu) residue from each methionine residue (XXIII).

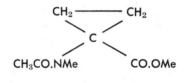

(XXI)

BOC-Lys(BOC)Phe-pF-Phe-Gly-Leu-Met-NH$_2$

(XXII)

Ac-MeLys(Ac)-MePhe-MepF-Phe-MeGly-MeLeu-MeAbu-NMe$_2$

(XXIII)

4.5.3 Recognition of Natural N-Methyl Amino Acids

Until recently, it has not been possible to detect and locate with certainty the presence of an N-methyl amino acid residue in a peptide by mass spectrometry because the mass number as well as the elemental composition of a given amino acid and its N-methylated lower homolog (*e.g.*, leucine or N-methylvaline) are the same. This problem can be solved, however, by the use of the mass spectrometric shift technique.

A comparative study of the mass spectra of an N-acyl oligopeptide methyl ester and its permethylated derivative provides information on the presence as well as the location of originally existing N-methyl amino acid residues in oligopeptide derivatives.[95] The increase in molecular weight after complete N-methylation indicates the number of methyl groups introduced in the molecule and thus the number of peptide -CONH- groups (together with -OH groups, if present) in the oligopeptide derivative before methylation. Each amino acid residue which has gained a methyl group reveals itself by an increase of 14 mass units recognizable from the peaks arising by peptide bond cleavage. An amino acid residue with its peptide nitrogen originally methylated or fully substituted shows no change in mass number after methylation (unless it contained other methylatable functional groups), and thus its position can be determined.

This procedure cannot, however, be used for oligopeptide derivatives which do not give mass spectra before methylation—for example, the heptapeptide derivative (XIII).[84] In such cases, the N-acyl peptide methyl ester may be treated with trideuterio methyl iodide to obtain the corresponding perdeuteriomethylated product. Comparison of the mass spectra of the permethylated and the perdeuteriomethylated derivatives would indicate a shift of three mass units (CH_3 versus CD_3) for each N-deuteriomethylated amino acid residue. Such a shift would not be observed if the amino acid residue did not originally contain any methylatable hydrogen.

This approach was illustrated by studying the mass spectrum of perdeuterio methyl fortuitine[95] and has been used for the study of the structure of stendomycin (*vide infra*).[96]

4.6 Mass Spectrometric Sequence Determination of Permethylated Peptide Derivatives

4.6.1 *Peptide Antibiotics*

4.6.1.1 The gramicidins

The gramicidins A, B and C are polypeptide antibiotics produced by *Bacillus brevis*, first described by Dubos and Hotchkiss.[97] Sarges and Witkop[32] have shown that they are N-formyl pentadecapeptides linked to a C-terminal ethanolamine residue; the complete structures as determined by these authors are shown in Table I.

The synthesis of gramicidin A has been described by Sarges and Witkop;[98] the synthetic compound was found to be identical in all respects with the natural one.

Mass spectrometry of these compounds showed some sequence peaks up to the seventh N-terminal residue (W. A. Wolstenholme, unpublished experiments). After permethylation, however, the sequence could be recognized *up to and including the twelfth residue*.[85]

As explained elsewhere in detail (Lederer, Reference 7), permethylation by the Kuhn method was accompanied by oxidation of Trp[9] in all gramicidin samples investigated (including a synthetic sample);

Table I
Structures of the Gramicidins A, B and C[32]

1	2	3	4	5	6	7	8	9	10	11	12	13	14	15

Valine–gramicidin A:
HCO-V*al*-Gly-Ala-leu-Ala-val-Val-val-Trp-leu-*Trp*-leu-Trp-leu-Trp-
                                                                    -NHCH₂CH₂OH

Isoleucine–gramicidin A:
  -*Ile*-   -*Trp*-
Valine–gramicidin B:
  -*Val*-   -*Phe*-
Isoleucine–gramicidin B:
  -*Ile*-   -*Phe*-
Valine–gramicidin C:
  -*Val*-   -*Tyr*-
Isoleucine–gramicidin C:
  -*Ile*-   -*Tyr*-
                NB: Val = ʟ-Val; val = ᴅ-Val; etc.

from a study of the mass spectra it was concluded that hydroxylation had occurred not on the indole nucleus nor on the β-carbon,* but on the α-carbon.

It was then found[89] that permethylation by other methods[90,93] did not lead to oxidation of the Trp[9] residue and gave a mass spectrum in complete agreement with the structure worked out by Sarges and Witkop.[32] In one case the sequence of as many as 14 amino acids from the N-terminus could be recognized (D. W. Thomas, unpublished results).

4.6.1.2. Stendomycin

The antifungal antibiotic stendomycin was isolated by Thomson and Hughes[99] from *Streptomyces antimycoticus*. Bodanszky *et al.*[100] showed that it is a cyclic acyl tetradecapeptide lactone (XXIV) resembling peptidolipin NA (IV) and that it is a mixture of homologous compounds which differ from one another with respect to their fatty acid constituents.

---

*Because of the absence of a peak at m/e 174; N-methyl- tryptophan gives a strong ion at m/e 144 corresponding to

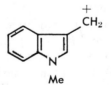

Me

Any methoxy tryptophan derivative having an OCH₃ on this part of the molecule would give an ion at m/e 174.

After acid hydrolysis they identified isomyristic acid and its lower homologs, as well as 14 moles of amino acids: Ala_1, Gly_1, Ser_1, Pro_1, Val_3, *allo*-Ile_2, *allo*-Thr_2, one mole of N-methyl threonine, one mole of "dehydrobutyrine" [= dehydro-α-aminobutyric acid = Δ-Abu[101]] and one mole of a "basic compound" B shown to be (XXV).[102]

13 12 11

aIle-Ser-aThr

"B" ↑

RCO-Pro-NMeThr-Gly-Val-aIle-Ala-Δ-Abu-aThr-Val-Val

1   2   3   4   5   6   7   8   9   10

(XXIV)

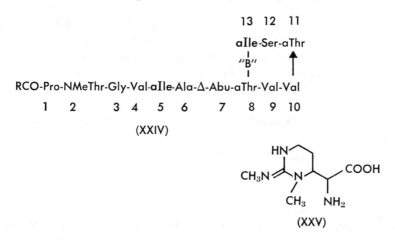

(XXV)

Permethylation of one mg of stendomycic acid, the "open form" obtained by mild alkaline hydrolysis of the antibiotic peptide, gave a product of sufficient volatility for mass spectral analysis—a rather satisfactory result considering the presence of four hydroxy amino acids in addition to a basic component expected to form a quaternary amine derivative of low volatility. The sequence of the first ten amino acids could be established from mass spectral analysis.[96]

Several problems not usually encountered with simple amino acids arose, involving determination of the location of the proline, threonine, N-methylthreonine, and "dehydrobutyrine" residues known to be present. All ions containing permethylated threonine usually lose methanol, giving ions containing N-methyldehydro-α-aminobutyric residues instead. The mass of these is the same as that of proline (97 m.u.); thus it is impossible to distinguish among each of the residues listed above without employing other techniques.

Use of CD_3I for the permethylation reaction allowed the division of these four residues into two groups: the masses of proline and N-methyl dehydro-α-aminobutyric acid (resulting from the N-methyl threonine) remained unchanged at 97 m.u., whereas the N-CD_3 group introduced in the threonine and Δ-Abu residues were detected by the shift of their masses to 100.

The sequence RCO-Pro-MeThr- is preferred over the inverse sequence RCO-MeThr-Pro-, because the latter would be expected to

give rise to a minor but nevertheless discernible mass spectral fragment, which would contain the threonine methoxyl group: RCO-
CH$_3$Thr- (m/e 343 for R = C$_{11}$H$_{23}$). This was not observed and thus
ÓCD$_3$
the assignment of a proline to position 1 is preferred although not
proven.

Positions 7 and 8 were defined by catalytic hydrogenation of permethylated stendomycic acid; the dihydro derivative (Abu) thus obtained
showed an increment of two mass units at the 7 position.

The partial structure (XXVI) can thus be written for the open form
of stendomycic acid:

RCO-Pro-NMeThr-Gly-Val-*allo*-Ile-Ala-Δ-Abu-*allo*-Thr-Val-Val-

R = C$_{13}$H$_{27}$ (53%), C$_{12}$H$_{25}$ (30%), C$_{11}$H$_{23}$ (17%)

(XXVI)

Sequence-determining peaks in the mass spectra beyond the tenth
amino acid could not be detected above the noise and background
level.[96]

A comparison of the mass spectral and chemical data[96] indicated
that an identical sequence had been assigned independently by each
method.

A tetrapeptide *allo*-Thr-Ser-*allo*-Ile-"B" obtained by partial hydrolysis of stendomycin and representing residues 11–14 was also submitted to mass spectrometry after acylation and permethylation. The
result was in agreement with the proposed structure.

4.6.1.3   Esperin

Esperin is an antibiotic produced by *Bacillus mesentericus* for which
structure (XXVII) was proposed.[103, 104] Alkaline hydrolysis of this
peptide lactone produced a linear peptide derivative, esperinic acid
(XXVIII).[104] However, on the basis of differences of the properties of
this acid and of a synthetic sample corresponding to structure (XXVIII),
the correctness of structures (XXVII) and (XXVIII) was questioned.[105]

C$_{10}$H$_{21}$CHCH$_2$CO.Glu

O ——— Asp.Val.Leu.Leu.OH   (XXVII)

C$_{10}$H$_{21}$CHOHCH$_2$CO.Glu.Asp.Val.Leu.Leu.OH   (XXVIII)

A reinvestigation[106] of the structure of esperin and esperinic acid using the permethylation techniques of Hakomori and Vilkas,[90, 91] and of Coggins and Benoiton[93] had led quite unequivocally to the structure (XXIX) for esperin.

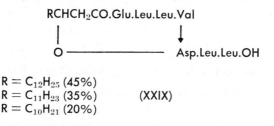

$$RCHCH_2CO.Glu.Leu.Leu.Val$$

O ——————— Asp.Leu.Leu.OH

R = $C_{12}H_{25}$ (45%)
R = $C_{11}H_{23}$ (35%)   (XXIX)
R = $C_{10}H_{21}$ (20%)

The fatty acid constituents of esperin were shown by mass spectrometry to consist of three principal homologs: C_{13}, C_{14} and C_{15} acids. The mass spectrum of the methyl ester of a hydrolysis product of esperin revealed a major peak at m/e 103 corresponding to the ion $\overset{+}{C}HOHCH_2COOCH_3$, which established the location of the hydroxyl group at position 3 of each fatty acid constituent.

The mass spectrum of permethylated esperinic acid clearly revealed the homologous fatty acid constituents and the entire amino acid sequence. Mass spectral fragmentation occurred exclusively at the peptide CO-NMe bonds; the resulting peaks as outlined in structure (XXX) delineated the sequence without the need for a detailed interpretation.

RCHOMeCH₂CO.MeGlu(OMe)	MeLeu	MeLeu	MeVal	MeAsp(OMe)	MeLeu	MeLeu	OMe
R = $C_{12}H_{25}$	412	539	666	779	922	1049	1176	1207
R = $C_{11}H_{23}$	398	525	652	765	908	1035	1162	1193
R = $C_{10}H_{21}$	384	511	638	751	894	1021	1148	1179

(XXX)

Use of deuterio methyl iodide verified that all *O,N*-methyl groups of (XXX) were introduced by the permethylation reaction; thus (XXXI) was proposed as the structure of esperinic acid.

RCHOHCH₂CO.Glu.Leu.Leu.Val.Asp.Leu.Leu.OH   (XXXI)

R = $C_{12}H_{25}$ (45%)

R = $C_{11}H_{23}$ (35%)

R = $C_{10}H_{21}$ (20%)

The mass spectrometric molecular weight (1063 and lower homologs) of esperin dimethyl ester is in agreement with a lactone structure formed by the hydroxyl group of the fatty acid residue at the N-terminus and one of the three carboxyl functions of the esperinic acid (XXXI). The lactone linkage as represented by formula (XXIX) was confirmed by successive hydrazinolysis and permethylation of esperin; the aspartic acid residue of the product contained an N-methylhydrazide ($-CONHNHCH_3$) which was recognized by mass spectrometry.

4.6.2   Oligopeptides Obtained by Degradation of Proteins

4.6.2.1   Heptapeptide from phospholipase A

A heptapeptide obtained in Utrecht by Dr. de Haas from the zymogen of phospholipase A purified by two successive Sephadex G-25 runs, as well as preparative paper chromatography, was permethylated by employing MeI–NaH in dimethylformamide.[109] The mass spectrum of the permethylated product was not very "clean" but did provide sufficient evidence that the N-terminal residue is a pyroglutamic acid; the presence of the latter is also indicated by an intense peak at m/e 98 which could originate from a pyroglutamic acid residue. The following sequence could be suggested from the mass spectrum.

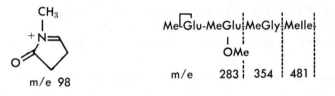

4.6.2.2   Octapeptide from silk fibroin

Morris et al.[107] have determined the sequence of an octapeptide (after permethylation) isolated from a chymotryptic digest of silk fibroin by means of ion-exchange chromatography. They noted that peptides thus obtained can be contaminated by a mixture of polymers extracted from the resin, the presence of which can render interpretation of the sequence peaks very difficult.

4.6.2.3   Octadecapeptide from immunoglobulin

An octadecapeptide isolated by Franek et al.[108] from the tryptic hydrolysate of pig immunoglobulin λ-chains was examined[92] by mass spectrometry after acetylation at N-terminus followed by permethylation. Up to ten amino acid residues from the N-terminus could be determined by the interpretation of the major peaks in the mass spectrum. The results so obtained summarized in Table II were in agreement with those independently available by sequential degradation of

Table II
Octadecapeptide of Pig Immunoglobulin. Partial N-Terminal Sequence as Determined by Mass Spectrometry

Position	1	2	3	4	5	6	7	8	9	10
	Ac.MeAla	MeAla	MeLeu[a]	MeThr OMe	MeLeu[a]	MeThr OMe	MeGly	MeAla	MeGln NMe$_2$	MeAla.
m/e	128	213	340	469	596	725	796	881	1051	1136
−MeOH					564	693	764	849	1019	

[a]MeLeu or MeIle.

213

the peptide. The mass spectrum also indicated that the peptide contained an approximately equimolecular mixture of alanine and threonine at the second N-terminal residue.

4.6.2.4 Docosapeptide from immunoglobulin

The mass spectrum obtained after permethylation of an N-terminal peptide from pig immunoglobulin containing 22 amino acid residues (isolated by Dr. F. Franek, Prague) revealed it to be a mixture of two peptides (A and B) having the partial sequence shown in Table III.[109]

Table III

N-Terminal Docosapeptide from Pig Immunoglobulin:
Partial Amino Acid Sequence by Mass Spectrometry

		MePyroglu	MeThr OMe	MeVal	MeLeu	MeGlu NMe$_2$	MeGlu ... OMe
A	m/e		255	368	495	665	822
(75%)	−MeOH		223	336	463	633	790
		MePyroglu	Pro	MeVal	MeVal	MeGlu NMe$_2$	MeGlu ... OMe
B	m/e		223	336	449	619	776
(25%)							

4.6.3 Mass Spectrometry as a Tool for Checking the Purity of Synthetic Peptides

In the course of mass spectrometric sequence determination a large number of synthetic peptide derivatives has been studied. Some of these were examined before, some after permethylation; most of them gave the expected mass spectra. In several cases, however, the mass spectrum either showed the presence of several impurities, or the mass spectra were entirely incompatible with the theoretical structures.

Sometimes it was evident that one or another of the protecting groups had not been removed, or that a coupling reaction had not succeeded. A special case might be described in more detail: (Lederer, Reference 7).

A hexapeptide related to eledoisin—H-Ala-Phe-Ile-Gly-Leu-Met-NH$_2$ prepared by Schröder et al.[110] was acetylated and then permethylated. The mass spectrum of the resulting product was in agreement with the expected N-acetyl permethylated hexapeptide derivative, with sequence peaks visible up to but *not* including methionine (see above). The spectrum, however, clearly showed the presence of another compound containing an additional oxygen atom in the N-terminal residue or its protecting group (confirmed by a mass measurement of m/e 144 (Table IV). Furthermore, methylation of the initial peptide

Table IV
Detection of an Unexpected Impurity in a Synthetic Peptide
Ala–Phe–Ile–Gly–Leu–Met–NH_2; according to Lederer[7]

(The mass spectrum of the acetylated and permethylated peptide shows two series of peaks with a mass difference of 16 m.u.)

Theoretical structure	128	289	446	487	614	
	CH_3CO-MeAla	MePhe	MeIle	MeGly	MeLeu	MeMet-NMe_2
"Impurity"	144	305	432	503	630	—
	CH_3OCO-MeAla	—	—	—	—	

Mass measurement of the peak at m/e 144. Calculated for $C_6H_{10}NO_3 = 144 \cdot 0661$. Observed: $144 \cdot 0665$.

Experiments by B. C. Das and D. W. Thomas

without prior acylation yielded a spectrum of this "oxygen homolog" in pure form.

It was concluded that part of the original peptide fraction must already be acylated and contain a methoxycarbonyl protecting group (CH_3OCO-Ala). This quite unexpected result was puzzling to Dr. Schröder and his colleagues, but it could finally be explained after analysis of a batch of t-butoxy carbonyl chloride used for the synthesis of the hexapeptide. It was found that it contained a certain amount of methoxycarbonyl chloride, which had thus been incorporated into the peptide and then had escaped "deprotection."

4.7 Scope and Limitations of Mass Spectrometry in Peptide Chemistry

The upper limit for mass spectrometry is of course an important question that interests all biochemists performing sequence determination of peptides. The situation can be summed up as follows: without permethylation, N-acyl oligopeptide methyl esters can be mastered up to hepta- or octapeptides, depending on the amino acid residues they contain. After permethylation this limit is increased, allowing in the most favorable case sequence determination of the first 12 N-terminal amino acid residues. Thus after permethylation, the decapeptide de-

Ac-Leu-Ala-Lys(TFA)-Val-Ala-Tyr-Val-Tyr-Lys(TFA)-Pro-OH

(XXXII)

rivative (XXXII) gives a mass spectrum[85] having the expected molecular ion at m/e 1580 (Figure 8) and all specific cleavages showing the entire amino acid sequence. Only a few "non-sequence" peaks occur, mainly in the molecular weight region and do not represent a serious handicap for a correct sequence determination.

Partial sequences have been obtained after permethylation with the tetradecapeptide derivative stendomycic acid,[96] the pentadecapeptide derivatives gramicidins A and B (Reference 7, Lederer, and Reference 89), and with an octadecapeptide[92] and even a docosapeptide[109] isolated from pig immunoglobulin; the sequences of only 10 to 12 N-terminal amino acid residues could be deduced. This is apparently due to the fact that in the high mass region the intensity of each successive sequence peak is lower than the preceding (frequently 10% or less) so that the ions beyond 10 to 12 residues become a very small percentage of the total ionization. Thus they fall below the sensitivity limits of the now available mass spectrometers.

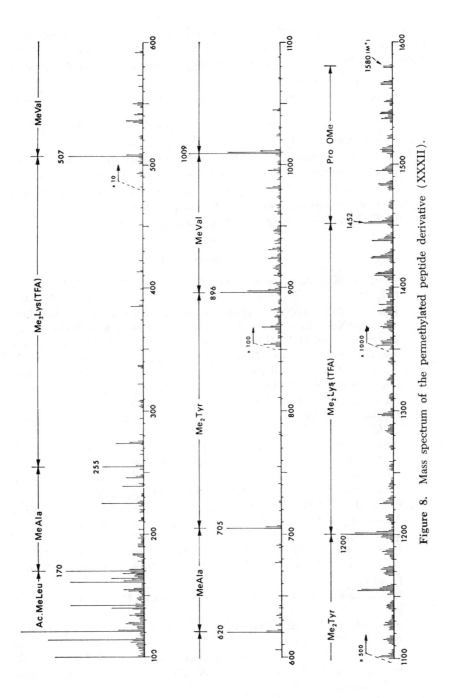

Figure 8. Mass spectrum of the permethylated peptide derivative (XXXII).

217

With the peptide derivatives studied up to now, the upper mass limits for the detection of significant peaks seem to be at m/e 1500–1600. It is clear that instrumental improvements will allow further advances in this field.

Mass spectrometry is particularly useful in the following cases:

1) For peptides having no free N-terminal residue (N-acetylated peptides and pyroglutamic acid peptides).
2) For the location of amides of aspartic acid and glutamic acid.
3) For finding and determining unexpected amino acid residues.
4) For branched-chain peptides, e.g. bacterial cell wall peptides.
5) For checking dubious chemical sequence determinations.
6) For checking the purity of synthetic peptides.

References

1. Biemann, K.   *Mass Spectrometry* (New York: McGraw-Hill, 1962); Hill, H. C.,   *Introduction to Mass Spectrometry* (London: Heyden and Son Ltd., 1966); McLafferty, F. W.   *Interpretation of Mass Spectra* (New York: W. A. Benjamin, Inc., 1966); Budzikiewicz, H., C. Djerassi, and D. H. Williams.   *Mass Spectrometry of Organic Compounds,* (San Francisco: Holden-Day, Inc., 1967); and Beynon, J. H. *Mass Spectrometry and its Applications to Organic Chemistry.* (Amsterdam: Elsevier, 1960).
2. Junk, G., and H. Svec.   J. Amer. Chem. Soc., **85**, 839 (1963).
3. Biemann, K., and J. A. McCloskey.   J. Am. Chem. Soc. **84**, 3192 (1962).
4. Biemann, K., J. Siebl, and F. Gapp.   J. Am. Chem. Soc. **83**, 3795 (1961); Andersson, C.-O., R. Ryhage, S. Stalberg-Stenhagen, and E. Stenhagen.   Arkiv. Kemi. **19**, 405 (1962); and Biemann, K.   *Mass Spectrometry* (New York: McGraw-Hill, 1962), p 260.
5. Heyns, K., and H. F. Grützmacher.   Ann. Chem. **667**, 194 (1963).
6. Andersson, C.-O., R. Ryhage, and E. Stenhagen.   Arkiv. Kemi. **19**, 417 (1962).
7. Jones, J. H.   Quart. Rev. **22**, 302 (1968); Shemyakin, M. M.   Pure Appl. Chem. **17**, 313 (1968); Lederer, E.   Pure Appl. Chem. **17**, 489 (1968).
8. Biemann, K.   *Mass Spectrometry* (New York: McGraw-Hill, 1962), p 284.
9. Senn, M., R. Venkataraghavan, and F. W. Lafferty.   J. Am. Chem. Soc. **88**, 5593 (1966).
10. Kiryushkin, A. A., Y. A. Ovchinnikov, M. M. Shemyakin, V. N. Bocharev, B. V. Rosinov, and N. S. Wulfson.   Tetrahedron Letters **1966**, 33.
11. Stenhagen E.   Z. Anal. Chem. **181**, 462 (1961).

12. Manusadzhyan, V. G., A. M. Zyakoon, A. V. Chuvilin, and Y. M. Varshavskii. Izv. Akad. Nauk Arm. SSR, Ser. Khim. **17,** 143 (1964).
13. Shemyakin, M. M., Y. A. Ovchinnikov, A. A. Kiryushkin, E. I. Vinogradova, A. I. Miroshnikov, Y. B. Alakhov, V. M. Lipkin, Y. B. Shetsev, N. S. Wulfson, B. V. Rosinov, V. N. Bocharev, and V. M. Burikov. Nature **211,** 361 (1966).
14. Heyns, K., and H. F. Grützmacher. Tetrahedron Letters **1963,** 1761.
15. Andersson, G.-O. Acta Chem. Scand. **12,** 1353 (1958).
16. Weygand, F., A. Prox, H. H. Fessel, and K. K. Sun. Z. Naturforsch. **20b,** 1169 (1965).
17. Andersson, B. A. Acta Chem. Scand. **21,** 2906 (1967).
18. Bricas, E., J. Van Heijenoort, M. Barber, W. A. Wolstenholme, B.C. Das, and E. Lederer. Biochemistry **4,** 2254 (1965).
19. Alpin, R. T., J. H. Jones, and B. Liberek. J. Chem. Soc. (C) **1968,** 1011.
20. Biemann, K., C. Cone, B. R. Webster, and G. P. Arsenault. J. Am. Chem. Soc. **88,** 5598 (1966).
21. Kiryushkin, A. A., Y. A. Ovchinnikov, M. M. Shemyakin, V. N. Bocharev, B. V. Rosinov, and N. S. Wulfson. Tetrahedron Letters **1966,** 33.
22. Flikwert, J. P., W. Heerma, H. Copier, G. Rijkstra, and J. F. Arens. Rec. Trav. Chim. **86,** 293 (1967).
23. Penders, T. J., W. Heerma, H. Copier, G. Dijkstra, and J. F. Arens. Rec. Trav. Chim. **85,** 879 (1966).
24. Heyns, K., and H. F. Grützmacher. Fortschr. Chem. Forsch. **6,** 536 (1966).
25. Van Heijenoort, J., E. Bricas, B. C. Das, E. Lederer, and W. A. Wolstenholme. Tetrahedron **23,** 3404 (1967).
26. Shemyakin, M. M., Y. A. Ovchinnikov, E. I. Vinogradova, M. Yu. Feigina, A. A. Kiryushkin, N. A. Aldanova, Y. B. Alakhov, V. M. Lipkin, and B. V. Rosinov. Experientia **23,** 428 (1967).
27. Vetter-Diechtl, H., W. Vetter, W. Richter, and K. Biemann. Experientia **24,** 340 (1968).
28. Biemann, K. *Mass Spectrometry* (New York: McGraw-Hill, 1962), p 153; Beynon, J. H. *Mass Spectrometry and its Applications to Organic Chemistry* (Amsterdam: Elsevier, 1960), p 251.
29. Barber, M., P. Jollès, E. Vilkas, and E. Lederer. Biochem. Biophys. Res. Commun., **18,** 469 (1965).
30. Vilkas, E., A. Rojas, B. C. Das, W. A. Wolstenholme, and E. Lederer. Tetrahedron **22,** 2809 (1966).
31. Senn, M., R. Venkataraghaven, and F. W. McLafferty. J. Am. Chem. Soc. **88,** 5593 (1966).
32. Sarges, R., and B. Witkop. Biochemistry **4,** 2491 (1965).
33. Vilkas, E., A. M. Miquel, and E. Lederer. Biochim. Biophys. Acta **70,** 217 (1963).
34. Lederer, E. Pure Appl. Chem. **7,** 247 (1963); Angew Chem. **76,** 214 (1964); Angew. Chem. Intern. Ed. Engl. **3,** 393 (1964).

35. Lanéelle, G., and J. Asselineau. Biochim. Biophys. Acta 59, 731 (1962).
36. Lanéelle, G., J. Asselineau, W. A. Wolstenholme, and E. Lederer. Bull. Soc. Chim. France 1965, 2133.
37. Guinand, M., G. Michel, and E. Lederer. Compt. Rend. Acad. Sci. 259, 1267 (1964).
38. Guinand, M., and G. Michel. Biochim. Biophys. Acta. 125, 75 (1966).
39. Barber, M., W. A. Wolstenholme, M. Guinand, G. Michel, B. C. Das, and E. Lederer. Tetrahedron Letters 1965, 1331.
40. Guinand, M., M. J. Vacheron, G. Michel, B. C. Das, and E. Lederer. Tetrahedron, Suppl. 7, 1966, 271.
41. Guinand, M., G. Michel, B. C. Das, and E. Lederer. Vietnamica Chim. Acta. 1966, 37.
42. Kiryushkin, A. A., V. M. Burikov, and B. V. Rosinov. Tetrahedron Letters 1967, 2675.
43. Kingston, D. G. I., Lord Todd, and D. H. Williams. J. Chem. Soc. 1966, 1669.
44. Vining, L. C., and W. A. Taber. Can. J. Chem. 40, 1579 (1962).
45. Wolstenholme, W. A., and L. C. Vining. Tetrahedron Letters 1966, 2785.
46. Kiryushkin, A. A., Y. A. Ovchinnikov, and N. S. Wulfson. Khim. Prirodn. Soedin., Akad. Nauk. Uz. SSR 2, 203 (1966); Chem. Abstr. 65: Biochemistry, 10, 15697-f (1966).
47. Briggs, L. H., B. J. Fergus, and J. S. Shannon. Tetrahedron, Suppl. 8, part I, 1966, 269.
48. Smith, D. W., H. M. Randall, A. P. MacLennan, R. K. Putney, and S. V. Rao. J. Bacteriol. 79, 217 (1960).
49. MacLennan, A. P., D. W. Smith, and H. M. Randall. Biochem. J. 80, 309 (1961).
50. Smith, D. W., and H. M. Randall. Am. Rev. Respirat. Diseases 92, 34 (1965).
51. Smith, D. W., H. M. Randall, A. P. MacLennan, and E. Lederer. Nature 186, 887 (1960).
52. MacLennan, A. P. Biochem. J. 82, 394 (1962).
53. Chaput, M., G. Michel, and E. Lederer. Biochim. Biophys. Acta 63, 310 (1962).
54. Vilkas, E., A. Rojas, B. C. Das, W. A. Wolstenholme, and E. Lederer. Tetrahedron 22, 2809 (1966).
55. Vilkas, E., C. Gros, and J. C. Massot. Compt. Rend. Acad. Sci. 266, Ser. C, 837 (1968).
56. Ikawa, M., E. E. Snell, and E. Lederer. Nature 188, 558 (1960).
57. Lanéelle, G. Compt. Rend. Acad. Sci. 263, Ser. C, 502 (1966).
58. Lanéelle, G. Thèse de Doctorat ès Sciences, Faculté des Sciences, Université de Toulouse (1967).
59. Jollès, P., F. Bigler, T. Gendre, and E. Lederer. Bull. Soc. Chim. Biol. 43, 177 (1961).

60. Chaput, M., G. Michel, and E. Lederer. Biochim. Biophys. Acta **78**, 329 (1963).
61. Vilkas, E., A. Rojas, and E. Lederer. Compt. Rend. Acad. Sci. **261**, 4258 (1965).
62. Lanéelle, M. A., G. Lanéelle, P. Bennet, and J. Asselineau. Bull. Soc. Chim. Biol. **47**, 2047 (1965); Lanéelle, G., and J. Asselineau. Eur. J. Biochem. **5**, 478 (1968).
63. Das, B. C., S. D. Géro, and G. Lanéelle. Unpublished experiments.
64. Harbon, S., G. Herman, B. Rossignol, P. Jollès, and H. Clauser. Biochem. Biophys. Res. Commun. **17**, 57 (1964).
65. Anderson, B., N. Seno, P. Sampson, J. G. Riley, P. Hoffman, and K. Meyer. J. Biol. Chem. **239**, PC 2716 (1964).
66. Bruneteau, M., and G. Michel. Compt. Rend. Acad. Sci. **267**, Ser. C, 745 (1968); Mester, L., A. Schimpl, and M. Senn. Tetrahedron Letters **1967**, 1697.
67. Heyns, K., and H. F. Grützmacher. Ann. Chem. **669**, 189 (1963).
68. Millard, B. J. Tetrahedron Letters, **1965**, 3041.
69. Prox, A., and F. Weygand. *Peptides.* Ed. by H. C. Beyerman, A. van de Linde and W. Massen van den Brink (Amsterdam: North-Holland Publ. Co., 1967), p 158; Wieland, T., G. Lüben, H. Otten-heym, J. Faesel, J. X. de Vries, A. Prox, and J. Schmid. *Angew. Chem. Intern. Ed. Engl.* **7**, 204 (1968).
70. Tschesche, R., R. Welters, and H.-W. Fehlhaber. Chem. Ber. **100**, 323 (1967).
71. Païs, M., X. Monseur, X. Lusinchi, and R. Goutarel. Bull. Soc. Chim. France **1964**, 817.
72. Zbiral, E., E. L. Menard, and J. M. Müller. Helv. Chim. Acta **48**, 404 (1965).
73. Warnhoff, E. W., J. C. N. Ma, and P. Reynolds-Warnhoff. J. Am. Chem. Soc. **87**, 4198 (1965).
74. Marchand, J., M. Païs, X. Monseur, and F.-X. Jarreau. Tetrahedron **25**, 937 (1969).
75. Wulfson, N. S., V. A. Puchkov, B. V. Rozinov, Yu. D. Denisov, V. N. Bochkarev, M. M. Shemyakin, Y. A. Ovchinnikov, A. A. Kiryushkin, E. I. Vinogradova, and M. Y. Feigina. Tetrahedron Letters, **1965**, 2805.
76. Wulfson, N. S., V. A. Puchkov, V. N. Bochkarev, B. V. Rozinov, A. M. Zyakoon, M. M. Shemyakin, Y. A. Ovchinnikov, V T. Ivanov, A. A. Kiryushkin, E. I. Vinogradova, M. Y. Feigina and N. A. Alda-nova. Tetrahedron Letters **1964**, 951.
77. Wulfson, N. S., V. A. Puchkov, B. V. Rozinov, A. M. Zyakoon, M. M. Shemyakin, Y. A. Ovchinnikov, A. A. Kiryushkin and V. T. Ivanov. Tetrahedron Letters **1965**, 2793.
78. McDonald, C. G., J. S. Shannon, and A. Taylor. Tetrahedron Letters **1964**, 2087.
79. Russel, D. W., C. G. McDonald, and J. S. Shannon. Tetrahedron Letters **1964**, 2759.

80. Bertaud, W. S., M. C. Probine, J. S. Shannon, and A. Taylor. Tetrahedron 21, 677 (1965).
81. McDonald, C. G., and J. S. Shannon. Tetrahedron Letters 1964, 3113.
82. Hassall, C. H., and J. O. Thomas. Tetrahedron Letters 1966, 4485.
83. Kuhn, R., I. Löw, and H. Trischmann. Chem. Ber. 90, 203 (1957).
84. Das, B. C., S. D. Géro, and E. Lederer. Biochem. Biophys. Res. Commun. 29, 211 (1967).
85. Thomas, D. W., B. C. Das, S. D. Géro, and E. Lederer. Biochem. Biophys. Res. Commun. 32, 199 (1968).
86. Biemann, K., C. Cone, N. R. Webster, and G. P. Arsenault. J. Am. Chem. Soc. 88, 5598 (1966).
87. Barber, M., P. Powers, M. J. Wallington, and W. A. Wolstenholme. Nature 212, 784 (1966).
88. Agarwal, K. L., R. A. W. Johnstone, G. W. Kenner, D. S. Millington, and R. C. Sheppard. Nature 219, 498 (1968).
89. Thomas, D. W. Biochem. Biophys. Res. Commun., 33, 483 (1968).
90. Vilkas, E., and E. Lederer. Tetrahedron Letters 1968, 3089.
91. Hakomori, S. I. J. Biochem. 55, 205 (1964).
92. Franek, F., B. Keil, D. W. Thomas, and F. Lederer. FEBS Letters 2, 309 (1969).
93. Coggins, J., and L. Benoiton. Abstracts of Papers, BIOL 18, 156th ACS National Meeting, 1968, Atlantic City, New Jersey.
94. Thomas, D. W., B. C. Das, S. D. Géro, and E. Lederer. Biochem. Biophys. Res. Commun. 32, 519 (1968).
95. Das, B. C., S. D. Géro, and E. Lederer. Nature 217, 547 (1968).
96. Thomas, D. W., E. Lederer, M. Bodanszky, J. Izdebski, and Muramatsu. Nature 220, 580 (1968).
97. Hotchkiss, R. D., and R. J. Dubos. J. Biol. Chem. 132, 791 (1950).
98. Sarges, R., and B. Witkop. J. Am. Chem. Soc. 87, 2020 (1965).
99. Thomson, R. Q., and M. S. Hughes. J. Antibiotics 16, 187 (1968).
100. Muramatsu, I., and M. Bodanszky. J. Antibiotics 21, 68 (1968).
101. Bodanszky, M., I. Muramatsu, and A. Bodanszky. J. Antibiotics 20, 384 (1967).
102. Bodanszky, M., J. Izdebski, I. Muramatsu, and A. Bodanszky. Peptides 1968: Proceedings of the 9th European Peptide Symposium, Orsay (Amsterdam: North-Holland Publ. Co., 1968), p 306.
103. Ogawa, H., and T. Ito. J. Agricult. Chem. Soc. Japan 24, 191 (1951); 26, 432 (1952).
104. Ito, T., and H. Ogawa. Bull. Agricult. Chem. Soc. Japan 23, 536 (1959).
105. Ovchinnikov, Y. A., V. T. Ivanov, P. V. Kostetsky, and M. M. Shemyakin. Tetrahedron Letters 1966, 5285.
106. Thomas, D. W., and T. Ito. Tetrahedron 25, 1985 (1969).
107. Morris, H. R., A. J. Geddes, and N. Graham. Biochem. J. 111, 38P (1969).
108. Franek, F., B. Keil, and F. Sorm. Abstracts of the 5th Meeting of FEBS, Prague, 1968, 173.

109. de Haas, G. H., F. Franek, B. Keil, D. W. Thomas, and E. Lederer, FEBS Letters **4**, 25 (1969).
110. Lübke, K., R. Hempel, and E. Schröder. Experientia **21**, 84 (1965).
111. Thomas, D. W. FEBS Letters **5**, 53 (1969).

ADDITIONAL REFERENCES

Amino acids and derivatives

Sur la structure de la nopaline métabolite abnormal de certaines tumeurs de Crown gall. Arlette Goldmann, D. W. Thomas, and G. Morel. Compt. Rend. Acad. Sci. Paris, **268**, 853 (1969).

Combined gas chromatography-mass spectrometry of amino acid derivatives. E. Gelpi, W. A. Koenig, J. Gilbert, and J. Oro. J. Chromatog. Sc., **7**, 604 (1969).

Isolation of lysinonorleucine from collagen. M. L. Tanzer and G. Mechanic. Biochem. Biophys. Res. Commun. **39**, 183 (1970).

Gas-liquid chromatography–mass spectrometry of carbon 13 enriched amino acids as trimethylsilyl derivatives. W. J. A. Vandenheuvel and J. S. Cohen. Biochim. Biophys. Acta **208**, 251 (1970).

Massenspektrometrische Identifizierung der Phenylthiohydantoin-Derivat von Aminosäuren. H. Hagenmaier, W. Ebbighausen, G. Nicholson, and W. Vötsch. Z. *Naturforsch.* **25b**, 681 (1970).

Gas liquid chromatography and mass spectrometry of deuterium containing amino acids as their trimethylsilyl derivatives. W. J. A. Vandenheuvel, J. L. Smith, I. Putter, and J. S. Cohen. J. Chromatog. **50**, 405 (1970).

Chemical ionization mass spectrometry of complex molecules. IV. Amino acids. G. W. A. Milne, T. Axenrod, and H. M. Fales. J. Am. Chem. Soc. **92**, 5170 (1970).

Synthetic model peptides

Sequence analysis of arginyl peptides by mass spectrometry. J. Lenard and P. M. Gallop. Anal. Biochem. **29**, 203 (1969).

Peptide sequencing by low resolution mass spectrometry. I. The Use of acetylacetonyl derivatives to identify N-terminal residues. V. Bacon, E. Jellum, W. Patton, W. Pereira, and B. Halpern. Biochem. Biophys. Res. Commun. **37**, 878 (1969).

The rational use of mass spectrometry for amino acid sequence determination in peptides and extension of the possibilities of the method. M. M. Shemyakin, Yu. A. Ovchinnikov, E. I. Vinogradova, A. A. Kiryushkin, M. Yu. Feigina, M. A. Aldanova, Yu. B. Alakhov, V. M. Lipkin, A. I. Miroshnikov, B. V. Rosinov, and S. A. Kazaryan. FEBS Letters **7**, 8 (1970).

Intramolecular alkylation reactions in the mass spectrometry of peptide derivatives. G. W. A. Milne, A. A. Kiryushkin, Yu. A. Alakhov, V. M. Lipkin, and Yu. A. Ovchinnikov. Tetrahedron **26**, 299 (1970).

Field ionization mass spectrometry. Benzyloxycarbonyl and tert-butyloxy-carbonyl derivatives of some simple peptides. P. Brown and G. R. Pettit. Org. Mass Spectrom. **3**, 67 (1970).

Mass spectrometric determination of the amino acid sequence in peptides and proteins. A. A. Kiryushkin & al. Sovrem. Probl. Khim. Peptidov Belkov 17 (1969); Chem. Abstr. Bio. **72**, 38811 (1970).

Sequence analysis of microgram amounts of peptides by mass spectrometry. J. Lenard and P. M. Gallop. Anal Biochem. **34**, 286 (1970).

Determination of amino acid sequences in peptide mixtures by mass spectrometry. F. W. McLafferty, R. Venkataraghavan and P. Irving. Biochem. Biophys. Res. Commun. **39**, 274 (1970).

Mass spectrometric determination of amino acid sequences in peptides.
  -fragmentation of peptides containing monoamino di-carboxylic acid groups,
  -synthesis of derivatives of peptides containing monoamino di-carboxylic acid groups. M. M. Shemyakin & al. Zh. Obshch. Khim. **40**, 407 (1970).

Mass spectrometric determination of amino acid sequences in peptides. Fragmentation of peptides containing asparagine and glutamine groups. M. M. Shemyakin & al. Zh. Obshch. Khim. **40**, 407 (1970).

Mass spectra of N-adamantoyl peptides. I. Lengyel, R. A. Salomone, and K. Biemann. Org. Mass Spectrom. **3**, 789 (1970).

Study by mass spectrometry of amino acid sequences in peptides containing histidine. J. F. G. Vliegenthart and L. Dorland. Biochem. J. **117**, 31 (1970).

Mass spectrometric determination of amino acid sequences in peptides. XV. Fragmentation of peptides containing monoamino dicarboxylic acid groups. M. M. Shemyakin & al. Zh. Obshch. Khim. **40**, 443 (1970).

Natural peptides

Application of mass spectrometry to protein chemistry. I. Method for amino terminal sequence analysis of proteoins. W. R. Gray and U. D. del Valle. Biochemistry **9**, 2134 (1970).

Isolation and mass spectrometric identification of the peptide subunits of mycobacterial cell walls. J. Wietzerbin-Falszpan, B. C. Das, I. Azuma, A. Adam, J. F. Petit, and E. Lederer. Biochem. Biophys. Res. Commun. **40**, 57 (1970).

Sequence Analysis of complex protein mixtures by isotope dilution and mass spectrometry. T. Fairwell, W. T. Barnes, F. F. Richards, and R. E. Lovins. Biochemistry **9**, 2260 (1970).

The structure of pepstatin. H. Morishima, T. Takita, T. Aoyagi, T. Takeuchi, and H. Umezawa. J. Antibiotics **23**, 263 (1970).

Sulfur containing peptides

Mass spectrometric determination of the amino acid sequence in peptides: desulfurization of sulfur-containing peptides. A. A. Kiryushkin, V. A.

Gorlenko, B. V. Rosinov, Yu. A. Ovchinnikov, and M. M. Shemyakin. Experientia **25**, 913 (1969).

Mass spectrometry of cysteine-containing peptides. M. L. Polan, W. J. McMurray, S. R. Lipsky, and S. Lande. Biochem. Biophys. Res. Commun. **38**, 1127 (1970).

Determination of amino acid sequence in peptides by mass spectrometry. Desulfurization of sulfur-containing peptides. R. Toubiana, J. E. G. Barnett, E. Sach, B. C. Das, and E. Lederer. FEBS Letters **8**, 207 (1970).

Mass spectrometry of peptide derivatives. Temporary protections of methionine as sulfoxide during permethylation. P. Roepstorff, K. Norris, S. Severinsen, and K. Brunfeldt. FEBS Letters **9**, 235 (1970).

Reductive desulfuration and mass spectrometric sequencing of sulfur-containing peptides. M. A. Paz, A. Bernath, E. Henson, O. O. Blumenfeld, and P. M. Gallop. Anal. Biochem. **36**, 527 (1970).

Peptide antibiotics, cyclodepsipeptides

The revised structure of viscosin, a peptide antibiotic M. Hiramoto, K. Okada, and S. Nagai. Tetrahedron Letters **1970**, 1087.

Isolation and structure elucidation of three new insecticidal cyclodepsipeptides, destruxins C and D and desmethyldestruxin B, produced by *Metarrhizium anisopliae*. A. Suzuki, H. Taguchi, and S. Tamura. Agr. Biol. Chem. **34**, 813 (1970).

Peptide hormones

Feline Gastrin. An example of peptide sequence analysis by mass spectrometry. K. L. Agarwal, G. W. Kenner, and R. C. Sheppard. J. Am. Chem. Soc. **91**, 3096 (1969).

Structure moléculaire du facteur hypothalamique hypophysiotrope TRF d'origine ovine: mise en évidence par spectrométrie de masse de la séquence PGA-His-Pro-NH$_2$. R. Burgus, T. F. Dunn, D. Desiderio, and R. Guillemin. Compt. Rend. Acad. Sci. **269** D, 1870 (1970).

Structural elucidation of a Gonadotropin-inhibiting substance from the bovine pineal gland. D. W. Cheesman. Biochim. Biophys. Acta **207**, 247 (1970).

Chapter 5

Molecular–Sieve Chromatography of Proteins

by S. Hjertén

5.1 Introduction

As early as 1952 Deuel and Neukom[1] briefly described chromatographic molecular sieving experiments. These authors seem to be the first who considered this technique. They cross-linked locust bean gum with epichlorohydrin. The gels formed were cut into small pieces and dehydrated with alcohol. After swelling in water the gel grains were packed in a column tube. The authors state that "the retention or breakthrough volume may be a measure of molecular size . . . and an effective dialysis may be accomplished by mere percolation." To explain the separations they claim that "the smaller the dissolved molecules the more quickly they diffuse into these cross-linked particles . . . and very large molecules do not diffuse into the swollen particles."

In 1955 Lindqvist and Storgårds[2] briefly mentioned the usefulness of starch columns for the molecular sieving of peptides. In papers published in 1955 and 1956, Lathe and Ruthven[3, 4] presented chromatograms which clearly demonstrated the great potentialities that beds of starch grains offer for the separation of peptides and proteins according to molecular size. However, chromatographic fractionations with beds of molecular sieving properties did not come into general use until bed materials with good physical and chemical characteristics became commercially available. The first commercial product, Sephadex (dextran cross-linked with epichlorohydrin), was introduced after publication of a paper in 1959 by Porath and Flodin[5] on the application of dextran gels for desalting and for chromatographic molecular sieving of low-molecular-weight substances. These dextran gels (G-25, G-50 and G-75) permitted separation of substances having molecular weights up to 30,000–40,000.

There was, of course, also a need for bed materials capable of fractionating substances of a much larger molecular size. In the search for
such bed materials, Hjertén and Mosbach[6,7,8] found that columns of
polyacrylamide gels were able to separate proteins as large as R-phy-
coerythrin (mol wt 290,000) and R-phycocyanin (mol wt 135,000).
Also Lea and Sehon[9] have described a method for the preparation of
gel grains of cross-linked polyacrylamide as well as of cross-linked
polyvinylethyl carbitol and polyvinyl pyrrolidone. Using granulated
agar gels Polson[10] could distinguish substances ranging in molecular
weight from 13,000 to several millions. The acidic character of agar
gels makes them less convenient for molecular sieving. With the neutral agarose gels Hjertén extended the fractionation range to cover not
only high-molecular-weight proteins but also subcellular particles.[7,11]
The fractionation range of the dextran gels was broadened when Flodin introduced new gel types with higher porosity (G-100, G-150, G-
200) and effective in the molecular weight range 50,000–500,000.[12]
Molecular-sieve chromatography on beds of porous glass[13] and on gels
of cellulose[14] and tanned gelatin[15] have also been described.

Comprehensive reviews on molecular sieving on both hydrophilic
and hydrophobic gels have been written by Determann[16] and Fischer.[17]

5.2   Comments on Terminology

The procedure involving fractionation of substances according to
their molecular size on columns packed with gel grains is often called
"gel filtration." This term has been criticized because filtration in its
common sense means the separation of only two phases, for instance,
by a filter paper.[18,19] Because the procedure is a chromatographic one,
it seems logical to include the word "chromatography" in the term. For
this reason "gel chromatography"[16] is to be preferred to "gel filtration";
unfortunately, however, this makes one think involuntarily of adsorption chromatography on gels of silica and aluminum hydroxide.[20,21] The
word "exclusion chromatography"[22] has the drawback of relating to a
postulated but not proved separation mechanism, namely, that the
solute molecules can penetrate the gel grains to a different extent
according to their molecular size, with the very large molecules completely excluded. The same weakness is associated with the terms
"restricted-diffusion chromatography"[23] and "gel-permeation chromatography.[18]

The author prefers "molecular-sieve chromatography"[8] or "chromatographic molecular sieving" in referring to the procedure; these
terms do not refer to any separation mechanism but only to the fact
that separations according to molecular size are obtained.[24]

5.3 The Separation Mechanism

5.3.1 *The Exclusion Hypothesis*

This hypothesis states that only part of a gel grain is accessible to the solute. It is assumed that the larger the molecular size of the solute, the smaller the available space. Molecules above a certain size will not penetrate a gel grain and will thus be eluted with the void volume (which equals the mobile volume between the gel grains). The other sample molecules will be eluted later according to their sizes. This explanation of the separations obtained in chromatographic molecular sieving was first considered by Deuel and Neukom,[1] as is evident from the quotations given in the Introduction. Lathe and Ruthven have proposed the same separation mechanism.[3,4] More refined treatments have been published by Porath,[25] Flodin,[12] and Squire,[26] and by Laurent and Killander[19] among others; the physical models used by these authors differ mainly in the way in which the solute molecules are assumed to be excluded. In the mathematical treatments of these models the distribution coefficient K_n of a substance is defined as the fraction of the volume of the water in the gel particles which is accessible to molecules of this substance, *i.e.*:

$$K_D = \frac{V_e - V_o}{V_i} \tag{1}$$

where

V_e = the elution volume of the solute of interest

V_o = the volume outside the gel grains (the void volume)

and

V_i = the volume of the water imbibed by the gel grains (the inner volume).

Laurent and Killander[19] have a slightly different definition of the distribution coefficient, namely

$$K_{Av} = \frac{V_e - V_o}{V_t - V_o} \tag{2}$$

where V_t is the total volume of the column bed.

Some of the mathematical expressions obtained for different assumptions on the separation mechanism are summarized below (see also Reference 27). In these expressions M is the molecular weight, a is the effective hydrodynamic (Stokes) radius of the solute of interest, and $A, B,$ and D are constants having different values in the different expressions.

Porath[25] has proposed a model based upon diffusion into conical holes, which leads to the relation

$$K_D{}^{1/3} = A - B \cdot a = A - D \cdot M^{1/2} \tag{3}$$

Using a similar treatment, but assuming that the holes have different shapes and that $a \propto M^{1/3}$ (Porath assumes $a \propto M^{1/2}$), Squire[26] has arrived at the formula

$$\left(\frac{1}{R}\right)^{1/3} = \left(\frac{V_e}{V_o}\right)^{1/3} = A - B \cdot a = A - D \cdot M^{1/3} \tag{4}$$

The hypothesis of Laurent and Killander[19] is based upon the assumption that the gel can be regarded as a suspension of rigid rods. The space between these matrix rods available to solute molecules was calculated according to an expression derived by Ogston[28] and resulted in the following relation:

$$(-\log K_{Av})^{1/2} = A + B \cdot a \tag{5}$$

It should also be mentioned that Andrews[29] has found the empirical formula

$$\frac{1}{R} = \frac{V_e}{V_o} = A - B \cdot \log M \tag{6}$$

5.3.2   Restricted Diffusion Hypothesis

According to Ackers[23] the above exclusion hypothesis works for the more compact gels (for instance, Sephadex G-75 and G-100). For gels with a higher water-regain capacity he claims that another effect, restricted diffusion, plays a more important role. Ackers assumes that there are two types of restricted diffusion that must be considered. One type is based on the assumption that the smaller the molecule, the greater the probability of its entering a pore in a gel grain. The other type is associated with the assumption that the diffusion coefficient is lower for a molecule inside a pore than for one in free solution because a molecule in a pore encounters an increased hydrodynamic resistance to motion. From his hypothesis Ackers has deduced the following relation:

$$K_D = \left(1 - \frac{a}{r}\right)^2 \left[1 - 2.10 \left(\frac{a}{r}\right) + 2.09 \left(\frac{a}{r}\right)^3 - 0.95 \left(\frac{a}{r}\right)^5 \right] \tag{7}$$

where r is the effective pore radius within the gel grain and a is the Stokes radius.

5.3.3   The Hypothesis Based on the Axial Migration of Particles in Poiseuille Flow

It is well-known that particles (for instance, erythrocytes) suspended in a liquid undergoing laminar flow through a tube migrate toward the axis of that tube.[30] It has been proposed by Hjertén and Mosbach[8] and by Pedersen[22] that this effect could be operative in molecular-sieve chromatography, but experimental evidence is still lacking.

5.3.4 The Hypothesis Based on Boltzmann Distribution

In 1962 Hjertén and Mosbach[8] advanced different hypotheses to explain the molecular-sieving action of polyacrylamide gel columns. One of the hypotheses is mentioned in the previous section. Another was based on the assumption (not proved) that molecular-sieve chromatography is a partition chromatography technique wherein the stationary and mobile phases can be regarded as the two components of a phase system—*i.e.*, the Brønsted formula should obtain.[31,32] This formula can be written as follows:

$$K = e^{-\frac{\lambda}{kT}} \tag{8}$$

where

K = the ratio of concentrations of the solute in the two phases (observe that this distribution coefficient is defined in a way other than that given in the previous sections, although Equation 1 is valid also for this new definition)

λ = a parameter, which depends on the molecular size of the solute

k = the Boltzmann constant, and

T = the absolute temperature.

It has been stated that Equation 8 is not workable for molecular-sieve chromatography of proteins, because a plot of $-\log K$ against molecular weight does not give a straight line.[16] Such a relationship, however, is not to be expected; in contrast to the case of low-molecular-weight compounds, λ for macromolecules is not proportional to the molecular weight of the solute but to its area, as pointed out by Brønsted[31] and proved mathematically by Albertsson.[32] If the Brønsted formula is valid, one should expect the following relation to hold for spherical protein molecules:

$$-\log K_D = \text{constant} \cdot M^{2/3} \tag{9}$$

where M = the molecular weight of the protein.

Andrews has measured distribution coefficients on Sephadex G-200 for a series of proteins of different molecular weights.[29] The plot of $-\log K_D$ against $M^{2/3}$ in Figure 1 is based upon his data. The points do not deviate more from a straight line than in the plot of K_D against log M, used by Andrews. From the considerations in the next section it is evident that one cannot expect Equation 9 to be generally valid.*

5.3.5 Thermodynamic Considerations

Under certain assumptions it can be shown by a purely thermodynamic treatment that

$$RT \ln K_D + v \left[p_s - p_m - (p_s^o - p_m^o) \right] + A \left[\gamma_s - \gamma_m - (\gamma_s^o - \gamma_m^o) \right]$$
$$+ RT \ln \frac{f_s}{f_m} = 0 \qquad (10)$$

where

- R = the gas constant
- T = the absolute temperature
- v = the partial molal volume of the solute
- p_s = the pressure in the stationary phase
- p_m = the pressure in the mobile phase
- A = the area of N solute molecules, where N = the Avogadro number
- γ_s = the interfacial tension for the face between the solute and the stationary phase
- γ_m = the interfacial tension for the face between the solute and the mobile phase, and
- f_s and f_m = the activity coefficients for the solute in the stationary and the mobile phase, respectively.

The superscripts o refer to the standard state.

The proof of Equation 10 and a more detailed discussion of it will be given elsewhere.[33]

For solutes of similar structure (so-called "isochemical" substances to which proteins are assumed to belong), $\gamma_s - \gamma_m$ is constant for a given gel bed. Thus the molal area A, the partial molal volume v of the protein, and the activity coefficients determine the value of the distribution coefficient (for a given gel bed $p_s - p_m$ is also constant). If the solute molecules are spherical (and isochemical), Equation 10 can be written in the following form:

$$-\log K_D = C_1 \cdot M + C_2 \cdot M^{2/3} + C \qquad (11)$$

---

*Using the chromatographic data of other authors we have thus found many experiments with macromolecules on Sephadex G-200 and other gel beds where Equation 9 is not applicable.

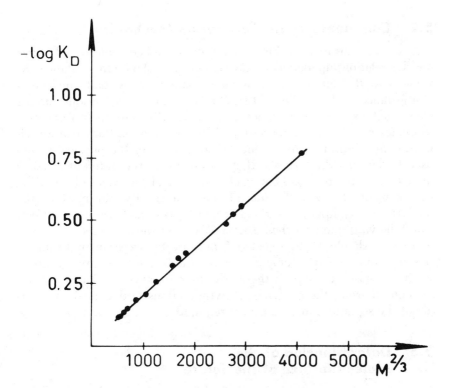

Figure 1. Plot of -log K_D against $M^{2/3}$ (K_D: the distribution coefficient, and M: the molecular weight) for different proteins.

where M is the molecular weight of the solute molecule and C a constant.*

When the first and third term to the right of the equality sign in Equation 11 can be neglected, the equation is reduced to the Brønsted formula in application to macromolecules (λ in Equation 8 is equal to $\lambda_o \cdot M^{2/3}$ where λ_o is a constant). In Reference 33 it is shown that for low molecular weight homologs Equation 10 can be written in the form $-\log K_D = \text{constant} \times M$ or $-\log K_D = \text{constant} \times \text{partial molal}$ volume. A similar relationship has previously been derived by ion-exchange theory for neutral low-molecular-weight substances,[34] and it has also been experimentally verified.[35]

---

*Because V_i in Equation 1 is difficult to determine accurately, it is more convenient in practice to use Equation 11 in the form

$$-\log(V_e - V_o) = C_1 M + C_2 M^{2/3} + C_3 \qquad (12)$$

where C_3 is a constant.

5.4 Comments on the Separation Mechanism

A series of formulas of different authors have been considered above for the relationship between chromatographic data (for instance, K-values and R-values) and parameters characterizing solute molecules (for instance, molecular weight and Stokes radius). The various authors claim that experimental values fit satisfactorily into their respective formulas. A rather common mistake is the assumption that such a verification of a formula is proof for the correctness of the physical model used in deriving the formula. If this were so, all the different physical models used for the mathematical treatment of the molecular sieving effects would be correct. But this is an absurdity, for most of the models are incompatible and cannot be transformed one into another. The following question then arises: Which of the proposed models is correct, or should all be rejected?* Because the experimental difficulties that must be overcome to answer these questions appear great, we have chosen to apply a thermodynamic treatment.[33] The great advantage of using thermodynamic considerations is that no hypothesis about the separation mechanism is required.

5.5 Determination of the Molecular Size of the Solute

From the different formulas presented, it is obvious that molecular-sieve chromatography offers the advantage of enabling molecular size of the solute of interest to be estimated. The solute need not be in a pure state; the only requirement is that it can be analyzed specifically, for instance, enzymatically. However, it should be stressed that the molecular size obtained must be regarded as tentative. In some cases the difference between the true molecular weight and that calculated from molecular-sieve chromatography is extremely great, particularly when the shape of the solute is unknown (most of the formulas given above apply only to globular macromolecules). For example, a polyethylene glycol (which is a linear molecule) with a molecular weight of 15,000 migrates approximately with the void volume on Sephadex G-100.[36] For a globular protein this elution volume should correspond to a molecular weight of about 100,000.

Many different opinions have been expressed about which of the proposed relationships between distribution coefficient and molecular

---

*There is no experimental evidence that macromolecules do penetrate the entire gel grain—an assumption in most of the hypotheses mentioned. Separations may very well take place on the surface of the gel grains although this hypothesis has not been experimentally verified.[45]

size of a solute to use when preparing an empirical calibration curve for estimation of the molecular size of a solute of interest. It is, of course, desirable to plot those parameters that will result in a straight line. However, not one of the proposed plotting methods will give a linear relationship in all cases and it is impossible to predict which of them will give it in any particular case. In practice, therefore, any plotting method can be used; in fact it is rather immaterial whether the calibration curve is a straight line or not because the molecular size obtained is always extremely approximate.

5.6 Factors Governing the Resolution of a Gel Column

The possibility of resolving two solute zones chromatographically depends partly on the distance ΔZ between the centers of the zones and partly on the width 2σ of the zones; σ = the distance between the center of a zone and a section through the zone where the concentration is $0.6\ c_o$, and c_o = the concentration in the center of the zone (see Figure 2). The resolution of two zones of the same width may conveniently be defined[37] by the relation

$$\rho = \frac{\Delta Z}{4\sigma} \qquad (13)$$

We will also introduce the height H of a theoretical plate defined by the equation

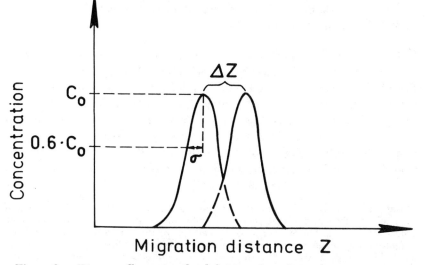

Figure 2. Diagram illustrating the definition of resolution between two zones.

$$H = \frac{\sigma^2}{L} \qquad (14)$$

where L is the length of the column bed.

The combination of Equations 13 and 14 gives

$$\rho = \frac{\Delta Z}{4\sqrt{HL}}$$

which can be transformed[37] to

$$\rho = \frac{1}{4}\sqrt{\frac{L}{H}} \cdot \frac{\Delta R}{R} \qquad (15)$$

where R is the ratio of zone velocity to mobile-phase velocity

$$R = \frac{V_o}{V_e}$$

Equation 15 shows that one should endeavor to attain low values of H in order to obtain a high resolution.

According to Giddings and Mallik,[38] the plate height H is determined by the following expression:

$$H = \frac{4D_m}{3Rv} + \frac{1}{20} R (1 - R) \frac{d_p^2 v}{D_m} + \sum \frac{1}{\frac{1}{2\lambda_i d_p} + \frac{D_m}{\omega_i d_p^2 v}} \qquad (16)$$

where

D_m = the diffusion coefficient in the mobile phase
v = the flow velocity of the mobile phase
d_p = the diameter of the spherical gel particles, and
λ_i and ω_i are geometrical factors.

Theoretical plate heights calculated with this formula are often much smaller than those obtained experimentally. (A contributing factor to this finding may be that the formula is derived for considerably smaller column diameters than are employed in molecular-sieve chromatography.)[38] Therefore it will be used mainly for a qualitative discussion. The three terms to the right of the equality sign correspond respectively to zone spreading caused by (a) ordinary diffusion, (b) eddy diffusion,* and (c) local nonequilibrium. Insertion of Equation 16 into

---

*Eddy diffusion occurs because certain parts of a zone take a more tortuous path than others. This means that some molecules move ahead of the average while others lag behind. The stream flow may also have different velocities in different regions of a chromatographic bed. In other words, eddy diffusion is ascribed to the local differences in *velocity* with which the different solute molecules migrate through the bed and to the different *distances* they have to migrate.

Equation 15 gives

$$\rho = \cfrac{\sqrt{L}\Delta R}{4\sqrt{\dfrac{4RD_m}{3v} + \dfrac{R^3(1-R)d_p^2 v}{20D_m} + R^2 \displaystyle\sum \cfrac{1}{\dfrac{1}{2\lambda_i d_p} + \dfrac{D_m}{\omega_i d_p^2 v}}}} \qquad (17)$$

We will now examine Equations 16 and 17 in a little more detail in order to show how the different parameters affect the resolution ρ.

5.6.1 The Bed Length L

From Equation 17 it is evident that relatively great increases in bed length L give only moderate increases in the resolution. Thus, if the bed length increases from 1 to 2 meters, the resolution will be only $\sqrt{2} = 1.4$ times greater.

5.6.2 The R- and ΔR-Values

The first and the third term under the square root sign in Equation 17 increase when R increases. By derivation it can be shown that the second term has a maximum for $R = 0.75$. We will therefore consider the following two cases.

(a) The second term is negligible in comparison with the first and third terms for all R-values up to 0.75. It will then also be negligible for $R > 0.75$. Consequently the resolution in this case will be higher the lower the R-value. However, such gel beds should not, of course, be selected so that the R-values of the solutes of interest correspond to the lower limit of the separation range of the bed (see Table I).

(b) The second term is predominant for R-values up to 0.75. In the range $0 < R < 0.75$ the resolution decreases with increasing R, just as in case (a). If the second term is predominant also in the range $0.75 < R < 1$ the resolution increases with increasing values of $R;$ consequently it may happen in certain experiments that we get higher resolution on a gel bed where the pertinent solutes migrate with R-values approaching 1 than on a bed where R-values are smaller than 0.75.*

As it is difficult to assess correct values for the parameters λ_i and ω_i, the third term under the square root sign in Equation 17 can only

---

*Giddings states that "any value of R near unity is injurious."[37] This incorrect statement is based on the use of an expression (Equations 7.2–14 and 7.2–24 in Reference 37) in which the fact that the distribution coefficient K approaches zero as R approaches one is ignored.

be estimated very roughly. Such an estimation together with a calculation of the first and third term indicate, however, that the second term might be predominant in molecular sieving of proteins only in the rare cases when an experiment is performed under extremely high flow rates on gel spheres of large diameter.

Table I
Separation Ranges of the Commercial Gel Products[*]

Trade name		Separation range for globular materials (mol wt)
Sephadex	G-10†	−700
	G-15	−1500
	G-25	1,000–5,000
	G-50	1,500–30,000
	G-75	3,000–70,000
	G-100	4,000–150,000
	G-150	5,000–400,000
	G-200	5,000–800,000
Bio-Gel	P-2	200–2,600
	P-4	500–4,000
	P-6	1,000–5,000
	P-10	5,000–17,000
	P-30	20,000–50,000
	P-60	30,000–70,000
	P-100	40,000–100,000
	P-150	50,000–150,000
	P-200	80,000–300,000
	P-300	100,000–400,000
Sepharose	6B	-2×10^6
	4B	$300,000 - 20 \times 10^6$
	2B	$2 \times 10^6 - 25 \times 10^6$
Bio-Gel A	0.5 m	$<10,000 - 500,000$
	1.5 m	$<10,000 - 1.5 \times 10^6$
	150 m	$1 \times 10^6 - >150 \times 10^6$
Sagavac	10	10,000–250,000
	9	25,000–500,000
	8	25,000–700,000
	7	$50,000 - 1.5 \times 10^6$
	6	$50,000 - 2 \times 10^6$
	5	$50,000 - 7 \times 10^6$
	4	$200,000 - 15 \times 10^6$
	3	$500,000 - 50 \times 10^6$
	2	$500,000 - 150 \times 10^6$
Gelarose	2%, 4%, 6%, 8%, 10%	

[*]Data are taken from the manufacturers' technical information material and should be considered extremely approximate.
†The practical operating range for Sephadex gels does not extend to molecular weights as low as those indicated.

In the above discussion we have not considered the fact that the ΔR-value for two given solutes is related to an "inherent" property of the gel and accordingly may vary with its composition. ΔR is thus a function of R. As we have considered ΔR as constant, the results derived are approximate and not always applicable.

As one very seldom knows how ΔR varies with R and also because Equation 17 is complicated and not exact, it is of very little practical use for the calculation of the R-values for which the resolution is optimal. Therefore, experimental investigations of the resolution on a series of gel beds of different compositions should be performed for each particular separation problem. The importance of varying the gel composition is illustrated in Figure 3 with serum as sample solution. If the aim of an experiment is to have the three main peaks separated as well as possible from each other, a gel of a total concentration $T = 6\%$ should be used. On the other hand, a 5% gel should be chosen if the interest is centered on the components corresponding to the first peak. From Figure 3 it is evident that even small alterations in gel composition will have a strong influence on the elution pattern. Although illustrated here with polyacrylamide gels, this finding is true also for dextran and agarose gels. Sometimes the steps in composition of the commercial gels are too large to give the desired optimum resolution. In such cases one may be forced to prepare the gel grains himself. Methods for the preparation of spherical gel grains of polyacrylamide[39] and agarose[40, 41, 42] have been described in detail. Gel grains of an irregular size—the same type as the commercial agarose product Saga-vac—can easily be prepared in the laboratory by pressing the gel through a metal net.[43]

In the brochures on commercial gels one can find empirical curves relating R-values (or K-values) to the molecular size of the solute. From such curves one can obtain the ΔR-values for two solutes of known molecular weight. Equation 17 shows that the resolution is a function of both ΔR- *and* R-values. Therefore, the gel that gives the highest ΔR-value will not necessarily give the highest resolution—a rather common misinterpretation, sometimes also found in the manufacturers' brochures.

5.6.3   *The Diameter d_p of the Gel Particles*

Equation 16 shows that the theoretical plate height H decreases with decreasing d_p, *i.e.*, the smaller the diameter of the gel spheres the higher the resolution (narrower zones). In practice one cannot use extremely small gel spheres because the resulting low flow rate means runs of long duration. It is also evident from Equation 16 that ordinary diffusion, corresponding to the first term to the right of the equality sign,

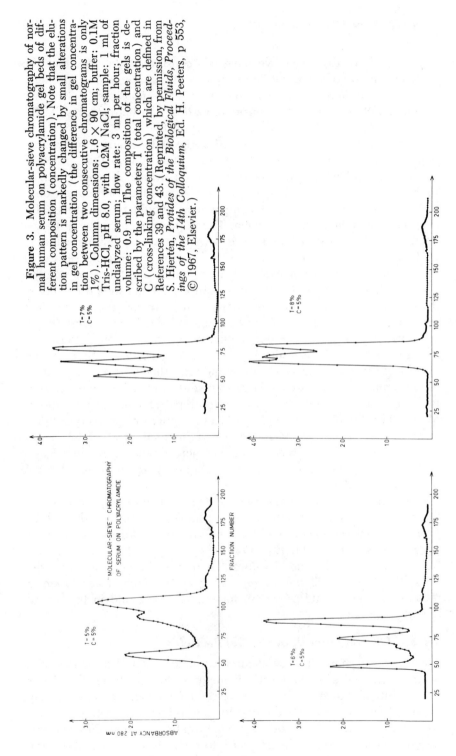

Figure 3. Molecular-sieve chromatography of normal human serum on polyacrylamide gel beds of different composition (concentration). Note that the elution pattern is markedly changed by small alterations in gel concentration (the difference in gel concentration between two consecutive chromatograms is only 1%). Column dimensions: 1.6 × 90 cm; buffer: 0.1M Tris-HCl, pH 8.0, with 0.2M NaCl; sample: 1 ml of undialyzed serum; flow rate: 3 ml per hour; fraction volume: 0.9 ml. The composition of the gels is described by the parameters T (total concentration) and C (cross-linking concentration) which are defined in References 39 and 43. (Reprinted, by permission, from S. Hjertén, Protides of the Biological Fluids, Proceedings of the 14th Colloquium, Ed. H. Peeters, p 553, © 1967, Elsevier.)

240

may contribute to increasing the plate height considerably if the gel grains chosen are so small that they cause a very large decrease in the flow rate v. However, as far as the resolution of proteins is concerned, many experiments have shown that there is very little risk that the diameter of the gel particles is too small—as long as the flow rate is satisfactory. For fractionation of high-molecular-weight proteins gel spheres of agarose are often preferable to those of dextran and poly-acrylamide, because the gel spheres of the former are much more rigid and therefore can be made smaller (with attendant increase in resolution power) without any risk of low flow rates.[40]

5.6.4. The Flow Rate v

Equation 16 shows that zone spreading caused by ordinary diffusion (corresponding to the first term to the right of the equality sign in Equation 16) diminishes when the flow rate increases. The reverse is true for zone spreading caused by eddy diffusion (second term) and local nonequilibrium (third term). There is thus only one flow rate for each experiment which will give optimum resolution. In cases where the first term can be neglected, Equation 16 indicates that it is favorable to use as small flow rates as possible. As a general rule one can say that the flow rate should lie in the range 1–3 ml per hour cm² in order to get satisfactory resolution.

5.7 Experimental Technique
5.7.1 Selecting the Gel Type

The following gel materials are commercially available: (a) cross-linked dextran, manufactured by Pharmacia Fine Chemicals AB, Uppsala, Sweden, under the trade name Sephadex; (b) cross-linked polyacrylamide, manufactured by Bio-Rad Laboratories, Richmond, California, USA, under the trade name Bio-Gel P; (c) agarose, manu-factured by Pharmacia Fine Chemicals AB, Uppsala, Sweden, under the trade name Sepharose; by Bio-Rad Laboratories, Richmond, Cali-fornia, USA, under the trade name Bio-Gel A; by Seravac Laboratories Ltd, Maidenhead, England, under the trade name Sagavac; and by Litex, Glostrup, Denmark, under the trade name Gelarose. All of these products except Sagavac are manufactured as spherical gel particles.

Some technical data on the different gels are listed in Table I. These data are given by the manufacturers, who will supply brochures con-taining further information on the mesh size of the gel grains and their physical and chemical properties, etc. The reader is requested to read these brochures, some of which also contain experimental advice.

From Table I it is evident that beds of Sephadex and Bio-Gel P have a separation range whose upper limit corresponds to substances with a molecular weight of about 500,000. For substances of higher molecular weights the agarose gels (Sepharose, Bio-Gel A, Sagavac, and Gelarose) offer the only alternative. It should be observed that the agarose gels cover part of the separation range of the dextran and polyacrylamide gels. Sepharose, Bio-Gel A, and Gelarose are manufactured in bead form and can therefore be expected to give narrower zones than Sagavac which can be obtained only as grains of an irregular size.

The quality of the first commercially available polyacrylamide gels was inferior to that of the dextran gels. However, since the introduction of a new suspension polymerization procedure developed by the author, the Bio-Gels P have been characterized by very good chemical and physical properties. In most cases Bio-Gel P and the corresponding type of Sephadex will give similar chromatograms.

5.7.2 Column Construction

For the harder gels the construction of the column tube is not so critical as for the softer gels (Sephadex G-150 and G-200, and Bio-Gels P-200 and P-300). To get satisfactorily high flow rates for the softer gels, it is important that the bed support be such that its pores will not become clogged by the gel particles, a condition which increases the resistance to flow. A porous sheet of polyethylene seems to give the best result in this respect. (Columns with such a bed support are commercially available from several companies).

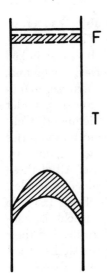

Figure 4. Drawing illustrating how an originally flat zone F will spread and become curved as it migrates down the column. The curvature of the zone is due to the higher flow rate at the wall of the column tube T than at its axis. This "wall" effect, which diminishes the resolution, is often less pronounced for column tubes of glass than for those of Plexiglas and still less for glass tubes coated with methyl cellulose.[8] [Reprinted, by permission, from S. Hjertén and R. Mosbach, Analytical Biochemistry 3, 109 (1962). © Academic Press.]

When chromatographing colored proteins on a gel column, one often observes that the originally flat zone will become more and more curved as it migrates down the column.[8] This is illustrated in Figure 4. Such a deformation of a zone, which diminishes the resolution, is due to a tendency of the buffer to flow faster in a layer near the wall of the tube. This "wall" effect is less pronounced for glass tubes than for Plexiglas tubes and still less for glass tubes treated with methyl cellulose as described in Reference 8.

5.7.3 Packing Procedure

To pack a column, the hydrated gel grains in the form of a rather diluted suspension are poured into the chromatographic tube and allowed to sediment. Fine material in the suspension should first be removed by repeated decantations. When the gel grains are of the soft type, the packing should be performed under reduced pressure, *i.e.*, the orifice of a narrow tubing, connected to the outlet of the column tube, is placed 10–30 cm below the upper liquid surface in the reservoir containing the gel suspension. This procedure allows only slight compression of the gel grains thus permitting proper flow rates; the subsequent elution is also performed under reduced pressure. A soft gel bed should be protected at its top by a filter paper or a 2-cm layer of a harder gel; otherwise there is a risk that the bed will be stirred up when the sample is subsequently applied.

To get a high resolution it is, of course, important that the gel bed be homogeneously packed. This can be checked by running a solution of a colored protein through the column. If the zone is strongly distorted, the column must be repacked. Pharmacia Fine Chemicals AB recommends the use of Blue Dextran 2000 as a test substance for judging the quality of the packing. This substance is not ideal, however, because in some cases we have observed that it adsorbs on the bed material.

To avoid bacterial and fungal growth the buffers should contain a bacteriostatic agent, for instance, sodium azide (0.02%).

5.7.4. Application of the Sample

For downward elution the sample is applied by layering on the top of the gel bed, which requires that the density of the sample be higher than that of the buffer. If this is not the case, sucrose, sodium chloride, or another easily soluble compound should be added. The syringe (or the pipet or polyethylene tubing) used for the application should have a bent tip to eliminate the risk of stirring up the top of the gel bed. If the column is to be used for upward elution, the sample is introduced from the bottom through the same tubing as will be used for the subsequent elution.

A high viscosity of the sample solution causes irregularities in the flow in the sample zone, which results in a crenelated appearance of the zone. To decrease this disturbance the protein concentration in a zone should in general not exceed 4%. Higher concentrations can be tolerated if the sample consists of substances which separate rapidly on the column. (Serum, which has a protein concentration of about 7%, can thus be applied undiluted; see Figure 3.)

5.7.5   Elution

In order to diminish the tendency of a gel bed to compress by gravity during elution, it is preferable to have the buffer stream upward through the bed.[44] This elution technique is not necessary to use for the harder, less compressible gels. In cases when constant flow rate is necessary, the use of a peristaltic pump is recommended; in most experiments, however, a Mariotte flask will provide a sufficiently constant flow rate.

5.7.6   Determination of the Material Distribution in the Eluate

As far as proteins are concerned, the material distribution is most often measured by absorption measurements at 280 nm. If the protein concentrations are very low, measurements at shorter wavelengths are recommended, for instance, at 230 nm where the light absorption is about ten times higher than at 280 nm. In such cases one has to choose buffers with a low absorption at these short wavelengths; phosphate, Tris–hydrochloric acid, and borate buffers are examples. When the background absorption is not constant, one should also measure the absorption at a wavelength where the proteins do not absorb, at 310 nm, for instance. If the difference in absorption at 280 and 310 nm is plotted against fraction number (or fraction volume) the background absorption can be expected to fluctuate only slightly, provided the absorption that originates from impurities in the buffer or on the cuvette surface is approximately the same at these wavelengths. The measurements at the two wavelengths should be made for each fraction without touching the cuvet; one should thus *not* measure all fractions first at 280 nm and then at 310 nm. The reference wavelength should lie close to but not coincide with the wavelength where the sample absorbs.[46]

5.7.7   Restoring the Flow Rate

When a column is used repeatedly, the flow rate will often become very low. Sometimes it may be necessary to repack the column. This can often be avoided, however, by stirring up and removing the upper 2-cm layer of the gel bed (the channels in this gel layer may be clogged by dust or other particles present in the buffer or the sample). In fact it is preferable to renew the upper layer of the gel bed after each run because clogging of the channels causes a deformation of the zones with attendant loss in resolving power of the column.

If the renewal of the gel bed at the top of the column will not restore the flow rate, one should force one or two ml of buffer into the column in a direction opposite to that used at the elution.[43] This "back flow" may release particles in the channels of the gel bed or in the bed support.

5.8 Applications

Molecular-sieve chromatography is widely used for the fractionation of proteins. Most issues of biochemical journals contain descriptions of experiments in which this technique has been employed. Therefore it might be sufficient to conclude by referring to Figure 3, which shows a molecular sieving of human serum.

ACKNOWLEDGMENTS

The author is much indebted to Mrs. Irja Blomqvist for skillful technical assistance. The work has been supported by the Swedish Natural Science Research Council.

REFERENCES

1. Deuel, H., and H. Neukom. Paper presented at the 122nd meeting American Chemical Society, September 1952; published in Nat. Plant Hydrocolloids **11**, 51 (1954).
2. Lindqvist, B., and T. Storgårds. Nature **175**, 511 (1955).
3. Lathe, G. H., and C. R. J. Ruthven. Biochem. J. **60**, xxxiv (1955).
4. Lathe, G. H., and C. R. J. Ruthven. Biochem. J. **62**, 665 (1956).
5. Porath, J., and P. Flodin. Nature **183**, 1657 (1959).
6. Tiselius, A. Experientia **17**, 433 (1961).
7. Hjertén, S., and R. Mosbach. Quarterly Report No. 3 to European Research Office (9851 DU), US Department of the Army, March 22, 1961.
8. Hjertén, S., and R. Mosbach. Anal. Biochem. **3**, 109 (1962).
9. Lea, D. J., and A. H. Sehon. Can. J. Chem. **40**, 159 (1962).
10. Polson, A. Biochim. Biophys. Acta **50**, 565 (1961).
11. Hjertén, S. Arch. Biochem. Biophys. **99**, 466 (1962).
12. Flodin, P. *Dextran Gels and their Application in Gel Filtration* (Halmstad, Sweden: Meijels Bokindustri, 1962).

13. Haller, W.   Nature **206**, 693 (1965).
14. Determann, H., H. Rehner, and T. Wieland.   Makromol. Chem. **114**, 263 (1968).
15. Polson, A., and W. Katz.   Biochem. J. **108**, 641 (1968).
16. Determann, H.   *Gelchromatographie* (Berlin, Heidelberg, New York: Springer-Verlag, 1967).
17. Fischer, L.   *Laboratory Techniques in Biochemistry and Molecular Biology*, Vol. 1, Ed. by T.S. Work and E. Work (Amsterdam, London: North-Holland Publishing Co., 1969), p 157.
18. Moore, J. C.   J. Polymer Sci., part A **2**, 835 (1964).
19. Laurent, T. C., and J. Killander.   J. Chromatog. **14**, 317 (1964).
20. Tiselius, A.   Arkiv. Mineral. Geol. **26B**, 1 (1948).
21. Shepard, C. C., and A. Tiselius.   Discussions Faraday Soc. **7**, 275 (1949).
22. Pedersen, K. O.   Arch. Biochem. Biophys., Suppl. 1 **1962**, 157.
23. Ackers, G. K.   Biochemistry **3**, 723 (1964).
24. Anderson, D. M. W., J. F. Stoddart.   Anal. Chim. Acta **34**, 40 (1966).
25. Porath, J.   J. Pure Appl. Chem. **6**, 233 (1963).
26. Squire, P. G.   Arch. Biochem. Biophys. **107**, 471 (1964).
27. Joustra, M. K.   *Protides Biol. Fluids. Proc. 14th Colloq., Bruges, 1966*. Ed. by H. Peeters (Amsterdam, London, New York: Elsevier, 1967), p 533.
28. Ogston, A. G.   Trans. Faraday Soc. **54**, 1754 (1958).
29. Andrews, P.   Brit. Med. Bull. **22**, 109 (1966).
30. Fåhraeus, R.   Acta Med. Scand. **161**, 151 (1958).
31. Brønsted, J. N.   Z. Physiol. Chem., A **1931**, 257.
32. Albertsson, P. Å.   *Partition of Cell Particles and Macromolecules* (Uppsala, Sweden: Almqvist & Wiksells Boktryckeri AB, 1960).
33. Hjertén, S.   J. Chromatog. **50**, 189 (1970).
34. Helfferich, F.   *Ion Exchange* (New York, San Francisco, Toronto, London: McGraw-Hill, 1962), p 95.
35. Ginzburg, B. Z., and D. Cohen.   Trans. Faraday Soc. **60**, 185 (1964).
36. Ryle, A. P.   Nature **206**, 1256 (1965).
37. Giddings, J. C.   *Dynamics of Chromatography: Part I, Principles and Theory* (London: Edward Arnold Publishers Ltd., and New York: Marcel Dekker, Inc., 1965).
38. Giddings, J. C. and K. L. Mallik.   Anal. Chem. **38**, 997 (1966).
39. Hjertén, S.   *Methods in Immunology and Immunochemistry*, Vol. 2, Ed. by C. A. Williams and M. W. Chase (New York: Academic Press, 1968), p 142.
40. Hjertén, S.   Biochim. Biophys. Acta **79**, 393 (1964).
41. Bengtsson, S., and L. Philipson.   Biochim. Biophys. Acta **79**, 399 (1964).
42. Hjertén, S.   *Methods in Immunology and Immunochemistry*, Vol. 2, Ed. by C.A. Williams and M.W. Chase (New York: Academic Press, 1968), p 149.
43. Hjertén, S.   Arch. Biochem. Biophys. Suppl. 1 **1962**, 147.

44. Porath, J., and H. Bennich.   Arch. Biochem. Biophys. Suppl. 1 **1962**, 152.
45. Tiselius, A.   Bull. Soc. Chim. Biol. **50**, 2201 (1968).
46. Hjertén, S., S. Höglund, G. Ruttkay-Nedecky.   Acta Virol. Prague **14**, 89 (1970).

Chapter 6

Thin–Layer Gel Filtration and Related Methods

by B. G. Johansson

The development of the gel filtration method with the use of gels of cross-linked dextran (Sephadex) constitutes a new dimension for separation of water-soluble substances, mainly according to their molecular size (Porath and Flodin, 1959).[1] The separating ability of Sephadex gels now available covers a molecular-weight range of less than 100 to about 300,000. The polyacrylamide gels introduced for gel filtration by Hjertén and Mosbach[2] appear to have separation characteristics similar to those of the dextran gels. Extension of the separation range to higher molecular weight has been made possible by the preparation of pearl-condensed agarose gels of different porosity[3] (see section 5.7.1). The gel filtration method is now widely used especially in biochemistry (for a review, see Determann[4]).

Ordinary column gel filtration enables chromatography of substances varying in quantity from less than one milligram up to hundreds of grams. The method has also been used on an industrial scale. In gel filtration on a microscale thin-layer chromatographic techniques have proved very useful.

Thin-layer gel filtration has so far been utilized mainly for separation of proteins. The method has been further developed to involve a second electrophoretic or immunoprecipitation step. This chapter concerns technical details of thin-layer gel filtration and some applications of the method as well as thin-layer gel filtration combined with electrophoresis or immunoprecipitation. The great interest focused on the separation of proteins will be reflected by the examples on applications of the thin-layer gel filtration method. This does not mean, however, that the field of application is restricted to this group of substances.

6.1 Thin-Layer Gel Filtration Proper

6.1.1 *Technique of Solvent Flow*

The first descriptions of thin-layer gel filtration techniques were given by Determann[5] and Johansson and Rymo.[6] Determann used

Sephadex G-25, which was partially dried on the plate in order to allow the ascending technique by capillary force commonly used in thin-layer chromatography. He succeeded in separating an artificial mixture of tyrosine, a pentapeptide, a condensation product of the pentapeptide (plastein), and horse serum albumin. The technique of Johansson and Rymo[6] differed from that of Determann in that the plate was used wet. This necessitated a descending buffer flow secured by a simple arrangement, with the plate supporting the gel layer resting at an incline of about 20° to the horizontal plane. A solvent flow was obtained with the aid of a thick filter-paper wick ensuring contact between the gel layer and a buffer reservoir. Separation was performed on Sephadex G-25 and Sephadex G-75.

The development of new types of gels with higher water-regain capacity rendered the ascending technique impossible; therefore the alternative technique with descending buffer flow will be described in some detail. This latter method was soon elaborated to allow the use of Sephadex G-100 and G-200 in thin-layer gel filtration.[7-10] Gel suspensions are preferably prepared from Sephadex "Superfine" with a particle-diameter range of 10–40 μm (available from Pharmacia Fine Chemicals, Sweden). Gels with larger particles have a substantially lower resolution power, and it would be of interest to try to increase the resolution with gels having particle diameters of less than 20 μm. Polyacrylamide gels (Bio Gel P 60 and P 300 with particle sizes 400 mesh obtained from Bio-Rad Laboratories, California) were successfully employed by Radola.[66]

6.1.2 Consistency of the Gel Slurry

The proper consistency of the gel slurry is very important. There are two main ways to prepare such gels. The gel powder may be allowed to swell in amounts of buffer just large enough to secure the suitable

Table I

Proportions of Gel Buffer in Swelling of Various
Sephadex Gels Used for Thin-Layer Gel Filtration

Type of Sephadex (Superfine)	Amount of buffer (ml) to be added to 1 g gel powder
G-25	5
G-50	11
G-75	15
G-100	19
G-150	22.5
G-200	25

consistency.[8,9] Approximate relative amounts of gel powder and buffer for different types of gel are given in Table I. Optimal proportions may vary slightly, however, from one gel batch to another. A gel of suitable consistency can also be obtained by allowing the gel powder to swell in an excess amount of buffer and to settle. As much buffer excess as possible is then gently sucked off through a narrow glass tube.

Both these methods seem to give satisfactory results, but it should be stressed that the optimal gel consistency can be obtained only by trial and error. If the gels are too dry, migration of the fractions will be irregular. Once the gel is spread, the plate should have a glassy surface but if the gel is too watery, it will run when the plate is tilted. The gel powder should be allowed to swell for at least 24 hours; the G-200 powder, for several days. With Bio-Gel P-300 it has been found that swelling must proceed 10–14 days to obtain reasonable running times.[66]

6.1.3 Spreading of the Layer

Spreading of the plates is simple and can be performed with one or another of a variety of equipments used for thin-layer chromatography. Adhesion of the gel to the glass surface does not require the presence of binders. As a matter of fact, no expensive equipment at all is necessary for spreading of the plate. The simple procedure of Baron and Economidis[11] gives quite satisfactory results.[12] According to this technique, several layers of masking tape are fixed along the edges of the long sides of the plate. The gel is then poured along one of the free edges of the plate and spread by means of a glass rod that is held firmly against the tape and drawn in a single, steady sweep along the plate. The writer uses a Perspex block with a groove of 0.5 mm for spreading the plates. A suitable thickness of the gel layer seems to be 0.5 mm,[7,8] but satisfactory results have also been reported with 0.25-mm[13] and even 0.9-mm gels.[9] Judging from personal experience, layers thinner than 0.5 mm are more difficult to spread satisfactorily, whereas layers thicker than 0.5 mm give somewhat indistinct gel filtration patterns.

6.1.4 Arrangement for Gel Filtration

No expensive apparatus is necessary for gel filtration on plates, but the commonly used thin-layer chromatographic equipment is generally unsatisfactory due to the necessity of descending buffer flow. A simple home-built apparatus, rather similar to that used by Determann[10] is shown in Figure 1. Along the long edges of the glass plate supporting the gel layer are two plastic rods (0.5 cm thick) on which is placed a glass coverplate. The two glass plates are held together by two clips,

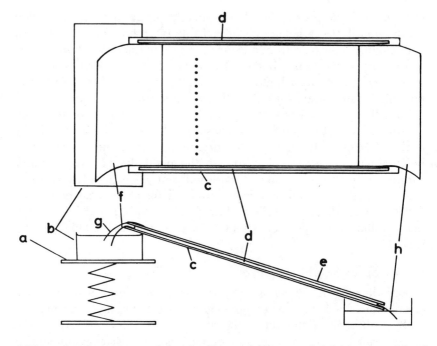

Figure 1. Apparatus used for thin-layer gel filtration: (a) laboratory table of adjustable height; (b) buffer reservoir; (c) glass plate supporting the gel layer; (d) plastic rods; (e) cover glass plate; (f) filter paper wick; (g) plastic film covering the filter paper wick; (h) filter paper wick draining the gel.

one on each side. A filter paper wick (Whatman 3 MM, preferably double-layer), which should be covered with plastic film to avoid evaporation, ensures conduction of solvent from the buffer reservoir to the gel layer; a second filter paper on the lower end of the plate drains the buffer from the gel. This equipment can easily be adjusted to allow the use of plates of varying size from 10×20 cm to 30×50 cm. The large plates, which are somewhat less easy to handle, have two advantages: longer separation distances and the possibility of applying a large number of samples to the same plate. Plates broader than 30 cm should not be used because of the risk of an irregular solvent flow. After the plates have been spread, they should be prerun overnight before application of samples.

6.1.5 Sample Application, Marker Substances, and Flow Rate

The samples, which should contain less than 100 μg of substances to be separated, are applied as round spots 2–4 mm in diameter, corresponding to sample volumes of about 1–3 μl. Samples must be applied very carefully in order to avoid damage of the soft gel layer, which can result in distortion of the zones. Linear application is also possible,[66, 67] but requires much more skill. Applicators used in paper electrophoresis may be useful for linear application. During application of samples the plate can be kept in the inclined position if the procedure is performed within less than 10 minutes.

The absence of a visible solvent front during the run necessitates the use of marker substances in order to determine and regulate the flow rate. Naturally pigmented proteins, such as hemoglobin and cytochrome c, have been used for this purpose.[9] Proteins rendered fluorescent by coupling with fluorescein isothiocyanate are also suitable.[8] But the most convenient and distinct marker substance seems to be bovine or human serum albumin, colored with Amido Black B.[14] Dextran Blue used in column gel filtration for determination of void volumes is not a satisfactory marker substance because of its tendency to trail. Watching the migration rate of the marker substances should enable adjustment of the inclination angle of the plate to give a suitable flow rate. When Sephadex G-200 is used, substances totally excluded from the gel phase should have a migration rate of 2 cm per hour at most, whereas Sephadex gels with a higher degree of cross-linking can be run faster. It is not possible to predict the most suitable angle of inclination for a given type of gel, because this angle depends on various factors such as properties of the gel batch and consistency of the gel slurry spread on the plate. Migration distances of unretarded substances should be at least 15 cm before the run is stopped. Longer runs of up to 30–40 cm give better resolution without any appreciable spreading of the zones.

6.1.6 *Detection of Separated Components*

Since it is difficult to dry the gel layer on the plate after the run without the appearance of minute cracks on the gel surface and a tendency for the gel layer to loosen from the plate, the easiest way of detecting the separated components is to transfer them to a filter paper sheet and perform the detection procedure on the paper.[8, 9] Whatman 3 MM seems to be the most suitable paper for this purpose[9] but Macherey and Nagel 212 may also be used[10] and for G-100 and G-75, Schleicher and Schüll 2043b.[8] The following transfer technique is recommended: a filter paper of exactly the same dimensions as the gel layer on the plate is applied to the gel, with the application being started at the end of the plate. The filter paper is applied to the gel by a rapid sweep of a plastic ruler in order to expel all air between the filter and the gel. After 5 minutes the separated components are almost completely transferred to the filter paper. The paper is then removed from the plate—often together with part of the gel, which is of minor importance—and dried in an oven at 100°C. Transfer to cellulose acetate membranes has also been suggested.[15-17] This technique gives a replica with very distinct zones and might be used for subsequent electrophoresis (see Section 6.3) or immunodiffusion (see Section 6.4.1).

For visualization of proteins on the paper replica any of the dyeing procedures employed in paper electrophoresis may be used. An advantage of dyeing with Lissamine Green (saturated solution in 5% acetic acid) compared with dyeing with Amido Black B is that all excess dye can easily be removed from the paper (Johansson and Rymo)[8] with the result that the background is completely unstained. Morris[9] suggests using a solution of 0.01% Nigrosine WS in methanol-glacial acetic acid–water (50:40:10) followed by washing in the same solvent when high sensitivity is required. Coomassie Brilliant Blue R-250, which was introduced by Fazekas de St. Groth et al.[18] for protein detection after cellulose acetate electrophoresis, is also useful as a very sensitive stain for visualizing proteins on the paper replica. A 0.1% solution of Coomassie in 5% sulfosalicyclic acid is recommended. Both Nigrosine and Coomassie Brilliant Blue leave considerable background color on the paper. Determann and Michel[10] used spraying of the paper replica with Pauly's reagent. This detection method seems to have no advantages over conventional protein staining methods. Protein-bound carbohydrates have been detected by the periodic acid-Schiff reagent (PAS reaction) after gel filtration on Bio-Gel P-60 and P-300; however, this staining method was unsuccessful in prints of Sephadex thin layers.[67] Enzymatic reactions as well as spraying with ninhydrin for detection of peptides and amino acids may, of course, also be performed on the paper replica.[8]

The detection of separated components without a preceding transfer to a filter paper (or cellulose acetate) replica is somewhat problematic, as mentioned above. Andrews[7] avoided drying of the plate by placing the wet gel plate in a closed chamber containing iodine vapor. Proteins appeared as brown spots. This technique was found to be less sensitive by Roberts,[19] who therefore used a modification of the chlorination technique described by Rydon and Smith.[20] The plates were partially dried, placed in an atmosphere of chlorine for 5–10 minutes and freed of chlorine in an air stream for 15 minutes. They were then sprayed with a solution of 20% w/v ammonium sulfate and 5% w/v sodium carbonate in water. After the plates were allowed to stand for 15 minutes, they were sprayed with a water solution containing 1% w/v starch and 1% w/v potassium iodide. By these means the N-chloroderivatives formed were visualized as blue spots on a colorless background. This technique enables the detection of proteins, mucopolysaccharides and amino sugars.

If detection of components directly on a dried plate is desired, careful drying of the plate is necessary. Drying at 50°C in an oven lined with wet paper to secure a slow and uniform drying process was suggested by Determann.[4] With this procedure almost homogeneous films of Sephadex G-100 and G-200 gels were obtained in a few hours. The dry plates were allowed to swell in methanol–glacial acetic acid–water (75:5:20) for 10 minutes and were then stained for 5 hours in a bath containing a saturated solution of Amido Black 10-B in the same solvent. The plates were washed afterwards for 2 hours in the dye-free solvent before they were dried. A rather similar procedure for dyeing dried G-75 plates was used by Johansson and Rymo.[6]

According to the experiences of Williamson and Allison[13] thick layers, substance overloading, and overheating may lead to cracking of the surface of G-200 plates during the drying process. With a layer thickness of 0.25 mm the gels could be carefully dried in 15–30 minutes at 50°C to a hard film, which could be stained with a saturated solution of Amido Black 10-B in methanol–glacial acetic acid–water (50:40:10) for 5–10 minutes and subsequently destained in two successive baths of the dye solvent. This procedure is claimed to be much quicker than the paper transfer technique and to allow densitometric scanning of the plates. According to the author's experience, however, this direct drying technique requires great care and is not at all so simple for routine protein work as the paper or cellulose acetate transfer technique.

Isolation of components, *e.g.*, proteins, on a microscale, is also possible by thin-layer gel filtration. Linear application of samples is then advisable. Localization of the components to be recovered may be performed by different techniques. One way is to determine the position of the desired component by means of a visible marker substance with the same migration rate, run on the same plate. Wollheim and Snigurowics[21] used a fluorescent conjugated macroglobulinemic serum for determination of the position of the serum macroglobulin fraction. Fluorescent substances in the gel buffer may also be included to visualize proteins without destroying them. Maier[22] detected proteins as dark areas against a bright orange background upon the exposure to ultraviolet light of thin-layer gels containing 0.2 mg Lissamine Rhodamine B-200 per gram of gel slurry. This method seems to be very promising for preparative work.

6.2 Applications

6.2.1 *Outside Protein Biochemistry*

Thin-layer gel filtration has been used little outside the field of protein biochemistry. Thin-layer gel filtration on Sephadex G-25 was applied by Grossman and Wagner[23] for characterizing components of Nitro Blue tetrazolium chloride preparations used in histochemical work. Stickland[24] employed thin-layer chromatography on Dowex-1 and Sephadex G-25 for removing interfering substances (polysaccharides and fluorescent material found in biological samples) before the separation of 5'-ribonucleotides on thin-layers of DEAE-cellulose. The potentialities of separating low-molecular-weight substances on Sephadex G-10 and G-15 by thin-layer gel filtration have not yet been extensively studied. Dutta et al. found[68] that the separation of various antibiotics on Sephadex G-15 depends on adsorption effects. Nor have less hydrophilic gels, e.g., Sephadex LH-20, been tried in thin-layer chromatography.

6.2.2 Within Protein Biochemistry

In investigations of proteins gel filtration deserves special attention as a simple analytical tool for several purposes such as the investigation of molecular size distribution of biological fluids, comparison of molecular sizes of proteins, and molecular weight determination on a microgram scale. The possibilities of running up to 30 samples in a single experiment is a definite advantage in comparative studies.

The good resolving power of thin-layer gel filtration for protein separation is demonstrated by the model experiment illustrated in Figure 2. The components of an artificial mixture of α-lactalbumin, β-lactoglobulin and serum albumin, all of bovine origin, were separated on Sephadex G-75 ("Fine"). Separation of these components was

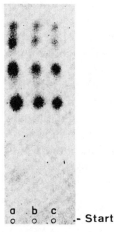

Figure 2. Gel filtration on Sephadex G-75, "fine." A mixture of bovine serum albumin (MW 69000), β-lactoglobulin (MW 35000), and α-lactalbumin (MW 15000) applied in a–c (from Johansson and Rymo).[6]

complete and what is more, the serum albumin was divided into a monomer and a dimer-polymer component. Gel filtration on Sephadex G-75 could not accomplish this division in columns of reasonable length (Johansson and Rymo).[6] Satisfactory resolution in the separation of blood serum proteins on Sephadex G-200 was shown by Johansson and

Rymo.[8] Proteins of normal human serum were separated into the three distinct fractions known from column gel filtration experiments (compare Figure 3 in Chapter 5). Monoclonal immunoglobulin (Ig) components of different classes (IgG, IgA and IgM) could be distinguished by their different migration rates. Furthermore monoclonal IgA was found to be resolved into several components owing to monomer-polymer equilibria. The monomer component of IgA was sometimes completely separated from IgG, again demonstrating the efficiency of the resolution compared with that of column gel filtration. Agostoni *et al.*[25] used thin-layer gel filtration for detecting Bence Jones proteins in serum from patients with myeloma. A spot behind the albumin on Sephadex G-200 was attributed to the presence of Bence Jones protein. On Sephadex G-75 there appeared two components, which presumably represented the monomeric and dimeric forms of light chains. It should be stressed, however, that with the exception of some few cases such as those described below, thin-layer gel filtration is hardly of any value in the diagnosis of diseases with plasma protein changes.

Thin-layer gel filtration of serum was used together with zone electrophoresis and immunoelectrophoresis by Hobbs[15] in a study of monoclonal immunoglobulin components. This additional analysis of molecular size, supplementary to the immunological classification, resulted in detection of a very interesting immunoglobulin component consisting of half-molecules of immunoglobulin G.[26] Purves[27] suggested the use of thin-layer gel filtration in the diagnosis of "heavy-chain disease" characterized by the presence of an abnormal component closely related to the heavy polypeptide chain of immunoglobulin G (or A) in serum and urine. Figure 3 exemplifies this use of the thin-layer technique on Sephadex G-200.

Preparative thin-layer gel filtration was used by Wollheim and Snigurowics[21] who studied the frequency of light chains of type κ and λ in monoclonal immunoglobulin M. After gel filtration on Sephadex G-200 that part of the gel containing the macroglobulin was transferred to test tubes and eluted with buffer. The eluates were concentrated and subjected to immunoelectrophoresis against anti-kappa and anti-lambda immune sera. This method was claimed to facilitate considerably the light-chain typing of macroglobulins, because it removed interfering immunoglobulins of other classes. It might be pointed out that immunogel filtration (see Section 6.4) could also solve the same problem in a simpler way.

The use of thin-layer gel filtration in investigations of proteinuria has recently been reported. The different patterns of proteinuria of glomerular, tubular and Bence Jones types on thin-layers of Sephadex G-75 were demonstrated by Davis *et al.*[28] Goddard and Hobbs[29] used filtra-

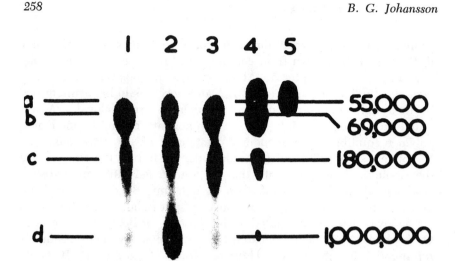

Figure 3. Thin-layer gel filtration of human serum on Sephadex G-200: (1) normal serum; (2) primary macroglobulinemia; (3) myelomatosis (IgG); (4) *predicted* pattern of serum in a case of heavy chain disease; (5) position of heavy chain (from Purves).[27]

tion on Sephadex G-75 for evaluation of the pattern of proteinuria in pyelonephritis. They found that the proteinuria was indistinguishable from the tubular type with an abnormally large amount of low-molecular-weight proteins. But gel filtration seems to offer no advantage over electrophoretic and immunologic methods as a routine procedure for the evaluation of different types of proteinuria. This conclusion also holds for the separation of myoglobin and hemoglobin on thin-layers of Sephadex G-50,[30] since simpler methods, *e.g.*, salting out, are available.

Thin-layer gel filtration has been utilized by the present author in studies of immunoglobulins. Separation of proteolytic fragments from immunoglobulin G by column gel filtration was first described by Hanson and Johansson[31] and the application of thin-layer gel filtration for this separation was suggested by Johansson and Rymo.[6] According to personal experience, thin-layer gel filtration is a convenient way to determine suitable conditions (enzyme–substrate ratio, pH, time) for maximal recovery of different fragments and to compare degradation products from different immunoglobulins.[32, 33] Thin-layer filtration through Sephadex G-200 was also found most useful for demonstration of the difference in molecular size between immunoglobulin A from serum and from colostrum (secretory IgA).[34]

Comparisons of fresh colostrum samples and the isolated colostral IgA by the same method clearly demonstrated that the IgA molecules are the same size. This was a strong indication that the large molecular size of colostral IgA is not due to polymerization during the preparation procedure. Colostral IgA, like IgA from most secretions, contains an additional component, a "secretory piece" not found in serum IgA. By thin-layer gel filtration on Sephadex G-150 this component is found to be of roughly the same molecular size as dimers of light chains of immunoglobulins.[35] In a study of proteins from swine colostrum, Porter *et al.*[70] used thin layer filtration on Sephadex G-200 in evaluating the changes in protein composition during the first 4 weeks of lactation.

Some other examples of thin-layer gel filtration used for characterization purposes deserve mention. Link[36] showed that it is possible to perform thin-layer filtration on Sephadex G-100 in 8M urea or 6.5M guanidine hydrochloride. With this method he demonstrated that reduced-alkylated β-trace protein isolated from cerebrospinal fluid migrated at the same rate as the untreated protein, indicating that no subunits of different molecular size were present. Thin-layer gel filtration has also been employed routinely together with electrophoresis as a convenient checking procedure in the preparation of aminoacylribonucleic acid synthetases.[37] Two fractions with valyl–RNA synthetase activity showing some distinct functional differences were found to have essentially the same molecular size when analyzed by thin-layer filtration on Sephadex G-200.[38] Various applications of thin-layer gel filtration have been reported by Radola,[67] including studies of human and animal serum proteins, soluble proteins from mammalian cells grown *in vitro,* and soluble leaf proteins from spinach.

6.2.3 Determination of Molecular Weight

Much interest has been focused on the possibilities of determining molecular weight by gel filtration. For a general discussion see Chapter 5 in this volume, and Determann.[4] Empirical correlations between elution volumes and molecular weights have been demonstrated by several authors.[7,39,40] On thin-layer gel filtration Andrews[7] found a linear relationship between the migration distances of a series of standard proteins and the logarithm of their molecular weights. The same relationship has been reported by other workers.[9,10,19] Although the empirical methods are proved to allow an estimation of molecular weights of a series of standard proteins, the anomalous behavior of some proteins (Figure 4) calls for strong caution in the evaluation of molecular weights by gel filtration alone. If the test protein differs in shape or density from the globular reference proteins commonly used, the method will give misleading information. The occasional presence of protein-gel interaction must also be considered.

Nevertheless, thin-layer gel filtration has been used by some workers for molecular weight determinations. Hygstedt and Jagenburg[41] de-

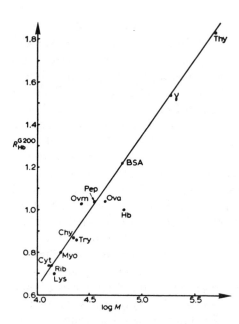

Figure 4. The relationship between the migration distances (expressed as $R_{hemoglobin}$) and logarithm of molecular weight on thin-layer gel filtration of standard proteins on Sephadex G-200. The anomalous behavior of lysozym (Lys), ovomucoid (Ovm), ovalbumin (Ova), and hemoglobin (Hb) is indicated (from Morris).[9]

termined the molecular weight of urinary β-glucuronidase as about 230,000 by thin-layer gel filtration on Sephadex G-200. The same gel was used for molecular-weight determination of the Fc-fragment, obtained from immunoglobulin A by reductive cleavage with sodium borohydride.[42] In a study of lactate dehydrogenase (LDH) a strong concentration dependence of the molecular weight of LDH-H₄ under acidic conditions was easily revealed by thin-layer gel filtration.[43] The molecular weight of lactoferrin from human milk could be calculated from its iron content after approximately the same molecular sizes of lactoferrin and serum transferrin were demonstrated by thin-layer gel filtration on Sephadex G-150.[44]

In order to avoid anomalous results on molecular weight determinations by gel filtration it is advisable to perform the chromatographic experiments under conditions where the proteins are denatured, i.e., by strong solutions of guanidine hydrochloride. (For a review see Reference 69.) This technique seems very promising and should of course be able to be adapted as a thin-layer gel filtration technique.

6.2.4 Future Extension of the Technique

Experience with thin-layer gel filtration has so far been confined to Sephadex gels G-25–G-200, whereas no information is available on the possibilities of using G-10 and G-15. Some experiences with polyacrylamide gels for thin-layer gel filtration have been reported by Radola[66, 67] who found a better resolution of ferritin on Bio Gel P-300 than on Sephadex G-200. The extension of the thin-layer technique to include the use of agarose would be especially welcome as it would broaden the range of application of the technique to embrace the separation of substances with molecular weights of more than 1,000,000. Preliminary experiments with pearl-condensed agarose gels with agarose concentrations ranging from 4–10% w/v indicate that no appreciable technical difficulties should impede thin-layer filtration on agarose gels.[45] As a matter of fact very distinct zones are found and furthermore the agarose gels are easier to dry to a homogeneous film on the plate than are Sephadex gels.

6.3 Thin-Layer Electrophoresis and Two-Dimensional Combination with Gel Filtration

Sephadex gels are excellent supporting media in column electrophoretic experiments (Gelotte *et al.*)[46] as its use is not restricted by any appreciable irreversible adsorption phenomena. Dose and Krause[47] tried horizontal electrophoresis on 0.2-cm thick layers of Sephadex G-25 and G-50 separating enzymes and low-molecular-weight substances such as peptides and amino acids. Thin-layer electrophoresis of amino acids and low-molecular-weight peptides was also tried out by Chudzik and Klein,[48] who succeeded in obtaining good differentiation between di- and tripeptides and their amino acid constituents by the combination of paper chromatography and thin-layer electrophoresis on Sephadex G-25. The absence of endosmotic flow in the Sephadex gels would make them suitable supporting media in isoelectric focusing experiments. Thin-layer isoelectric focusing on Sephadex G-75 has been described by Radola.[71] By this method horse radish peroxidase could be separated into a great number of components with different p I.[72]

Whereas ordinary thin-layer gel electrophoresis of proteins on Sephadex gels seems to be a method of less interest, the combination of thin-layer gel filtration with a second electrophoretic step should add two advantages to the thin-layer filtration method: a higher resolving power and electrophoretic characterization of the separated components. A procedure for two-dimensional thin-layer gel filtration-electrophoresis of proteins was elaborated by Johansson and Rymo[49] and described in further detail by Hanson *et al.*[14] In this technique thin-layer filtration is first run in 0.5-mm thick layers of Sephadex G-200 or G-100 on plates of the dimensions 30 × 30 cm. The plate is then turned 90° and electrophoresis is performed for 3–4 hours at 10 V/cm. The separations are done in 0.05M sodium barbital buffer, pH 8.6. The

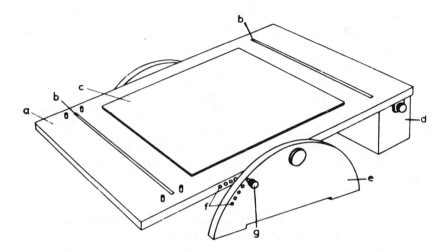

Figure 5. Schematic drawing of apparatus used for two-dimensional thin-layer gel filtration-electrophoresis with Perspex cover removed: (a) Perspex support; (b) slits for filter paper connections; (c) cooling plate of brass; (d) buffer reservoir; (e) Perspex stand; (f) holes in the Perspex support and Perspex stand for adjustment of inclination of the thin-layer plate. On electrophoresis a second buffer reservoir is connected (from Hanson et al.).[14]

method requires a cooling device for the electrophoresis. A relatively simple and suitable apparatus is shown in Figure 5. This apparatus is, of course, also useful for thin-layer gel filtration alone.

The two-dimensional method can separate serum proteins into at least 12 components. The localization of components indicates that molecular sieve effects are virtually absent in the electrophoretic step. The only marked anomalous electrophoretic migration was shown by α_2-macroglobulin, which was substantially retarded probably due to the frictional effect of the Sephadex gel. Vergani et al.[25] modified the technique by transferring the gel-filtrated proteins to cellulose acetate membranes on which the electrophoresis was then run. In this way they avoided the frictional retardation of the α_2-macroglobulin and other high-molecular-weight proteins. Their technique also has the advantage of not requiring any special equipment for thin-layer gel electrophoresis.

Thin-layer gel filtration-electrophoresis is a useful method for investigating materials available only in very small amounts, because the analysis is performed with samples containing 10–50 μg of proteins. In an investigation of human growth hormone[50] lyophilized preparations studied with this method showed pronounced heterogeneity (Figure 6) Porter et al.[73] employed thin-layer gel filtration-electrophoresis on Sephadex G-150 in a study of the interaction of heparin with plasma

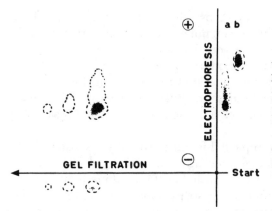

Figure 6. Two-dimensional thin-layer gel filtration-electrophoresis of lyophilized human growth hormone (HGH) on G-200, showing pronounced heterogeneity of the preparation. References in the electrophoretic direction: (a) lyophilized HGH; (b) human serum albumin (from Hanson et al.).[50]

proteins and demonstrated that bound heparin appeared predominantly with the low molecular weight fractions. Bergmann *et al.*[51] were able to demonstrate the low-molecular size of abnormal γ-globulins found on agar gel electrophoresis of cerebrospinal fluid from patients with a rare disease called subacute sclerosing leucoencephalitis. The method was also used for comparisons of degradation products of immunoglobulin G with trypsin and plasmin.[52]

6.4 Immuno-Gel Filtration

6.4.1 *Immunodiffusion*

The addition of an immunodiffusion step to the separation of proteins by thin-layer gel filtration constitutes a sensitive method analogous and supplementary to immunoelectrophoresis. This method is referred to here as immuno-gel filtration.* Several procedures for immuno-gel filtration have been employed.[13–15, 17, 53–55] Grant and Everall[54] performed immunodiffusion in the Sephadex layer. Although excellent immunoprecipitation patterns were demonstrated in their paper, this technique has several drawbacks, as later pointed out by the authors themselves.[56, 57]

Immunodiffusion in agar can be done by the procedure of Hanson *et al.*[14] This technique is essentially as follows. After thin-layer gel filtration on 10 × 20-cm plates, all Sephadex gel not containing protein is removed, leaving 3–5-cm broad strips containing the separated proteins. Molten agar cooled to 50°C is then applied to the plate, where it forms a gel about 1 mm thick. Careful technique enables the agar

---

*Other designations suggested are immunochromatography[17, 53] and gel immunofiltration.[54]

to flow over the Sephadex gel strips without any disturbing effects. With this technique satisfactory immunoprecipitation patterns can be obtained, but it should be observed that the procedure requires considerable technical skill. Immunodiffusion in agar gel can also be performed after transfer of the separated proteins from the thin-layer gel to cellulose acetate membranes[57] or filter paper strips.[13] These are afterward placed on the agar gel in which the immunodiffusion procedure is to take place.

A most useful technique employed is transferral of the separated proteins to cellulose acetate membranes after thin-layer gel filtration and then performance of the immunodiffusion in this medium by placing on the membranes two thin filter paper strips soaked in antiserum.[15, 17, 55] If two cellulose acetate membranes are simultaneously applied to the thin-layer gel, one of them may be stained for protein and the other one used for the immunodiffusion. Both strips may then be superimposed to reveal the separation as well as the immunoprecipitation pattern.[15, 17] The results obtained with these various procedures are difficult to compare, but the techniques involving transfer of the proteins to cellulose acetate membranes may be superior to the other techniques whether the immunodiffusion is then performed in this medium or in

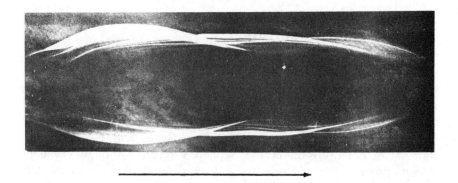

Figure 7. Immuno-gel filtration of a pathological human serum using immunodiffusion in cellulose acetate membrane against goat antihuman immune serum. The arrow indicates the flow direction during gel filtration. (From Agostoni *et al.*)[55]

agar. The immuno-gel filtration pattern of a human serum with the immunodiffusion performed on cellulose acetate membranes is shown in Figure 7.

6.4.2 *Immunological Quantitation*

Immunological quantitation of components after thin-layer gel filtration is possible by using principally a slight modification of the method

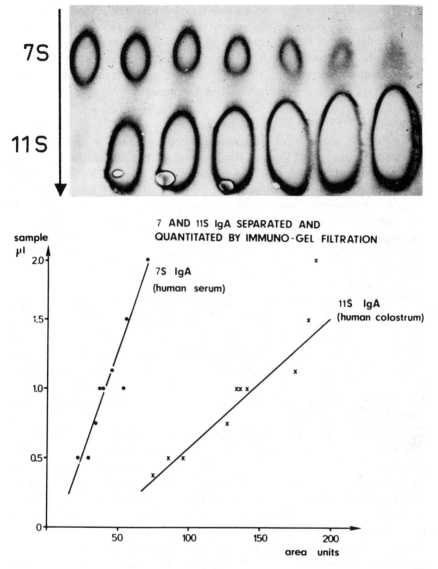

Figure 8. Quantitation of 7S and 11S immunoglobulin A by single immunodiffusion in agarose gel after thin-layer gel filtration on Sephadex G-200. Above: immunodiffusion patterns (the arrow indicates flow direction in gel filtration); below: standard curves.

of Mancini *et al.*[58] After thin-layer filtration immunodiffusion is allowed to proceed in antibody-containing agarose gel. The crucial step in this technique is to bring this gel into close contact with the Sephadex layer without distorting the separated components. This is quite possible, however, with a 1.5% w/v agarose gel of the dimensions 10 × 10 cm, which is carefully placed on the thin-layer gel. This method seems to be useful for quantitation of proteins showing monomer-polymer equilibria. The use of this method is exemplified in Figure 8, which demonstrates the possibility of quantitating high-molecular-weight secretory IgA in the presence of IgA monomer.[59]

6.4.3 Applications

Up to now immuno-gel filtration has been used only in a few investigations. Hanson et al.[50] used the method in a comparative study of lyophilized and nonlyophilized human growth hormone. The antigenic relationship between immunoglobulin G and some low-molecular-weight components occurring in cerebrospinal fluid from patients with subacute sclerosing leucoencephalitis was easily established by immuno-gel filtration[51] (see also Section 6.3). Thin-layer gel filtration on Sephadex G-75 followed by immunodiffusion is said to allow detection of Bence Jones proteins in serum much more clearly than immunoelectrophoresis.[60]

The method is also useful in studies of pathological immunoglobulins. Solomon[61, 62] detected high-molecular-weight immunoglobulin G in the sera of two patients with myeloma and was also able to show the occurrence of 7S immunoglobulin M in some patients with lymphocytic neoplasms. Furthermore monomeric and dimeric Bence Jones proteins were distinguished and even components containing either the variant or the invariant half of the light chain of immunoglobulins could be demonstrated.

The immunoprecipitation step after thin-layer gel filtration might be considerably refined by application of Laurell's method, i.e., electrophoresis of antigen in antibody-containing agarose gels[63] as suggested by Williamson and Allison.[13] In their procedure the separated proteins are transferred to a filter paper strip which is placed on the antibody-containing agarose gel and subjected to electrophoresis. As many as 20 components could be detected in mouse serum by this means.[13, 64] Another alternative is to remove the surplus of Sephadex gel, as described by Hanson et al.[14] and then carefully transfer the whole antibody-containing agarose gel to the Sephadex plate. Figure 9 illustrates such an experiment and shows the molecular-size distribution of α-lipoprotein of normal serum.[65]

The examples cited obviously show that thin-layer gel filtration combined with immunoprecipitation is a simple and convenient method which may be used as a supplement to immunoelectrophoresis for characterization of proteins in biological fluids or in the preparation of

Figure 9. Thin-layer gel filtration combined with electrophoresis in antibody-containing agarose gel. Separation of human serum by thin-layer gel filtration on Sephadex G-200 and subsequent electrophoresis into agarose gel containing anti-α-lipoprotein immune serum. The gel filtration direction as well as position of albumin (Alb) and β-lipoprotein (Lipid) are indicated in the figure.

proteins. If necessary the sensitivity as well as the resolving power may be increased by employing electrophoresis in antibody-containing agarose gels after thin-layer gel filtration.

REFERENCES

1. Porath, J., and P. Flodin.   Nature **183**, 1657 (1959).
2. Hjertén, S., and R. Mosbach.   Anal. Biochem. **3**, 109 (1962).
3. Hjertén, S.   Biochim. Biophys. Acta **79**, 393 (1964).
4. Determann, H.   *Gel Chromatography* (Berlin, Heidelberg, New York: Springer-Verlag, 1968).
5. Determann, H.   Experientia **18**, 430 (1962).
6. Johansson, B. G., and L. Rymo.   Acta Chem. Scand. **16**, 2067 (1962).
7. Andrews, P.   Biochem. J. **91**, 222 (1964).
8. Johansson, B. G., and L. Rymo.   Acta Chem. Scand. **18**, 217 (1964).
9. Morris, C. J. O. R.   J. Chromatog. **16**, 167 (1964).
10. Determann, H., and W. Michel.   Z. Anal. Chem. **212**, 211 (1965).
11. Baron, D. N., and J. Economidis.   J. Clin. Pathol. **16**, 484 (1963).
12. Maier, C. L., Jr.   J. Chromatog. **32**, 577 (1968).
13. Williamson, J., and A. C. Allison.   Lancet **2**, 123 (1967).
14. Hanson, L. Å., B. G. Johansson, and L. Rymo.   Clin. Chim, Acta **14**, 391 (1966).

15. Hobbs, J. R.   *In the Scientific Basis of Medicine: Annual Reviews* (London: Athlone Press, 1966), p 106.
16. Vergani, G., R. Stabilini, and A. Agostoni.   J. Chromatog. **28**, 135 (1967).
17. Kohn, J.   Immunology **15**, 863 (1968).
18. Fazekas de St. Groth, S., R. G. Webster, and A. Datyner.   Biochim. Biophys. Acta **71**, 377 (1963).
19. Roberts, G. P.   J. Chromatog. **22**, 90 (1966).
20. Rydon, H. M., and P. W. G. Smith.   Nature **169**, 922 (1952).
21. Wollheim, F., and J. Snigurowicsz.   Scand. J. Haematol. **4**, 111 (1967).
22. Maier, C. L.   Chromatog. **32**, 577 (1968).
23. Grossman, H., and H. Wagner.   J. Chromatog. **35**, 301 (1968).
24. Stickland, R. G.   Anal. Biochem. **10**, 108 (1965).
25. Agostoni, A., C. Vergani, and E. Cirla.   *Protides Biol. Fluids, Proc. 14th Colloq.*, Bruges, 1966. Ed. by H. Peeters (Amsterdam, London, New York: Elsevier, 1967).
26. Hobbs, J. R., and A. Jacobs.   Clin. Exptl. Immunol. **5**, 199 (1969).
27. Purves, L. R.   Lancet **2**, 186 (1965).
28. Davis, J. S., F. W. Flynn, and H. S. Platt.   Clin. Chim. Acta **21**, 357 (1968).
29. Goddard, P. F., and J. R. Hobbs.   Proc. Roy. Soc. Med. **61**, 335 (1968).
30. Stránsky, Z., and M. Srch.   J. Chromatog. **28**, 146 (1967).
31. Hanson, L. Å., and B. G. Johansson.   Nature **187**, 599 (1960).
32. Cederblad, G., B. G. Johansson, and L. Rymo.   Acta Chem. Scand. **20**, 2349 (1966).
33. Johansson, B. G.   "Studies on Human Immunoglobulins With Special Reference to Immunoglobulin A in Colostrum," Thesis, University of Lund (1967).
34. Axelsson, H., B. G. Johansson, and L. Rymo.   Acta Chem. Scand. **20**, 2339 (1966).
35. Hanson, L. Å., B. G. Johansson.   *Gamma Globulins: Structure and Control of Biosynthesis,* Nobel Symposium 3, Ed. by J. Killander (Stockholm: Almqvist & Wicksell, 1967), p 141.
36. Link, H.   *Immunoglobulin G and Low Molecular Weight Proteins in Human Cerebrospinal Fluid.* Acta Neurol. Scand. **43**, Suppl. 28. (Copenhagen: Munksgaard, 1967).
37. Rymo, L.   Personal communication.
38. Lagerkvist, U., and J. Waldenström.   J. Biol. Chem. **242**, 3021 (1967).
39. Whitacker, J. R.   Anal. Chem. **35**, 1950 (1963).
40. Wieland, T., P. Duesberg, and H. Determann.   Biochem. Z. **337**, 303 (1963).
41. Hygstedt, O., and R. Jagenburg.   Scand. J. Clin. Lab. Invest. **17**, 565 (1965).
42. Yakulis, V., N. Costea, and P. Heller.   J. Immunol. **102**, 488 (1969).
43. Determann, H., in *Protides Biol. Fluids, Proc. 14th Colloq.*, Ed. by H. Peeters (Amsterdam, London, New York: Elsevier, 1967), p 563.

44. Johansson, B. G.   Acta Chem. Scand. **23**, 683 (1969).
45. Hjertén, S., and B. G. Johansson.   Unpublished results.
46. Gelotte, B., P. Flodin, and J. Killander.   Arch. Biochem. Biophys. Suppl. No. 1, 319 (1962).
47. Dose, K., and G. Krause.   Naturwiss. **49**, 349 (1962).
48. Chudzik, J., and A. Klein.   J. Chromatog. **36**, 262 (1968).
49. Johansson, B. G., and L. Rymo.   Biochem. J. **92**, 5P (1964).
50. Hanson, L. Å., P. Roos, and L. Rymo.   Nature **212**, 948 (1966).
51. Bergmann, L., S. J. Dencker, B. G. Johansson, and L. Svennerholm. J. Neurochem. **15**, 781 (1968).
52. Hanson, L., and B. G. Johansson.   Intern. Arch. Allergy **31**, 380 (1967).
53. Carnegie, P. R., and G. Pacheco.   Proc. Soc. Exptl. Biol. Med. **117**, 137 (1964).
54. Grant, G. H., and P. H. Everall.   J. Clin. Pathol. **18**, 654 (1965).
55. Agostoni, A., C. Vergani, and B. Lomanto.   J. Lab. Clin. Med. **69**, 522 (1967).
56. Grant, G. H., and P. H. Everall.   Proc. Soc. Anal. Chem. **4**, 143 (1967).
57. Grant, G. H., and P. H. Everall.   Lancet **2**, 368 (1967).
58. Mancini, G., A. O. Carbonara, and J. F. Heremans.   Immunochemistry **2**, 235 (1965).
59. Hanson, L. Å., J. Holmgren, B. G. Johansson, and C. Wadsworth.   Unpublished results.
60. Vergani, C., and A. Agostoni.   Clin. Chim. Acta **16**, 326 (1967).
61. Solomon, A.   J. Lab. Clin. Med. **70**, 876 (1967).
62. Solomon, A.   Federation Proc. **26**, 529 (1967).
63. Laurell, C.-B.   Anal. Biochem. **10**, 358 (1965).
64. Williamson, J.   Personal communication.
65. Johansson, B. G.   Unpublished results.
66. Radola, B. J.   J. Chromatog. **38**, 61 (1968).
67. Radola, B. J.   J. Chromatog. **38**, 78 (1968).
68. Zuidweg, M. H., J. G. Oostendorp, and C. J. K. Bos.   J. Chromatog. **42**, 552 (1969).
69. Ackers, G. K.   Advan. Protein Chem. **24**, 343 (1970).
70. Porter, P., D. E. Noakes, and W. D. Allen.   Immunology **18**, 245 (1970).
71. Radola, B. J.   Biochim. Biophys. Acta **194**, 335 (1969).
72. Delincée, H., and B. J. Radola.   Biochim. Biophys. Acta **200**, 404 (1970).
73. Porter, P., M. C. Porter, and J. N. Shanberge.   Biochemistry **6**, 1854 (1967).

Chapter 7

Microelectrophoretic Determination of Protein and Protein Synthesis in the 10^{-7} to 10^{-9} Gram Range

by H. Hydén and P. W. Lange

Most organs in the body are composed of several types of cells some of which may dominate in the expression of the function of the organ. The heterogeneity of the tissue composition requires the use of extraordinary methods of sample extraction and analysis.

The requirements are usually met by any method which allows the analysis of substances in amounts of 10^{-7}–10^{-9} g in a volume of 10^5 μm^3 provided pure samples of the various cell types can be obtained.

During the last five years the advantage of using such micromethods for separation of protein has been recognized. In 1965, Grossbach[1] described a micromodification of the disc electrophoretic method using glass capillaries filled with polyacrylamide gel for determination of protein in the range of 10^{-9} g. In 1966, we described a micromethod for the separation of 10^{-7}–10^{-9} g of protein on polyacrylamide gel in 200-μ diameter glass capillaries.[2] Felgenhauer[3] described a similar method in 1967 for the separation of 10–30 μg of protein with a resolution of 1–3 μg of a single protein. Gofman[4] used a Plexiglas chamber instead of capillaries for microelectrophoresis of protein.

Neuhoff[5] increased the resolution of the electrophoretic pattern of protein by using higher concentration of acrylamide gel with the addition of 0.5% hydantoin. Neuhoff and Schill[6] eluated protein microfractions obtained by this method and showed that they could be used

Most of this article is taken from J. Chromatog. **35**, 336–351 (Courtesy of Elsevier & Company, Amsterdam, The Netherlands).

271

for immunoprecipitation. Felgenhauer[7] achieved great sensitivity, 10^{-9} g, in the detection of a single protein by using his modification of an immunoprecipitation following microelectrophoresis of protein on polyacrylamide gel. Recently a review of disc electrophoresis, also describing microtechniques, was published by Maurer.[8]

The present paper describes how micro-disc electrophoresis of labeled proteins in brain cells, nerve cells or glia can be used to assess the type and turnover of soluble proteins in comparable loci of the same brain or in identical structures of different brains. The specific activities obtained are corrected for variations in the local concentration of the precursors. The new procedure is a further development of a method for micro-disc electrophoresis which we have previously published.[2]

7.1 Sampling of Cell Material

7.1.1. Isolation of Nerve Cells and Glia

The animals are killed rapidly and bled and the brain is immediately removed. A section is made through the part of the brain which is of interest. This section, 2–4 mm in thickness, is placed in a solution of $0.25M$ sucrose or Krebs-Ringer solution, and on the cooling stage of a stereo microscope. This is provided with $20 \times$ oculars and an auxiliary $2 \times$ front lens, giving a magnification of $20 \times -160 \times$. The light sources are provided with infrared filters to remove heat.

The brain section is carefully kept under solution and flooded with a $0.01M$ solution of Methylene Blue for 20 seconds. It is then washed and kept under pure sucrose or salt solution. This treatment will render the nerve cell bodies slightly blue for at least 5 minutes. After that time, the staining may have to be renewed. Even if this is not done, the cells are still recognizable to the trained eye against the glistening white glia because of their yellowish appearance and because the synaptic knobs surrounding the cell border keep the stain. This has been described previously.[9]

The cells, together with part of the surrounding glia, are lifted from the section by inserting a pointed stainless steel wire (Nicrothal L, AB Kanthal, Hallstahammar, Sweden) 18–28 μm diameter, below the cell. Each cell is lifted out of the section and placed in Krebs-Ringer solution. This is done by free hand manipulation. The stereo-microscope should be provided with hand supports.

The nerve cell bodies plus a considerable part of the dendrites, free swimming, are freed from the adhering glia by gentle manipulation from below. Figure 1 shows six isolated nerve cells photographed in incident light. Note the absence of glia and the length of the dendrites.

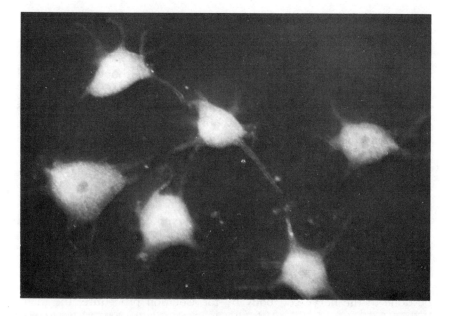

Figure 1. Six nerve cell bodies including the first part of the dendrites, unstained and freed from surrounding glia by free hand dissection and photographed in incident light. Magnification 100 ×.

When these become too thin they break off, but evidently repair the holes immediately (see below). A careful study of the time required to isolate ten nerve cells already placed in an isotonic solution from the surrounding glia showed it to take around 2 minutes.

When the glia are removed from the cell body plus processes, they have a tendency to stick together, especially in an isotonic ion solution. It is then easy to shape the glia into a clump containing 7–8 nuclei of about the same volume as that of the nerve cell in the case of the large Deiters' neurons. The dry weight per volume unit is the same for both the nerve cells and the surrounding glia;[9] this has been checked by quantitative X-ray microradiography.[10–12] Hence, results obtained from the same volume of both types of cells can be compared.

A pertinent question has been whether cells obtained by rapid free-hand dissection constitute suitable material for biochemical analysis, or if, in spite of the short time required for sampling, the cells have undergone postmortem changes. In isolation methods utilizing centrifugation, the shearing forces break glial cells and are likely to damage nerve cells. The small force used for the manipulation on microdissection does not, by comparison, invoke such shearing forces. Possible mechanical impacts are many orders of magnitude smaller here, which is also demonstrated by our results.

It is to be noted that it takes 5 minutes for a trained investigator to kill the animal, place the brain section in solution on the cooling stage and isolate 10 nerve cells.

Such cells have been placed in an ion solution without substrate and the oxygen consumption determined by micromanometry for 2 hours.[3] At this time the rate of oxygen consumption in $\mu l \ O_2 \cdot 10^{-4}/cell$ was close to zero. Glucose or β-hydroxybutyrate was then added, and the respiration was immediately resumed and rose within 10 minutes to $10 \ \mu l O_2 \cdot 10^{-4}/cell$, the dry weight of which was $2 \cdot 10^{-8}$ g. The isolated nerve cells also had the capacity to phosphorylate.[14] Isolated nerve cells were, furthermore, placed in a Krebs-Ringer bicarbonate–glucose saline, and membrane potentials were recorded by inserting micro-electrodes under visual inspection.[15] The average membrane potential at 23° of 120 cells was 40 mV. When a 95% N_2–5% CO_2 atmosphere was introduced, the membrane potentials dropped to around 20 mV, but rose again when 95% O_2–5% CO_2 was reintroduced.

The results from these experiments show that whatever damage may be done to the nerve cell body by the isolation procedure, the membrane of the processes and the perikaryon seem to have self-repaired to the extent that no appreciable amount of substance can have leaked out. It is also possible to cultivate such mammalian nerve cells in the presence of glia.[16]

From the data available, it seems clear that nerve cells from a defined area of the brain and isolated by free hand dissection from a fresh brain section, are excellent material for biochemical studies.

7.1.2   Isolation Samples from Homogeneous Cell Regions

Certain areas in the brain, like the granular layer of the retina or the pyramidal nerve cell layer of the hippocampus, have a relatively uniform cell population. These areas are accessible to the same kind of sampling technique as described in the previous paragraph. Stainless steel tools, forged under a stereo-microscope, are used for removing sections consisting of 100–1000 cells. The cell material is homogenized as described below, and protein concentration determined according to Lowry.[17] An example of analysis of the pyramidal cell layer of hippocampus will be given (Figure 3).

7.1.3   Homogenization

The cell material is homogenized in a glass capillary, diameter 400 μm. Water alone dissolves 39% of the total protein of nerve cell material (from the hippocampus rat). When a Tris solution, pH 7.4, containing 0.1% of Triton X-100 was used, 72% of the total protein was extracted. At electrophoresis 92% of this protein went into the polyacrylamide gel, as measured by radiometry. When in addition to the Triton X-100, 0.1% of SDS was included into the extraction fluid, 92% of the total protein was extracted. More than 90% of this material went into the gel at electrophoresis. The homogenization is performed by introducing into the solution containing the cells a loop of steel wire 28–70 μm thick, fastened to a Teflon tube. This is attached to a dental drill driven at 12,000 rpm.[18] The homogenization proceeds for 1–3 minutes under visual inspection through a microscope. The solution is then centrifuged in a Hematocrit centrifuge at 11,000 rpm for 5–10 minutes. This leaves a clear supernatant.

7.2 Preparation of Capillaries and Polymerization of Gels

Glass capillaries of different width are available from Drummond Science Company Broomall, Pennsylvania. We have routinely used capillaries of 33 mm length, and 215 μm and 440 μm diameter, respectively. The capillaries are boiled in a detergent solution, RBS 25 (Kistner, Göteborg), rinsed in hot water, distilled water, 95% ethanol and acetone. A solution of Siliclad (Clay-Adams, Incorporated, New York) diluted 1:100 is sucked up, after which the capillaries are rinsed in distilled water and dried.

Recrystallized monomers according to Loening[19] were used for the preparation of gels. The various solutions described in our previous publication have been slightly changed according to the modification recommended by Neuhoff[5] (see Section 7.2.1). The modification consisted essentially in adding 0.5% hydantoin for the preparation of 25% polyacrylamide gel. The hydantoin increases the time for polymerization tenfold, increases the pore size and stabilizes the gel. The capillaries were filled by capillary action with the acrylamide solution to half their volume. The polymerization proceeds in a nitrogen atmosphere overnight. In order to remove possible artifacts caused by the ammonium persulfate[20] a small volume of 20 μmole of Tris, pH 8.4, thioglycolate was placed on top of the gel at the electrophoresis.

Before each run a 5% upper gel was carefully applied by a micropipet to a length of 2 mm. This gel was polymerized by light for 5–10 minutes. Before applying the sample, superfluous water on top of the gel was sucked off and a small amount of Stock B solution (see Section 7.2.1) diluted 1:8 was applied for some minutes and removed. Then 0.5 μl–1.5 μl of the sample was applied with a micropipet. Thus, the small volume 200-μm diameter capillary contained: sample 0.5 μl, upper gel 0.07 μl, running gel 0.5 μl. The 400-μm diameter capillary contained: sample 1.7 μl, upper gel 0.3 μl, running gel 2.5 μl.

7.2.1   Polymerization of the Lower Gel

Stock A. pH 8.8: 860 mg, Tris (The British Drug Houses, specially purified) + 0.063 ml N,N,N',N'-tetramethylethylenediamine + 3.6N H_2SO_4 + H_2O to 10 ml and a pH of 8.8.

Stock B. pH 6.7: 2.85 g Tris + 1M H_3PO_4 to pH 6.7 and H_2O to 50 ml.

Stock C. 20 g recrystallized acrylamide (Eastman Organic Chemicals) + 200 mg N,N'-methylenebisacrylamide + 3.75 mg $K_3Fe(CN)_6$ + H_2O to 37.5 ml.

Stock D. 70 mg $(NH_4)_2S_2O_8$ + 25 ml 2% Triton X-100 + H_2O to 50 ml.

Stock E. 5.98 g Tris + 0.46 ml N,N,N',N'-tetramethylethylenediamine + 1M H_3PO_4 + H_2O to 100 ml and a pH of 6.7.

Stock F. 200 mg $(NH_4)_2S_2O_8$ + 5 ml 2% Triton X-100 + H_2O to 10 ml.

Preparation. One ml of (0.5 ml Stock A + 1.5 ml Stock C) + 1.0 ml Stock D, gives a 20% solution. To this is added 100 mg acrylamide + 10 mg hydantoin in order to get a final 25% polyacrylamide gel.

For homogenization. 20 μmole sodium thioglycolate is added to the sample in a 0.25M sucrose + 0.5% Triton X-100 solution, buffered by Stock B to pH 6.7.

7.2.2   Polymerization of the Upper Gel

One ml of (0.5 ml Stock E + 1.5 ml Stock C) + 1 ml H_2O. One ml of this solution is added to 1 ml Stock E. One ml of this solution is added to 0.8 ml H_2O + 0.2 ml Stock F. The final concentration of this upper gel will be 5%.

7.3   Electrophoresis

Electrode buffer. pH 8.5: 3.0 g Tris + 14.4 g glycine + H_2O to 500 ml. Fluorescein: one drop of a saturated solution.

The capillaries are inserted through a 1.5-mm thick rubber stopper that fits in one end of a glass tube (5 mm diameter) containing electrode buffer. The other end of the capillary is located in a small jar containing the same buffer. The electrodes are 0.5-mm platinum wires. Protein samples in five capillaries are run at the same time. The electrophoresis proceeds in a refrigerator or the capillaries are cooled by a fan. The apparatus for microseparation consists of a rack for five capillary runs (Figure 2). The jars for electrode buffer are placed on a vertically movable stand. The capillary holders are adjustable. Both voltage and current can be measured during the electrophoresis. The voltage during the run is kept constant, as is stressed by Neuhoff.[5]

Figure 2. Rack for five-capillary gel electrophoresis.

For separation on the 200-μm diameter gels the voltage is set to 75 V and the current is around 4 μA which, however, continuously falls. After separation for 20 minutes, the state of "steady stacking" has usually been achieved, and the protein is entering the running gel. This is controlled by the fluorescent dye. At this stage, the capillary is removed from its holder and 13 mm of the upper part of the capillary is broken off and the capillary reinserted in its holder. This will decrease the resistance which is reflected in the increase of the current to around 4 μA. Protein separation occurs thus in a polyacrylamide gel of 1.2 μl with a diameter of 200 μm. The run takes around 10 minutes.

For the 400-μm diameter gels, the corresponding details are: 60 V for the first 20 minutes under which period the protein is concentrated in the upper 5% gel and the current falls continuously from around 20 μA to 5 μA. At the stage of "steady stacking," 13 mm of the upper part of the capillary is cut. The separation of protein then proceeds on a 3-μl gel with a diameter of 400 μm.

Certain factors have to be considered as possible sources of error in micro-disc electrophoresis as in any other type of electrophoresis using supporting media. It is of utmost importance to avoid electroendosmosis in microelectrophoresis. It can always be assumed that a layer of fluid exists between the outer part of the gel and the wall of the capillary in which molecular streaming can occur. Hjertén[21] has found that if the glass wall is coated with a methyl cellulose solution with a viscosity of 100 centipoises, electroendosmosis is completely avoided. Therefore, the capillaries used in our procedure have been coated with a thin layer of methyl cellulose (Methocel, Dow Chemical Company, Midland, Michigan).

Another factor of importance is the temperature gradient from the center of the gel to the outside. The temperature in the middle of the gel must not exceed 40°C. Any excessive cooling of the outside of the capillary relative to the inner part of the gel should be avoided.[22] If not, the excessive gradient will produce a bending of the fractions in the direction of the run. In the examples given in this paper (see Figures 4 and 5) it can be seen that the stained protein discs are evenly horizontal. In polyacrylamide gels of larger diameter, 5–10 mm, Gandini and Dravid[22] have pointed out that the glass wall of the tubes should increase when the maximally favorable current is determined in relation to the thermal conductivity of the glass and the external cooling system. This is less critical in microelectrophoresis since the thermal capacity of the thin capillaries including the gel is small and an equilibrium easily obtained as is demonstrated by our results.

7.3.1 *Extrusion of the Gel and Protein Staining*

After separation, the gel is removed by inserting a tightly fitting wire in the anode end of the capillary. The gel is extruded into 80% ethanol. For staining, 0.5% Amido Black in 7.5% acetic acid is used for 5 minutes.

Figure 3 gives a schematic outline of the procedure used and Figures 4 and 5 show photographs of the 200-μm and 400-μm diameter polyacrylamide gels on which 10^{-7} g of brain protein have been separated for 25 minutes and stained by Amido Black.

7.4 Radiometric Determination of the Protein Fractions

The following description concerns the determination of the radioactivity in protein microfractions which have incorporated ^{3}H-labeled amino acids. A method for determination of specific radioactivities of ^{3}H-purines and ^{3}H-pyrimidines has been described by Koenig and Brattgard.[23] The principle is to measure tritium radioactivity after the tritium has been converted to gas. By this means, the factor of self-absorption is removed or greatly reduced. In the protein analysis, we have followed the same principle and used the method with some modifications.

The protein fractions in the micro-gels are cut out under a stereomicroscope. By using a strictly standardized procedure, the various fractions could be localized by taking the fifth, stained gel as a refer-

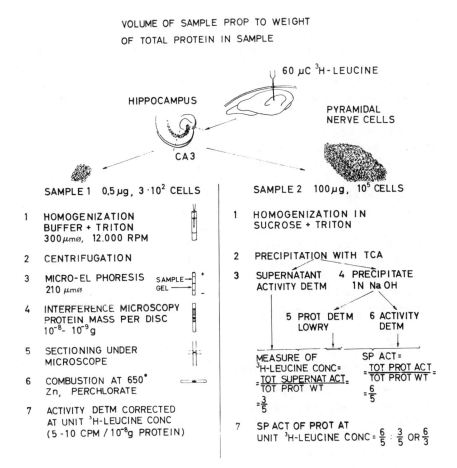

VOLUME OF SAMPLE PROP TO WEIGHT
OF TOTAL PROTEIN IN SAMPLE

60 μC ^{3}H-LEUCINE

HIPPOCAMPUS

PYRAMIDAL
NERVE CELLS

CA3

SAMPLE 1   0,5 μg, 3·10^2 CELLS

1 HOMOGENIZATION
   BUFFER + TRITON
   300 μmø, 12.000 RPM

2 CENTRIFUGATION

3 MICRO-EL PHORESIS   SAMPLE→ +
   210 μmø   GEL → −

4 INTERFERENCE MICROSCOPY
   PROTEIN MASS PER DISC
   10^{-8}- 10^{-9} g

5 SECTIONING UNDER
   MICROSCOPE

6 COMBUSTION AT 650°
   Zn, PERCHLORATE

7 ACTIVITY DETM CORRECTED
   AT UNIT ^{3}H-LEUCINE CONC
   (5-10 CPM / 10^{-8}g PROTEIN)

SAMPLE 2   100 μg, 10^5 CELLS

1 HOMOGENIZATION IN
   SUCROSE + TRITON

2 PRECIPITATION WITH TCA

3 SUPERNATANT   4 PRECIPITATE
   ACTIVITY DETM   1N NaOH

                  5 PROT DETM   6 ACTIVITY
                    LOWRY   DETM

MEASURE OF   SP ACT=
^{3}H-LEUCINE CONC=
$=\dfrac{\text{TOT SUPERNAT ACT}}{\text{TOT PROT WT}}$ $=\dfrac{\text{TOT PROT ACT}}{\text{TOT PROT WT}}=$
$=\dfrac{3}{5}$ $=\dfrac{6}{5}$

7 SP ACT OF PROT AT
   UNIT ^{3}H-LEUCINE CONC $= \dfrac{6}{5} : \dfrac{3}{5}$ OR $\dfrac{6}{3}$

Figure 3.   Outline of procedure.

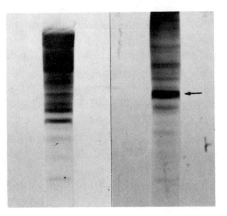

Figure 4 (far left). Brain protein separation pattern on micropolyacrylamide gels, 200 μm diameter, stained with Amido Black. Magnification 25 ×.

Figure 5 (left). Brain protein separation pattern on micropolyacrylamide gels, 400-μm diameter, stained with Amido Black. Albumin added, fraction at arrow. Magnification 25 ×.

ence. A small correction has to be made for the 10% linear swelling which this gel undergoes in the acetic acid stain mixture. The cut-out gel segment containing the protein, 50–500 μm thick, is inserted into a Supramax glass capillary, i.d. 600 μm and 1.5 cm in length, together with powdered $KClO_4$ and Zn particles. The capillary is closed and heated at 650° for 45 minutes. This ensures oxidation to water and reduction to gas. The capillary is then introduced in a specially constructed counting tube provided with a mechanism for crushing the capillary, and letting the ^3H gas escape into the carrier gas of the tube system which is a 99% helium–1% isobutane gas mixture.

The counting tube used by us is a Philips tube PW 4340, the mylar window of which has been removed. The radioactivity is counted with an anticoincidence probe (Philips PW 4149) which has been adapted to a proportional spectrometer, Baird Atomic, model 563. The results are recorded by a Baird Atomic Printer 620. The efficiency of this combustion method was determined in the following way: one-mm per test piece were combusted and counted as described. Thirty determinations were performed and the result was 15 ± 1.5 cpm per sample. The efficiency of the method was thus 40%.

7.5   Interference Microscope Technique for Protein Determination

The calculation of the specific activity of the protein fractions after the micro-gel separation requires a determination of the amount of protein in the single discs (around 10^{-8} g). A direct determination of the weight of a protein fraction after its separation from the gel does not seem to be possible since the separation cannot be made quantitatively. The problem is therefore to determine exceedingly small amounts of protein in presence of large amounts of gel material. We have found that a   possible way to solve the problem is to use an interference microscope technique. For this purpose a Leitz interference microscope was used ("Leitz Interferenz-Mikroskop für Durchlicht").

The objects to be investigated are thus protein discs in a micro-polyacrylamide gel with circular cross section (diameter around 200 μm) or with a form of the cross section which has been more or less deformed. The refractive index of the gel material has been measured interferometrically and found to be 1.485 ± 0.005. The refractive index of proteins lies around 1.59. The micro-gel is embedded in a mixture of glycerol and benzyl alcohol with a refractive index of 1.485. This mixture penetrates the gel completely which can easily be seen from the image of the gel in the interference microscope. The embedded gel is placed at right angles to the interference fringes. If the refractive index of the embedding medium is exactly the same as that of the gel, the interference pattern in the microscope field is not influenced by the gel. However, generally there is a small difference in refractive index and the fringes within the gel are therefore slightly displaced (Figure 6). This displacement (optical path difference of the gel relative to the embedding medium) is a function of the thickness of the gel and the (small) difference in refractive index between gel and the surrounding medium. Within the protein discs, the displacement of the

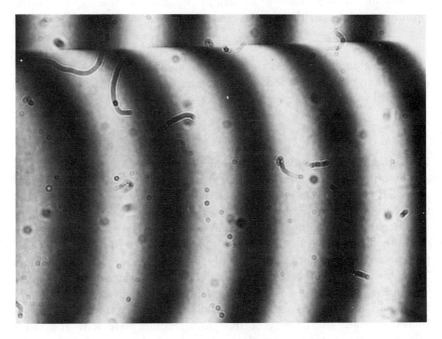

Figure 6. Interference micrographs of a 200 μ diameter gel embedded in benzyl alcohol–glycerol, $n = 1.485$. Magnification 210 ×.

fringes is the sum of the displacement due to the gel and that of the protein, the latter being, of course, in its turn a function of the amount of protein and the difference in refractive index between the protein and the embedding medium. Thus, the optical path difference for a point within a protein disc is $F = F_p + F_g$, where F_p is the optical path difference due to the presence of protein and F_g that of the gel. Further, as is well known,

$$F_p = d_p(n_p - n_m)$$

and

$$F_g = d_g(n_g - n_m)$$

where d_p is the effective thickness (*i.e.*, the compressed, pore-free thickness) of the protein, n_p the refractive index of the protein, and d_g and n_g the equivalent symbols for the gel. The refractive index of the embedding medium is n_m. Thus, the general expression for the displacement (optical path difference) within a protein disc is

$$F = d_g(n_g - n_m) + d_p(n_p - n_m)$$

from which we want to calculate d_p in order to determine the amount of protein. This is done most simply by first measuring F_g in the protein-free gel near the protein disc and then F within the disc. Thus

$$F_p = F - F_g = d_p(n_p - n_m)$$

If the value of n_p (often near 1.60) is known, d_p can be calculated. If not, the gel is embedded in two media with different refractive indices and we thus obtain:

$$F_{p1} = d_p(n_p - n_{m1})$$
$$F_{p2} = d_p(n_p - n_{m2})$$

from which both d_p and n_p can be calculated.

The optical path difference is measured practically by means of the so-called trace displacement method which is schematically illustrated in Figure 7. It should then first be observed that we do not have to measure the displacement of the protein disc relative to the embedding medium but relative to that of the gel. Figure 7 schematically shows part of a microgel with a protein disc in the interference microscope. Since the amount of protein is so small, the displacement of the fringes within the disc relative to that of the fringes of the gel is also relatively small and not easy to estimate visually. The following procedure is therefore used: the density of the photographic plate with the interference image of the gel with the protein disc is first recorded along the line A—B in the middle of the protein-free gel. This record is made on transparent paper. The second record is made along the same line from B to C where the protein disc is situated. When placing the first record (on transparent paper) above the second, it can easily be seen where the fringes are displaced owing to the presence of protein and thus the protein disc is located. It is also easy to measure the optical path difference within different parts of the disc (D_1, D_2 and D_3 in Figure 7). The optical path differences are in this case $D_1/D_0 \cdot \lambda$, $D_2/D_0 \cdot \lambda$ and $D_3/D_0 \cdot \lambda$, where D_0 is the length of separation between the fringes on the record and λ the wavelength used. If the thickness of the disc along the direction of the gel is constant and uniformly filled with protein, and if the gel is circular, the calculation of the protein content in the disc from a record along the A—B—C line is simple. However, for

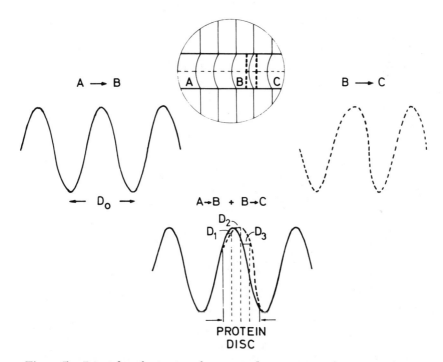

Figure 7. Principles of microinterferometric determination of protein amount in separated fraction.

a lens form of the disc and/or a noncircular cross section of the gel and with a nonuniform distribution of protein within the disc, several records (usually three or four) along the direction of the gel have to be made. We then obtain a distribution of d_p-values within a disc. Through simple graphic integration methods the amount of protein within a disc can be calculated.

The accuracy of the method is mainly dependent upon the accuracy of the displacement measurement. It would seem reasonable (with large fringe separation) that the error should not exceed 1/50 λ and under favorable conditions 1/100 λ. In our measurements, the optical path differences are in the region of 1/10–1/4 wavelength which means that the error would vary in the region of 8–20% or under favorable conditions 4–10%.

Figure 8 shows part of two records of the interference fringes in the gel outside and within a protein disc, situated above each other. The displacement of the fringes owing to the presence of a protein disc in the gel is easily seen.

The following practical example shows a simple case when 0.02 μg of albumin was analyzed. The pattern consisted of two bands, one

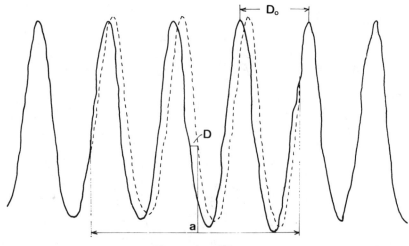

Protein Disc

Figure 8. Densitometric record of interference fringe displacement in protein disc. Phase difference in point $a = \dfrac{D}{D_0} \cdot \lambda$.

very strong, the other very weak. The diameter of the gel with circular cross section was 230 μm. The strong disc was plane parallel with a thickness of 110 μm. The separation of the recorded fringes was 79 mm. The displacement within the disc at the middle of the gel was 14 mm. The wavelength used was 546 nm or 0.546 μm. The optical path difference of the protein (through the gel diameter) is thus $14/79 \cdot 0.546$ μm = 0.097 μm. The refractive index of albumin is around 1.60 and that of the embedding medium is 1.485. Thus $0.097 = d_p (1.60-1.485)$ from which we obtain $d_p = 0.097/0.115 = 0.84$ μm. This means that every μm^3 in the disc contains $0.84/230$ μm^3 of protein. The volume of the disc is $\pi \cdot 115^2 \cdot 110$ μm^3. Thus the disc contains $\pi \cdot 115 \cdot 110 \cdot 0.84/230$ μm^3 of protein. If the density of the protein is set at 1.3, the amount of protein is

$$\pi \cdot 115^2 \cdot 110 \cdot \frac{0.84}{230} \cdot 1.3 \cdot 10^{-12}\, \text{g} \simeq 2.2 \cdot 10^{-8}\, \text{g}$$

The smallest amount of protein which can be measured in a disc by this method (gel diameter ≈ 200 μm) can be estimated as follows:

Assume $D_0 = 100$ nm and $D = 2$ nm. Further $n_p = 1.60$ and $n_m = 1.485$. The thickness of the protein disc $= 10$ μm and the density of protein ~ 1.3.
This gives

$$\frac{2}{100} \cdot 0.546 = d_p \cdot 0.115$$

$$d_p = \frac{2}{100} \cdot \frac{0.546}{0.115}$$

The weight of the protein in the band is thus

$$\pi \cdot 100^2 \cdot 10 \cdot \frac{2 \cdot 0.546}{100 \cdot 0.115 \cdot 200} \cdot 1.3 \cdot 10^{-12} \, g = 2 \cdot 10^{-10} \, g$$

The error will be around 50% for this extremely small amount of protein.

7.6 Correction for Variation in Precursor Concentration

When comparing the specific activity of the protein bands (*i.e.*, the incorporation of labeled amino acids) or that of unseparated proteins from corresponding regions in two brain halves or from identical brain areas of different animals, it should be observed that the *concentration* of available labeled amino acids may show great variations. These variations are a function of different circulation conditions, dilution through cerebrospinal fluid, etc., all factors which are difficult or impossible to control. It is, therefore, necessary to obtain a measure of the concentration of the labeled amino acids in the investigated brain region in order to be able to compare the incorporation of the labeled amino acids in the two brain hemispheres or in the brain of different animals.

It is obvious that higher concentrations of amino acids should result in a higher incorporation of them, up to a certain saturation limit. In order to compare quantitatively the incorporation of labeled amino acids (expressed, for instance, as specific activity of unseparated and fractionated proteins from the brain regions investigated) it is necessary to correct for the influence of the variations in the labeled amino acid concentration. This means that it is necessary to have information concerning the (mathematical) relation between the specific activity of the proteins and the concentration of the labeled amino acid in the cells from which the proteins are isolated. If this relation is known, the specific activity of the protein fractions in a microgel can be referred to the same amino acid concentration and a quantitative comparison of the incorporation of precursors between different brain regions is possible.

In order to investigate whether a linear relationship exists between the protein specific activity and the free amino acid concentration, and

the possible range of such a relation, the following experiments were
made (see scheme in Figure 3):

[3]H-Leucine was injected bilaterally in varying amounts (5–60 μl,
1 μC/μl) into the lateral vertricles of a total of 20 rats, each weighing
175–200 g. The animals were killed 1.5 hours after the injection. The
brain was rapidly removed, and the two CA3-regions of the hippocampus (selected as a brain region of interest in connection with a
behavioral experiment) were immediately dissected out and processed
as outlined in the right half of the scheme in Figure 3. About 100 μg of
pyramidal neurons from the sample was homogenized in 0.5% Triton
X-100. The solution was precipitated with TCA. The precipitate was
washed and its total radioactivity determined (cpm). It was then dissolved in 1M NaOH and the amount of protein (μg) in the solution
was determined according to Lowry's method.[17] In the supernatant,
the total radioactivity of ^{3}H-leucine was determined (cpm). From
these two latter determinations, a measure of the concentration of
[3]H-leucine in the cells investigated is obtained as the quotient between
the total ^{3}H-leucine activity and the amount of protein in the sample.
The prerequisite for this procedure is that the amount of protein
in the sample can be used as a measure of the sample volume. It seems
reasonable to assume that for identical regions in the two hemispheres
or in the brains from different animals, the amount of protein per unit
volume of that region should vary within narrow limits. It is therefore
obvious that the incorporation of amino acids can be compared only for
identical brain regions in the same or in different animals. Thus, the
protein content of different brain regions with different cell structures
cannot be compared with respect to the volume of the different regions.
It should further be observed that this method cannot be used for a
comparison of the volume of identical brain regions from pathological
and normal cases.

Figure 9 shows the results. The specific activities of the unseparated
proteins from the 100–μg samples of the CA3-region (cpm/μg protein)
are plotted against the concentration of ^{3}H-leucine (cpm of the supernatant per μg protein in the sample). A regression analysis shows that
a linear relationship exists between the variables with a high degree of
significance ($p = 0.01 - 0.001$). The correlation coefficient has the
same degree of significance. With a high degree of significance the
experiment has thus shown that within the concentration region up
to 50 units (as described above) the specific activity of the protein is
directly proportional to the concentration of the ^{3}H-leucine.

Since a linear relationship exists between specific activity and amino
acid concentration, it can finally be stated that the specific activity of
unseparated proteins from a 100-μg sample or of a protein band in a

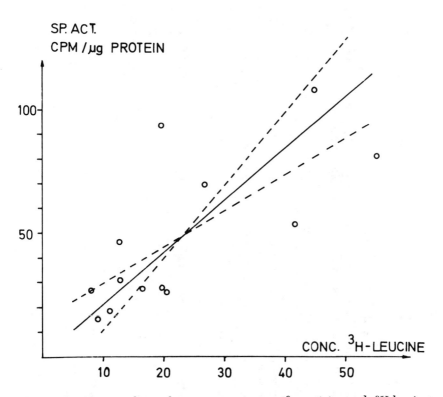

Figure 9. Linear relation between protein specific activity and ^{3}H-leucine activity.

microgel can be compared with the equivalent activities of samples from the identical brain region of the other hemisphere in the same animal or from different animals simply by dividing the specific activity by the ^{3}H-amino acid concentration. All values of the specific activity are thus compared at the same ^{3}H-amino acid concentration, here called unit ^{3}H-leucine concentration.

7.7 Application

The following example of the application of the microprocedure for protein analysis is given to illustrate its potentialities.

Rats were used in an experiment to induce a transfer of handedness in the retrieval of food and to study the effect in isolated nerve cells of a control area of the hippocampus. No surgical measure was undertaken to induce the animals to switch from the preferred paw to the other paw in taking small food pills from far down a narrow glass tube. A wall was simply arranged close to and parallel to this tube. Since the

Table I

Incorporation of ^{3}H-Leucine in Fractionated and Unfractionated Proteins of Pyramidal Nerve Cells in Hippocampus CA3 from the Rat, Expressed as Specific Activities, Corrected for ^{3}H-Leucine Concentration

Hippocampus CA3	Protein ($g \times 10^{-8}$)	^{3}H-protein activity (count/min)	Protein fractions 4 and 5			Unseparated proteins
			Uncorrected specific activity (count/min per 10^{-8} g protein)	^{3}H-Leucine concentration (count/min in supernatant/μg protein in sample)	Corrected specific activity (count/min per 10^{-8} g protein)	Corrected specific activity (count/min per 10^{-8} g protein)
Experimental ratos (18)						
Average right and left sides	2.02±0.15 (n = 14)	13.1±1.50 (n = 49)	6.50±0.89	13.0±1.83 (n = 17)	0.50±0.09	0.113±0.016
Control rats (4)						
Average right and left sides	2.02±0.15 (n = 7)	16.40±1.46 (n = 18)	8.12±0.96	30.37±2.80 (n = 8)	0.27±0.04	0.049±0.005

Numbers in parenthesis: number of rats; n: number of experiments.

animal cannot reach over cross-wise with the preferred paw, it tries to use the other paw and eventually learns. Two training periods per day were given. On the 4th–5th day the rats were still on the rising part of the performance curve. Nerve cells from that period were isolated from the so-called CA3-region in the hippocampus which contains medium-sized pyramidal nerve cells. One hour and a half before the sampling of material, 60×2 μl with 1 μC/μl of a ^{3}H-leucine solution was injected bilaterally in the lateral ventricles. Nerve cell bodies from the CA3 area of the hippocampus were sampled, homogenized and protein extracted with a salt solution containing Triton X-100 to solubilize the proteins, as described above under the Correction procedure. In the separation pattern of the proteins on the microgel, there are two prominent fractions on the acid half of the gel, designated 4 and 5 (Figure 4).

The sum of the specific activities in these two fractions was determined as well as the specific activity of the unseparated proteins in the sample of the CA3 region. The results are summarized in Table I.

From Table I it is seen that the concentration of ^{3}H-leucine of the hippocampal nerve cells (cpm in supernatant/μg of protein in the sample) is lower in the experimental animals than in the control animals. This could possibly be interpreted as a consumption of the ^{3}H-leucine during the training period, resulting in an underestimate of the available ^{3}H-leucine during the training. The corrected specific activity would then overestimate the amount of protein synthesized. However, such an underestimation of the ^{3}H-leucine concentration would imply a higher incorporation of ^{3}H-leucine in the protein of the experimental tissue (the CA3 region of the hippocampus) than in that of the controls, resulting in a higher uncorrected specific activity of the experimental tissue. The data observed (Table I) contradict this. Instead it is highly probable that the lower ^{3}H-leucine concentration in the experimental tissue is a result of a high consumption of ^{3}H-leucine also in other parts of the brain.

REFERENCES

1. Grossbach, U.   Biochim. Biophys. Acta **107**, 180 (1965).
2. Hydén, H., K. Bjurstam, and B. McEwen.   Anal. Biochem. **17**, 1 (1966).
3. Felgenhauer, K.   Biochim. Biophys. Acta **133**, 165 (1967).
4. Gofman, T. Y.   Biokhimiya **32**, 690, (1967).
5. Neuhoff, V.   Arzneimittel-Forsch. **18**, 35 (1968).
6. Neuhoff, V., and W.-B. Schill.   Hoppe-Seylers Z. Physiol. Chem. **349**, 795 (1968).
7. Felgenhauer, K.   Biochim. Biophys. Acta **160**, 267 (1968).

8. Maurer, H. R.   *Disk-Elektrophorese* (Berlin: Walter de Gruyter & Co., 1968; Ann Arbor, Michigan: Ann Arbor-Humphrey Science Publishers, 1971).

9. Hydén, H.   Nature 184, 433 (1959).

10. Hydén, H., and S. Larsson.   J. Neurochem. 1, 134 (1956).

11. Hydén, H., and B. Rosengren.   Biochim. Biophys. Acta 60, 6 (1962).

12. Rosengren, B. H. O.   Acta Radiol., Suppl. 1959, 178.

13. Hydén, H., and P. Lange.   Acta Physiol. Scand. 64, 6 (1965).

14. Hillman, H., and H. Hydén.   Histochemie 4, 446 (1965).

15. Hillman, H., and H. Hydén.   J. Physiol. (London) 177, 398 (1965).

16. Hansson, H., and P. Sourander.   Z. Zellforsch. Mikroskop. Anat. Abt. Histochem. 62, 26 (1964).

17. Lowry, O. H., N. J. Rosebrough, A. L. Farr, and R. J. Randall.   J. Biol. Chem. 193, 265 (1951).

18. Eichner, D.   Experientia 22, 620 (1966).

19. Loening, U. E.   Biochem. J. 102, 251 (1967).

20. Brewer, J. M.   Science 156, 256 (1967).

21. Hjertén, S.   Thesis, University of Uppsala (1967); Chromatog. Rev. 9, 122 (1967).

22. Gandini, S., and A. Dravid.   Proc. 1st Intern. Meeting Soc. Neurochem., Strasbourg (1967), p 77.

23. Koenig, E., and S.-O. Brattgård.   Anal. Biochem. 6, 424 (1963).

Chapter 8

Electrofocusing of Proteins
by C. W. Wrigley

8.1   Introduction*

Electrofocusing involves the migration of an ampholytic molecule in a pH gradient, under the influence of an applied voltage, to the region where the pH is equal to the isoelectric point of the molecule. This principle has been recognized for many years but until recently attempts to apply it to protein fractionation have been largely unsuccessful, mainly because of difficulty in establishing a satisfactory pH gradient. This problem was overcome several years ago by the introduction of synthetic carrier ampholytes for this purpose. As a result of this advance, electrofocusing has become established as an important method of protein fractionation. During 1967 and 1968, descriptions of the method appeared in about 60 publications—more than the number of articles on the use of earlier electrofocusing procedures during the previous 50 years.

This chapter deals with electrofocusing using synthetic carrier ampholytes. Background information and detailed instructions are given for the fractionation of proteins by electrofocusing on a preparative scale using density gradients to stabilize the pH gradient. Analytical fractionation in polyacrylamide gel is also described.

Density gradient electrofocusing has been reviewed by Haglund.[1] Vesterberg[2] summarized and discussed the contents of nine major publications on electrofocusing of proteins. Sessions devoted to electrofocusing were included in the Sixteenth and Seventeenth Colloquia on Protides of the Biological Fluids held at Bruges, Belgium in 1968[3] and 1969.[4] Recently, electrofocusing has been reviewed further by Haglund[128] and Vesterberg.[129]

---

*Literature review concluded April, 1969.

8.1.1   *Definitions*

Before describing the principle of isoelectric focusing, the meanings of certain terms should be clarified:

8.1.1.1   **Ampholytes**

An ampholyte or amphoteric electrolyte is a molecule which can behave as an acid and a base. Its net charge is negative, zero or positive, depending on the pH. Only ampholytic molecules can be fractionated by isoelectric focusing. Proteins are the compounds most amenable to fractionation by isoelectric focusing, but many other biological molecules of medium and low molecular weight are also suitable. The formation of a natural pH gradient for isoelectric focusing depends largely on the use of compounds with ampholytic properties. Ampholytes used for pH gradient formation will be referred to as "carrier ampholytes" to distinguish them from the "sample ampholytes" being fractionated.

8.1.1.2   **Isoelectric point (pI)**

The isoelectric point of an ampholyte is the pH at which its net charge is zero. At this pH the ampholyte does not migrate in an electric field. This operational property may differ from the more fundamental concept of the isoionic point (the pH of zero net charge with all bound ions other than H^+ removed) due to binding of buffer ions to the protein under the conditions of measuring the isoelectric point. For example, bovine serum albumin has a higher isoelectric point (5.4) in 0.15M NaCl than its isoionic point (4.9), due to binding of chloride ions.

See References 5, 6, and 7 for further discussion of the acid-base properties of proteins and Reference 7 for a list of isoelectric points of various proteins.

8.1.2   *The Principle of Electrofocusing*

Throughout this article the term "isoelectric focusing" will be used interchangeably with electrofocusing. Many other terms have been used: isoelectric fractionation, isoelectric condensation, stationary electrolysis, steady state electrolysis, isoelectric equilibrium electrophoresis, *la focalisation électrique, Elektrofokussierung.*

Consider the behavior of a protein with an isoelectric point of 6 in a tube of stabilized solution in which the pH progresses continuously from 3 at the top to 10 at the bottom (Figure 1). Through this column is passed a direct current, anode at top, cathode below. Molecules of the protein in a region where the pH is lower than its isoelectric point are negatively charged and migrate down the tube. Conversely, molecules in a region where the pH is higher than their isoelectric point move up the tube. The protein therefore becomes concentrated or focused in a zone where the pH equals the isoelectric point. The protein remains focused in the same position as long as the pH gradient is stable and the voltage is applied.

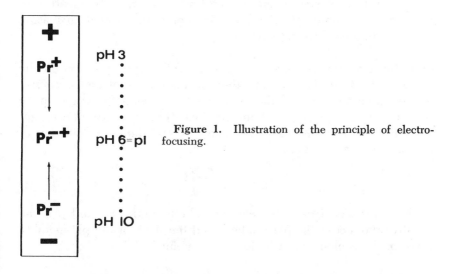

Figure 1. Illustration of the principle of electro-focusing.

In the same way, components of a heterogeneous mixture of proteins become arranged in a series of zones down the tube in order of their individual isoelectric points. Such an arrangement of proteins has been referred to by Kolin[8] as an "isoelectric spectrum." The distinction between isoelectric focusing and electrophoresis should be emphasized. A "mobility spectrum"[8] produced by electrophoresis represents the arrangement of proteins on the basis of their size and net charge at the pH of fractionation. An important advantage offered by isoelectric focusing compared with other methods of fractionation is that protein zones become sharpened as fractionation proceeds, since the forces producing the separation tend to minimize spreading due to diffusion.

8.1.3 *Quantitative Aspects*

The following equation (Svensson[9]) describes the concentration course of an ampholytic component focused as a zone in a natural pH gradient. It is obtained by equating the electric and diffusional mass flows for the component.

$$\frac{Cui}{q\kappa} = D \frac{dC}{dx} \tag{1}$$

In this equation

C is the concentration of the component.

u is its mobility (cm^2 $volt^{-1}$ sec^{-1}).

i is the electric current (amps).

q is the cross-sectional area (cm^2) of the electrolytic medium.

κ is the conductivity (ohm^{-1} cm^{-1}).

D is the diffusion coefficient (cm^2 sec^{-1}) of the component.

x is the coordinate along the direction of the current, increasing from anode to cathode.

From this equation, Svensson[9] derived the following expression for the distance (x_i) from the peak to either inflection point of the concentration distribution curve of the component, assuming the curve to be Gaussian and the pH gradient and conductivity to be constant throughout the zone. This interval is analogous to the standard deviation of the Gaussian distribution.

$$x_i = \pm \sqrt{\frac{q \kappa D}{pi}} \qquad (2)$$

In this expression $i/q\kappa$ is the electric field strength (E) and p is the negative product of the pH gradient and the slope of the pH-mobility curve of the component at its isoelectric point.

$$p = -\frac{du}{dx} = -\frac{du}{d(pH)} \cdot \frac{d(pH)}{dx} \qquad (3)$$

The quantities D and $du/d(pH)$ are properties of the sample being fractionated. It follows from equations 2 and 3 that sharp zones will be obtained for compounds with low diffusion coefficients and steep pH-mobility curves near their pI. The possession of these properties makes proteins especially suitable for isoelectric focusing. It is also clear that for good resolution, a high field strength and a shallow pH gradient should be used. It is therefore necessary to allow for the heating effects of a high electric field in designing suitable apparatus and to provide satisfactory means of producing a shallow, stable pH gradient having a reasonable conductivity.

8.1.4 Artificial pH Gradients

Procedures for pH gradient production have been classified[9] into the formation of artificial and natural pH gradients. An artificial pH gradient is formed at the diffused junction of two buffers of different pH and different density, layered one on top of the other. However when a voltage is applied to such a system, the charged buffer ions migrate and the course of the pH gradient changes rapidly. An artificial pH gradient is unsuitable for isoelectric focusing over a long period because of its instability in an electric field. Kolin[8, 10] used an artificial pH gradient in the isoelectric fractionation of hemoglobin, catalase and cytochrome c. Separation was recorded by direct photography after 4 minutes of electrolysis but could not be continued for a longer time.

8.1.5 Natural pH Gradients

In contrast to artificial pH gradients which are unstable in an electric field, natural pH gradients are produced and stabilized by applying a voltage. They are therefore much more suitable for isoelectric focusing than artificial pH gradients. The simplest example of a natural pH gradient is the electrolysis of a simple salt solution in a convection-free medium. As electrolysis proceeds, acid becomes concentrated near the anode and base near the cathode. A crude pH gradient is thus produced, with pH increasing from anode to cathode.

Consider next the electrolysis of a mixture of basic, neutral and acidic amino acids in a vessel divided into three compartments by two cross-membranes at right angles to the direction of current flow. Acidic amino acids become concentrated in the acid compartment, neutral ones in the center, and basic amino acids in the cathode compartment. This is the basis of a process, patented in 1912 by two Japanese chemists, for the production of glutamic acid from protein hydrolysates.[11]

A smoother pH gradient can be obtained by using an apparatus with more cross-membranes and with ampholytes having a better spread of isoelectric points than the amino acid mixture. A sample ampholyte introduced into such a system of carrier ampholytes migrates in the pH gradient to the point where its net charge is zero. A number of publications describing separations of this type have been reviewed by Svensson.[11] In general resolution was poor, limited by the number of partitions which themselves produced an undesirable endosmotic flow of electrolyte.

A fresh theoretical appraisal of natural pH gradients was made by Svensson[9, 12] in 1961 and 1962. He suggested replacing the multicompartment apparatus by a density gradient and outlined the properties desirable for carrier ampholytes suited to forming a stable natural pH gradient. Svensson recommended the use of "low molecular ampholytes with buffering capacity and conductance in the isoelectric state." A search[12, 13] showed that insufficient carrier ampholytes were available commercially, especially in the 4–7 pH range. Peptide mixtures, obtained from various proteins by partial hydrolysis, were reasonably suitable for pH gradient production but made the analysis of sample proteins difficult.[13] So far, the most satisfactory carrier ampholytes have been those synthesized especially for the purpose.

8.1.6 Synthetic Carrier Ampholytes

The carrier ampholytes synthesized by Vesterberg[2, 14] for pH gradient formation consist of a heterogeneous mixture of different poly-amino–polycarboxylic acids of reasonably low molecular weight. They can be represented by the following general formula, in which R represents H or $-(CH_2)_x COOH$, and m, n, p, and x are less than 5.

$$-CH_2-N-(CH_2)_n-N-(CH_2)_p-N-$$
$$\qquad\quad |\qquad\qquad\quad |\qquad\qquad\quad |$$
$$\qquad\quad R\qquad\qquad (CH_2)_m\qquad\quad R$$
$$\qquad\qquad\qquad\qquad\quad |$$
$$\qquad\qquad\qquad\qquad NR_2$$

Amino groups can be primary, secondary or tertiary. The synthesis is designed to produce a large number of homologs and isomers, with wide variation in the degree of substitution. Consequently a continuous spectrum of isoelectric points is obtained covering the range 3–10. Coarse isoelectric fractionation may be necessary so that the relative proportions of different species can be adjusted to produce an approximately linear pH gradient in the range required.

Such a mixture of carrier ampholytes with pI-values from 3 to 10, for example, has a pH of about 6.5. To produce a pH gradient with this mixture it is incorporated in an anticonvectant medium, such as a density gradient or a polyacrylamide gel, between the anode and cathode which make contact with the carrier ampholytes through solutions of acid and base, respectively. Carrier ampholytes near the acid solution of the anode became positively charged and are repelled from the anode to be replaced finally by the carrier ampholytes with the most acid pI. In the same manner the most basic ampholytes become concentrated close to the cathode solution, and carrier ampholytes with intermediate pI-values are arranged between the electrodes in order of their pI-values. The resulting pH gradient is stable as long as the voltage is applied.

8.2 Density Gradient Electrofocusing

Density gradients have been widely used to stabilize against mixing during electrofocusing. Following electrofocusing in such a medium, the density gradient solution can be drained into a fraction collector for analysis. Since fractionated components are actually separated in this way, the procedure is preparative as well as analytical.

The design of an apparatus for electrofocusing in a density gradient must provide for the escape of electrode gases without disturbing the gradient. The simplest apparatus is a glass U-tube. One arm contains the density gradient; one electrode is located above the gradient and another in the balancing solution of the other arm. The more complicated apparatus of Vesterberg and Svensson[24] has a central balancing tube and a surrounding annular column which contains the density gradient. Because the latter apparatus has been almost universally used for density gradient electrofocusing, Part 8.2 of this chapter is concerned mainly with its application to electrofocusing. Use of the U-tube apparatus and adaptation of a general-purpose electrophoresis column are described in Section 8.2.7.

8.2.1 *The Vesterberg and Svensson Electrofocusing Column*

Design of the column,[15,24] illustrated in Figure 2, is based on earlier designs by Svensson.[16] The column itself is available from LKB Produkter-AB, Stockholm, Sweden in two sizes having capacities of about 110 ml and 440 ml. The two columns are similar in length. Isoelectric

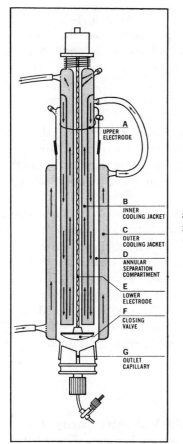

A
UPPER
ELECTRODE

B
INNER
COOLING JACKET

C
OUTER
COOLING JACKET

D
ANNULAR
SEPARATION
COMPARTMENT

E
LOWER
ELECTRODE

F
CLOSING
VALVE

G
OUTLET
CAPILLARY

Figure 2. Diagram of Vesterberg and Svensson column for electrofocusing in a density gradient (by courtesy of LKB-Produkter AB, Sweden).

focusing takes place in an annular separation compartment, D, which contains the density gradient and carrier ampholytes. The lower electrode solution makes contact with the electrode E in a central tube to allow the escape of electrolysis gas without disturbing the density gradient. Inner and outer cooling jackets, B and C, provide efficient thermostating to avoid convective disturbances due to heating. The inner jacket is located inside the outer at a tapered joint, just below the upper electrode A. Following an electrofocusing experiment, the portion of the lower electrode solution in the inner tube is sealed off from the density gradient by closing a valve, F, and the density gradient is drained through the outlet capillary G.

8.2.2 Commercially Available Synthetic Carrier Ampholytes

Carrier ampholytes synthesized according to Vesterberg[2,14] are available commercially from LKB Produkter-AB, Stockholm, Sweden. Their availability and specifications follow:

Available pH ranges: pH 3–10; 3–6, 5–8, 7–10; 3–5, 4–6, 5–7, 6–8, 7–9 and 8–10. Narrower ranges can be prepared from these in the laboratory using the density gradient electrofocusing column (Section 8.2.5.3). Carrier ampholytes are supplied as a 40% w/v solution in water. Unit pack is 25 ml.

Molecular weight range: generally 300–600, not larger than 1,000.

Solubility in water: greater than 50% w/v.

Storage stability: Carrier ampholytes may decompose and discolor if exposed to direct sunlight or temperatures above 20°C for long periods. Opalescence or sediment probably indicates microbial growth: remove by suitable filtration or centrifugation. Store solutions at 4°C or frozen.

Ultraviolet absorption: Ampholine carrier ampholytes have negligible absorbance at visible wavelengths. Ultraviolet absorption is reasonably low down to about 260 nm. It varies slightly from one batch of ampholytes to another, and within the same batch from ampholytes of one pI region to another (see Section 8.2.5.10). The absorbance of a 1% solution of mixed carrier ampholytes in a 1-cm cuvet is specified to be less than 0.30 at 254 nm, less than 0.05 at 280 nm and less than 0.08 at 365 nm. Figure 3 shows the absorption spectrum for a particular batch of mixed carrier ampholytes (pH range 3–10).

8.2.3 Ancillary Equipment

8.2.3.1 Power supply

The supply should cover the range 0–1,000 V, 3-watt capacity (6 watts for the large column). For certain applications, a range of 0–400 V is adequate. A meter, reading current down to 0.5 mA, is required.

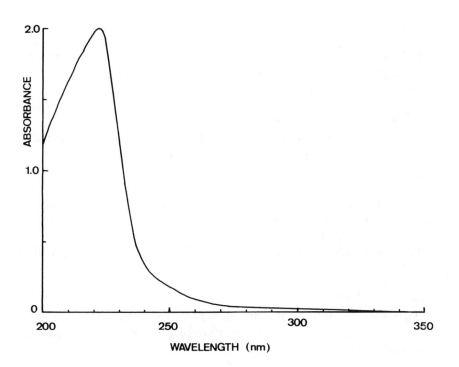

Figure 3. Absorption curve for a 1% solution of mixed, unfractionated synthetic carrier ampholytes. Path length, 1 cm; range of pH, 3–10. LKB Produkter batch number 11.

8.2.3.2 Gradient mixer

The density gradient can be formed manually by the sequential layering of fractions of decreasing density (Section 8.2.4.3). Alternatively, a continuous gradient can be formed automatically by using one of various devices reviewed by Svensson.[17] A device using series-coupled open vessels has been described[18] for the production of linear density gradients. It incorporates a specially shaped plunger to compensate for the density difference between the solutions in the two vessels. The device is available commercially from LKB Produkter-AB.

8.2.3.3 Temperature regulation

For accurate pI determinations, the temperature should be controlled during isoelectric focusing and during pH measurement. Furthermore, it is advisable to perform isoelectric focusing at reduced temperatures because of the risk of bacterial growth during the rather long time required for focusing. For these reasons, equipment is needed for refrigeration, temperature control, and circulation of cooling liquid.

8.2.3.4 Pump

Following electrofocusing the column can be drained by gravity. However, the use of a multichannel peristaltic pump is to be preferred for emptying and also for filling the column.

8.2.3.5 Fraction collection

A time-operated fraction collector combined with the metering pump is the most satisfactory system.

8.2.3.6 Flow analysis

Monitoring the column effluent may give a more detailed analysis of fractionation than examination of individual fractions. Equipment used for recording ultraviolet absorbance of the liquid stream[1, 19] should have a narrow-diameter flow cell to avoid mixing and steaming within the cell.[20]

8.2.3.7 pH Measurement

To realize the full capabilities of the technique the pH meter used should be accurate to 0.01 pH unit. Flow analysis of pH has been used by Valmet[21] and Jonsson et al.[53]

8.2.3.8 Removal of carrier ampholytes

For purification of protein components from fractions, carrier ampholytes and sucrose can be removed by any of the following procedures:
   (a) Gel filtration[123] on Sephadex dextran gel (AB Pharmacia, Uppsala, Sweden) or on BioGel polyacrylamide beads (Bio Rad Laboratories, Richmond, California).
   (b) Exhaustive dialysis against dilute salt solution for several days. Equipment for dialyzing a large number of samples has been described by Rust[22] and Tappan.[23]
   (c) Ultrafiltration[124] using Diaflo membranes (Amicon Corp., Lexington, Massachusetts).
   (d) Protein precipitation with ammonium sulfate.[125]

8.2.4 Procedure

Instructions are given below for performing an exploratory electro-focusing experiment in a sucrose density gradient using wide pH range carrier ampholytes in the 110-ml Vesterberg and Svensson column. Information so gained about the pI of a component of interest would normally lead to a more precise experiment using a shallower pH gradient and higher field strength to produce better resolution. The general conditions set out below are intended to be varied as necessary according to the recommendations in Section 8.2.5. A full description of an electrofocusing experiment is given in Reference 24.

8.2.4.1 Electrode solutions

	Cathode at top		Anode at top	
Upper electrode solution	1% ethanolamine	10 ml	0.5% sulfuric acid	10 ml
Lower electrode solution	sulfuric acid	0.2 ml	ethanolamine	0.4 ml
	water	14 ml	water	14 ml
	sucrose	12 g	sucrose	12 g

Concentrated sulfuric acid may be replaced by the same volume of concentrated phosphoric acid, and ethanolamine by ethylenediamine.

8.2.4.2 Density gradient solutions

The following volumes are based on the use of 1% carrier ampholytes for charging the 110-ml column:

Light solution:	40% carrier ampholyte solution	0.7 ml
	water	58 ml
	sucrose	3 g
Dense solution:	40% carrier ampholyte solution	2.0 ml
	water	40 ml
	sucrose	28 g

Carrier ampholytes are distributed unequally between the light and dense solutions to compensate for the initially lower conductivity of the dense solution due to its sucrose content. Dissolved air should be removed from solutions by degassing to avoid bubble formation during subsequent flow analysis.

8.2.4.3 Filling the column

Connect cooling water to jacket. Check to verify that the column is vertical. With valve F open, pour the lower electrode solution into the central tube through the upper nipple using a fine-tipped funnel and attached capillary tubing. The solution should be at least 1.5 cm above the bottom of the inner jacket. Check to verify that no bubbles are trapped around the valve by shutting and opening it several times. Drain some liquid through the outlet capillary.

Introduce the density gradient through the other nipple so that it flows smoothly down the wall of the separation compartment at up to 4 ml per minute. For stepwise gradient formation, measure 0, 0.1, 0.2, 0.3 ml . . . . 3.2 ml of light solution into 33 tubes labeled 0, 1, 2, 3 . . . .

32, respectively. Then mix in 3.2, 3.1, 3.0, 2.9 ml . . . . 0 ml of dense solution into tubes 0, 1, 2, 3 . . . . 32, respectively. The 3.3-ml portions are added to the separation compartment in the order 0–32 either by hand-pouring or with a peristaltic pump.

The following automatic procedure is presented as an alternative: light solution is pumped at 2 ml per minute into a beaker containing the dense solution. At the same time the solution is pumped from the beaker into the electrofocusing column at 4 ml per minute. Efficient mixing in the beaker is important.

Finally, apply the upper electrode solution as a layer over the gradient until the upper electrode is covered by about 1 cm of liquid.

Since the volume of the separation chamber varies slightly from column to column, use of the prescribed volumes may result in under- or overfilling. If a particular column is overfilled, remove liquid from the top of the density gradient so that the electrode contacts only the upper electrode solution. If underfilled, add more electrode solution.

8.2.4.4 Electrolysis

Connect leads to the appropriate terminals and turn on the power. Record voltage and current during the run. Follow recommendations in Section 8.2.5 concerning voltage adjustment and total time for electrofocusing.

8.2.4.5 Draining the Column

At the end of a run turn off the power and carefully close the valve at the bottom of the central tube. Insert a plastic tube (1–2 mm) through the upper nipple and remove the lower electrode solution. This precaution avoids the risk of contaminating the sucrose gradient solution in the event of valve leakage. It is also wise to remove the upper electrode solution (approximately the volume layered initially) since the acid or base quickly diffuses into the upper region of the density gradient. The solution should drain directly through a short length of tubing (up to 1 mm in diameter) into the flow cuvet of the UV monitor (flowing from top to bottom), through the peristaltic pump (at 1 ml per minute), and finally to the fraction collector. The suggested direction of flow through the UV monitor takes advantage of the density gradient stabilization but the possible trapping of air bubbles should be checked. The minimum size of fractions is generally dictated by the quantity needed for assay procedures. Generally 3–4 ml is a convenient fraction size.

For accurate pH determinations in the alkaline region, fractions should be collected and stored under an inert gas such as argon[24, 25, 26] to prevent the alteration of pH due to absorption of carbon dioxide. This precaution is not necessary if pH is measured in a flow cell or immediately after collection of fractions.

8.2.4.6 Large column

For the 440-ml electrofocusing column, the following alterations are necessary. Use four times the prescribed quantities of electrode solutions. Make up 215 ml of light and 215 ml of dense gradient solutions. For stepwise gradient formation, mix 46 fractions of 9 ml each, with light solution replacing dense solution in 0.2-ml increments. Apply a maximum power of 4–6 watts at similar or slightly lower voltages.

8.2.5 *Design of the Experiment*

8.2.5.1 Temperature

Vesterberg and Svensson[24] compared the behavior of myoglobin components during electrofocusing at 4° and 25°C. They found that the pH of the focused zone depended on the temperature of pH measurement and not on the temperature of focusing. Isoelectric point at a certain temperature can therefore be determined by measuring pH at this temperature, irrespective of the pH of electrofocusing. Generally pI decreases with increasing temperature.

The use of higher temperature increases the rate of focusing, but also increases the risk of loss of enzymic activity, of denaturation and of microbial growth in the sucrose gradient. Choice of temperature therefore depends largely on the lability of the sample protein. A compromise procedure sometimes adopted is to electrofocus at 4°C for much of the overall time and to conclude by electrofocusing for several hours at a higher temperature, say 20°C.[19, 29] Time is allowed to refocus any zones disturbed by the temperature change. The higher final temperature minimizes difficulties of bubble formation and fogging associated with flow analysis of cold solutions.

8.2.5.2 Concentration of carrier ampholytes

Comparison of experiments performed at different concentrations of carrier ampholytes suggests that the isoelectric point of a protein is independent of carrier ampholyte concentration.[24] The concentration used almost universally is 1% w/v, *i.e.*, the 110-ml electrofocusing column requires 1.1 g ampholyte or 2.75 ml 40% stock solution. Generally buffering capacity is inadequate with lower ampholyte concentrations, although 0.5% has been used. Concentrations over 1% are useful for increasing the solubility of proteins that tend to precipitate at the pI.[21]

8.2.5.3   pH Range

For best resolution, the pH range should be as narrow as possible, although a preliminary run using a wide pH range is usual when dealing with proteins of unknown pI. Electrofocusing in the range pH 1–3 has been performed using a mixture of acids.[53]

The narrowest ranges available commercially are 2 pH units wide. Use of ranges down to 0.5 pH unit wide is recommended between pH 4 and pH 9. It is not recommended at the extremes of the nominal pH range as there are relatively fewer individual carrier ampholytes having pI-values above 9 and below 4. Carrier ampholytes for narrow-range pH gradients must be prepared in the laboratory. Perform a normal electrofocusing run without sample using a higher concentration, say 4%, of carrier ampholytes. From the resulting fractions, carrier ampholytes of the required pH range can be chosen and bulked. Store frozen. This preparation is regarded as a 4% solution of ampholytes when it is used for electrofocusing; its sucrose content is not taken into account when making up the new density gradient. If ampholytes are required pure, use a high-molecular-weight solute for the density gradient. The carrier ampholytes can then be purified by dialysis.

As an alternative to the above procedure, the sample can be included in the initial run and is already prefractionated for further electrofocusing in the narrow-range experiment.

8.2.5.4   Protein loading

The sample can be applied either in a narrow zone in the density gradient or throughout the gradient. Fractionation is fastest and best if the point of application is close to the final position of focusing. Application in a zone can be performed by dissolving the sample in one or two of the middle fractions when producing a stepwise gradient, or in solution of intermediate density drawn from the mixing device when forming a continuous gradient. The density of the sample zone should not exceed that of the solution below it. These densities can be checked roughly with a pocket saccharimeter. For sample loading throughout the column, the protein is incorporated in the light gradient solution, or even in both light and dense solutions. In the latter case, the sample may be protected from upper and lower electrode solutions by nonprotein layers of light and dense solutions, respectively.

Sample application throughout the gradient is useful for dilute samples, but the salt content in the gradient should not exceed an overall concentration of 0.0005M. If necessary, dialyze sample against water, carrier ampholytes, or 1% glycine.[27]

The amount of protein to be loaded is largely determined by the heterogeneity and solubility of the sample and by the slope of the pH gradient. For the 110-ml column, 5 mg protein per zone is recommended for analytical fractionation with high resolution. Load up to 25 mg per zone for preparative runs. With overloading, it is possible to exceed zone carrying capacity of the density gradient so that drops of the concentrated protein fall below the level of the zone. This situation is most likely to occur at high field strength in a steep pH gradient. High sample loadings per zone are possible with shallower pH gradients since zones are wider. For satisfactory use of very high loadings, see Wadstrom[28] who applied 500 mg to a 110-ml column.

As mentioned earlier, some proteins which tend to precipitate at the pI can be kept in solution by raising the ionic strength with a higher concentration of carrier ampholytes. Alternatively, precipitation can be overcome by the inclusion of urea or dimethylformamide throughout the gradient.[28, 29] These reagents may, however, produce changes in tertiary structure. Cyanate, formed from urea, is a possible source of artifact formation.[100] Only fresh solutions of purified urea should be used but cyanate may form during a run, especially at high pII. IIigh concentrations of urea have been used to investigate ribosomal proteins[30] and subunits of α-crystallin.[31, 32, 33] In certain work nonionic detergents might prove useful.

8.2.5.5 Solute for density gradient production

Sucrose is by far the most widely used solute for preparing density gradients. It is generally harmless to proteins and can even have a protective effect. Its spectroscopic purity is discussed in Section 8.2.5.10. In certain cases it may be desirable to replace sucrose by another solute —if, for example, sucrose interferes in the assay procedure to be used. A satisfactory solute should be nonionic, have a high solubility and low viscosity in water, and should increase the density by at least 0.12 g/ml. Alternative solutes have included 50 or 60% glycerol[34, 35] and 75% ethanediol.[36] High-molecular-weight solutes such as dextran[37] and Ficoll[38] (AB Pharmacia Fine Chemicals, Uppsala, Sweden) have also been used but these cannot easily be separated from proteins following fractionation.

8.2.5.6 Polarity of electrodes

Electrode polarity should generally be chosen to place the carrier ampholytes of higher conductance in the lower part of the column where conductivity is otherwise reduced by the high sucrose concentration. Carrier ampholytes isoelectric in the range pH 6–7 have relatively lower conductivities. Thus, for pH ranges below 7 the cathode should be on top and for alkaline ranges the anode should be on top. For the fractionation of crude extracts it may be convenient to arrange polarity so that unwanted material is focused to the bottom of the column, especially if it contains precipitated matter. (If precipitates interfere with normal column draining, liquid can be removed through a fine tube carefully inserted from above; see Reference 24, page 832.)

8.2.5.7   Field strength

In theory, the highest possible voltage gives the best resolution (see Equation 2). However, a practical limit on field strength is set by the risk of convective disturbance of the gradient due to ohmic heating. A maximum power of 1 watt for the 110-ml column is observed by many workers although for most applications 2 watts is a reasonably safe limit. Satisfactory use of the column at 4 watts has been reported.[39] In actual use, an indication that power is too high is given by the appearance of eddy zones in the gradient, seen as disturbances in the refractive index of the medium. As focusing of the carrier and sample ampholytes proceeds, the electrical resistance of the column increases and the voltage may be gradually increased if necessary. (For example, start at 400 V and increase by 200 V each hour.) A final voltage of 400–700 V is generally adopted for wide pH ranges and up to 1000 V for narrow pH ranges. An additional limiting factor is the actual precipitation of protein at high field strength, reported for example by Jeppsson.[29]

8.2.5.8   Time

An indication that electrofocusing has reached equilibrium is given by dropping of the current to a constant level. In addition, focused bands of major components can sometimes be seen as refractile zones, especially against a crosshatched background. Neither of these criteria is a precise indication and it is generally advisable to continue electrofocusing for some hours beyond the time thus indicated. Within reason, resolution is not adversely affected by continued electrolysis beyond the minimum time. Formation of a pH 3–10 gradient takes about 8 hours at 400 V and 5°C. Focusing of protein zones takes longer. Overall time generally varies from 24–72 hours. Focusing is faster for wide pH ranges than for narrow, and at higher field strengths and higher temperatures.

8.2.5.9   Draining efficiency

Resolution of components is much better within the column than after draining. Loss of resolution during draining has been demonstrated by Vesterberg and Svensson for two myoglobin components[24] by photographing the zones in the column just before draining. Separation of the two peaks in a densitometric scan of the photo was much better than separation in the actual elution profile. To improve the efficiency of draining, various workers have silicone-treated the inner walls of the column with 0.1% aqueous Z-4141 (Dow Corning, Midland, Michigan)[25, 26] or with Repelcote (Hopkins and Williams Ltd., Chadwell Heath, Essex).[40] However this practice has not yet gained general acceptance.

8.2.5.10 Analysis of protein

A serious disadvantage of density gradient electrofocusing is the difficulty of detecting the sample ampholytes in the presence of carrier ampholytes and sucrose. The most satisfactory detection procedure is an enzyme assay, if appropriate. However, carrier ampholytes may complex the metal ions essential for enzyme activity. If this is the case, the metal should be added in excess. An antigenic assay is also very useful. In this case neither carrier ampholytes nor sucrose interfere.

The popular Lowry procedure for protein estimation[11] cannot be used without prior dialysis, since the carrier ampholytes react strongly. However its use should be satisfactory with TCA-precipitated and washed protein.

Neither carrier ampholytes nor sucrose interfere seriously in the Biuret procedure for protein estimation,[42] but the relatively low sensitivity of this method (0.5–3 mg protein/ml) makes it generally unsuitable. A microversion of this method may be more satisfactory.[43]

Ultraviolet fluorescence, a sensitive procedure for protein determination, involves little interference from ampholytes (P. E. Penner, personal communication).

The most frequently used procedure is estimation of absorbance at 280 nm. This method is reasonably sensitive, well-suited to flow analysis, and not subject to sample destruction by the procedure. However the carrier ampholytes themselves have significant absorbance at 280 nm and variation in the absorbance of individual carrier ampholytes can be mistaken for protein peaks. It is therefore essential to perform a blank run without sample to determine the base line for the batch of carrier ampholytes in use. Such a blank run is illustrated in Figure 4. Fluctuation in base line is not great for absorbance at 280 nm nor for the Biuret procedure. Background absorbance at 254 nm is considerable. Commercial flow analyzers based on the 254 nm line of mercury are thus poorly suited to analysis after electrofocusing.

Sucrose of low UV absorption should be used. Certain papers[24, 25] specify Mallinckrodt sucrose (Mallinckrodt Chemical Works, St. Louis,

Missouri). Satisfactory results have been obtained with certain other samples (analytical reagent grade). Earland and Ramsden have warned[44] that zones of protein-like material can be electrofocused from certain sucrose samples. The suggested blank run should reveal difficulties of this type.

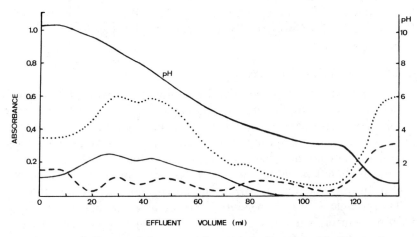

Figure 4. Blank run of density gradient electrofocusing with 1% carrier ampholytes (pH 3–10), without sample, in a 110-ml column at 5°C for 20 hours at 400 V. The anode was at the top. Fractions were analyzed for pH (solid line), absorbance at 280 nm (dashed line) and at 254 nm (dotted line), and absorbance after applying the Biuret procedure to 1-ml samples (solid line).

In summary, the best procedure for estimating total protein is the determination of absorbance at 280 nm. Fractions thus shown to be interesting can be further examined after removal of sucrose and carrier ampholytes by dialysis.

8.2.6   Results and Applications

The electrofocusing of a colored protein, such as native hemoglobin, provides an excellent demonstration of the phenomenon as it is happening and is also an instructive preliminary experiment for learning to use the apparatus. Electrofocused zones of hemoglobin are shown in Figure 5. The profile obtained by monitoring absorbance at 254 nm during the elution of this column is illustrated in Figure 6. The pH gradient is reasonably linear. The pH reading of the peak tube for each component indicates its isoelectric point.

Proteins fractionated by electrofocusing in a density gradient are listed in Table I. The colored metalloporphyrin proteins have been studied in some detail, probably because they are readily located. As mentioned earlier, enzymes are also amenable to electrofocusing since

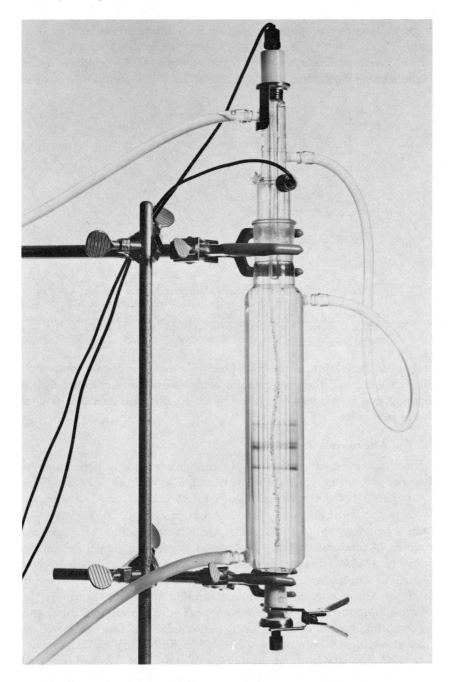

Figure 5. Fractionation of human hemoglobin in the Vesterberg and Svensson column for density gradient electrofocusing. Separation time, 65 hours (From Haglund[1].).

in general carrier ampholytes and sucrose do not interfere in their assay.

The literature contains very little mention of electrofocusing of whole serum. Isoelectric points obtained by other methods, however,

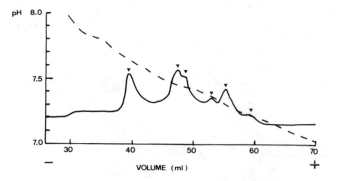

Figure 6. Elution profile (absorbance at 280 nm) of hemoglobin components from the column shown in Figure 5 and superimposed pH gradient (dashed line). Only the relevant portion is shown (From Haglund[1].). The pI-values of the components marked are 7.64, 7.46, 7.44, 7.36, 7.30 and 7.23.

have been listed for serum proteins.[74] Valmet[21] indicates that difficulties concerning the precipitation of serum proteins during electrofocusing can be largely overcome by increasing the concentration of carrier ampholytes.

8.2.7 Alternative Apparatus

Electrofocusing of vitamin B_{12}-binding proteins was performed by Flodh et al.[49] in the LKB Uniphor column electrophoresis system. This equipment is not designed specifically for electrofocusing but is meant to adapt to almost any one of the many forms of electrophoresis.

An extremely simple apparatus for electrofocusing in a density gradient of moderate volume has been used for enzyme fractionation by Weller et al.[60] The unit illustrated in Figure 7 has been made to their design. Electrofocusing takes place in an 11-ml density gradient in the right arm. Beneath the gradient and in the balance arm is the cathode solution containing 40% sucrose. To fill the unit, 14 ml of this dense cathode solution is placed in the U-tube; 11 1-ml fractions are layered on top of one another in the separation arm in the manner described earlier, with light solution replacing dense solution in 0.1-ml increments. The cathode is inserted into the left arm and the anode into 1–2 ml 2% sulfuric acid layered on top of the density gradient. Electrofocusing is complete after about 10 hours (final voltage, 500 V).

Table I
Applications of Density Gradient Isoelectric Focusing to Protein Fractionation

Protein	Origin	Reference
Metalloporphyrin protein		
Ferrimyoglobins	Horse	Vesterberg and Svensson, 24
Ferri- and ferromyoglobins	Horse	Vesterberg, 45
Myoglobin	Horse	Riley and Coleman, 78
Hemoglobin	Hagfish	Svensson, 16
Hemoglobin	Hagfish	Haglund, 1
Hemoglobin	Hagfish	Paleus and Vesterberg, 46
Hemoglobin	Hagfish	Quast and Vesterberg, 39
Cytochrome c	Beef heart	Flatmark and Vesterberg, 25
Lactoperoxidases	Cow's milk	Carlstrom and Vesterberg, 26
Vitamin B_{12}—binding proteins	Body fluids, leucocytes	Grasbeck, 47
Vitamin B_{12}—human intrinsic factor complex	Human gastric juice	Grasbeck, 34
Vitamin B_{12}—binding protein	Human saliva	Grasbeck and Visuri, 48
Vitamin B_{12}—binding proteins	Mouse gastric mucosa	Flodh *et al.*, 49
Vitamin B_{12}	Commercial preparation from bacteria	Aares *et al.*, 50
Enzymes		
Enzymes and toxins	*Staphylococcus aureus*	Vesterberg *et al.*, 15
Deoxyribonuclease	*S.aureus*	Wadstrom, 51
Hyaluronate lyase	*S.aureus*	Vesterberg, 52
Hemolysin	*S.aureus*	Wadstrom, 28; Mollby and Wadstrom, p 465 in 4
Invertase	Yeast	Vesterberg and Berggren, 35

Table I, continued

Protein	Origin	Reference
Laccase		Svensson, 13
Laccase A and B	*Polyporus versicolor*	Jonsson et al., 53
Cellulases and other glycosidases	*Aspergillus*	Ahlgren et al., 36
Cellulases and other glycosidases	Various fungi	Ahlgren et al., 54
Cellulase	Rot fungus	Bucht and Eriksson, 55
Glycosidases	Jack bean	Li and Li, 56, p. 455 in 4
N-Acetyl hexosaminidase	Pig kidney; calf, rat, and human brain	Sandhoff, 57
Phospholipase C	*Clostridium perfringens*	Bernheimer et al., 58
Sulfatase B	Ox liver	Allen and Roy, 59
Lactate dehydrogenase	Hamster skeletal muscle	Weller et al., 60
Carbonic anhydrase B	Ox	Jonsson and Pettersson, 19
Clostridiopeptidase B		Mitchell, 61
Protease		Rebeyrot and Labbe, 62, and p. 493 in 4
Bromelain	Pineapple	Berndt et al., 63
Lipoxidase	Soybean	Catsimpoolas, 64
Plasma proteins		
Whole serum, gamma globulin	Human serum	Valmet, 21
Immunoglobulins	Human serum	Howard and Virella, p. 449 in 4
Estradiol-binding β-globulin	Human serum	VanBaelen, et al., p. 695 in 3
Testosterone-binding globulin	Human pregnancy serum	VanBaelen et al., 65 and p. 489 in 4
Albumin	Human serum	Carlsson and Perlman, p. 439 in 4; Valmet, p. 443 in 4
Albumin	Bovine plasma	Kaplan and Foster, 66
Albumin	Serum	Reis and Wetter, 67
Apolipoproteins A and B	Serum	Blaton and Peeters, p. 707 in 3
Myeloma proteins		Eulitz, p. 481 in 4

Table I, continued

Protein	Origin	Reference
Clotting factors	Ox blood	Pechet and Smith, 68
Fibrin-stabilizing factor	Human plasma	Earland et al., p. 485 in 4
Transferrins	Human plasma	Jeppsson, 29
Miscellaneous		
Ovotransferrin	Hen egg	Wenn and Williams, 40
Ovalbumin	Chick, duck, turkey	Smith and Back, 69
α-Crystallin subunits	Bovine lens	Bjork, 31
		Bloemendal & Schoenmakers, 32
		Schoenmakers & Bloemendal, 33
Lens proteins		Bours et al., p. 475 in 4
Various protein hormones	Human origin	Moritz, p 701 in 3
Lactoglobulin		Kaplan and Foster, 66
Glucagon-like immunoactive components		Markussen and Sundby, p. 471 in 4
Venom	Gut	Toom et al., 70
Lymphocyte-stimulating leucoagglutinin	Snake	Weber and Grasbeck, 71
Ribosomal proteins	Red kidney beans	Brown et al., 30
Colicins E_2 and E_3	*Streptococcus fecalis*	Herschman and Helinski, 72
Fiber antigen	*Escherichia coli*	Pettersson et al., 27
Whey proteins	Adenovirus type 2	Catsimpoolas et al., 75
Trypsin inhibitor	Soybean	Catsimpoolas et al., 77
Plant pigments	Soybean	Svensson, 13
Plant pigments	Whortleberry	Jonsson and Pettersson, 73
	Red beet, various berries, tea	

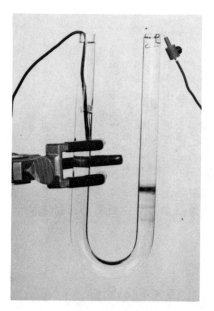

Figure 7. U-tube apparatus for electrofocusing. A preparation of hemoglobin has been fractionated in a density gradient in the right arm. The left arm contains a balancing solution of 4% ethanolamine in 40% sucrose. The tube has an internal diameter of 1 cm and volume of about 28 ml. Ampholyte pH range, 3–10.

The density gradient is emptied from the column through a fine tube attached to the top of the right separation arm by pumping 40% sucrose into the left arm. For temperature regulation the whole U-tube can be placed in a water bath.

Weller *et al.* also recommend the use of a larger U-tube which accommodates a 33-ml gradient. Presumably temperature regulation from the outside only would be inadequate for columns any larger. The system is attractive for pilot or analytical runs. However, for precise work too few fractions of useful volume would be obtained unless the usual analytical techniques were scaled down.

Valmet[4, 126] has described an apparatus for performing "zone convection electrofocusing" in free solution rather than in a density gradient. The fractionation bed is horizontal, and upper and lower surfaces of the trough are corrugated to prevent extensive horizontal mixing.

8.2.8 Resolving Power

Vesterberg and Svensson[24] derived an expression for the smallest difference (ΔpH) allowed between the isoelectric points of two proteins to enable them to be resolved by electrofocusing.

$$\Delta pH = 3.07 \sqrt{\frac{D(dpH/dx)}{-E(du/dpH)}} \qquad (4)$$

Vesterberg and Svensson[24] resolved myoglobins differing in pI by 0.06 in a 110-ml column with 1000 V applied using a pH range of 6.5–7.5.

Through use of the diffusion coefficient of myoglobin and values of other quantities from the experiment, a resolving power of 0.02 pH unit was obtained in the above equation. The authors made the point, on the basis of this result, that "if the pI difference between two proteins is great enough to be picked up by a good pH meter, the proteins can be resolved by steady-state electrolysis." This theoretical estimate of resolving power has been substantiated experimentally.[26, 45]

8.2.9 *Validity of pI-Values*

Determination of isoelectric point by electrofocusing embodies a great saving compared with classical procedures. Whereas in a single electrofocusing experiment pI-values can be determined for several proteins, estimation by moving-boundary electrophoresis requires about four experiments at each of four different pH-values for each protein.

Furthermore, isoelectric point determined by electrofocusing is likely to come close to the isoionic point, since the solution is salt-free and the ionic strength of the carrier ampholytes is very low. In one instance where a comparison can be made, the pI of a deoxyribonuclease from *S. aureus* determined by electrofocusing[15, 51] is very close to its isoionic point[76] after the difference in temperature between the two estimations has been corrected.[2] For hemoglobins[16] and myoglobins,[24] pI-values obtained by electrofocusing are higher than those obtained by moving-boundary electrophoresis due to the binding of anions to the proteins at the ionic strength (usually 0.1) used for electrophoresis. On the other hand, the two procedures give similar pI-values for cytochrome c which does not bind anions at its high pI.[25]

Although modification of pI by salts is unlikely with electrofocusing, the possible effects of sucrose and carrier ampholytes on proteins must be considered. A negative dependence of pI on sucrose concentration has been reported[25] for the basic protein, cytochrome c (pI of over 10). The difficulty was overcome by extrapolating from observed pI-values to zero sucrose concentration. The interference of sucrose at high pH is probably associated with alterations in the dissociation of protein amino groups because of the reduced dielectric constant at high solute concentrations. In some cases, pI has been shown to be independent of sucrose concentration.[53]

At present no direct evidence indicates that interference from carrier ampholytes produces anomalies in pI-values. The findings that pI-values are independent of carrier ampholyte concentration[24] and of the presence of 6M urea[28] in certain cases suggest that carrier ampholytes do not complex with proteins. Electrostatic interaction between protein and carrier ampholytes with similar pI-values should be minimal at the steady state of electrofocusing when both have zero net charge. Thus any *reversible* interaction of protein and carrier ampholyte at pH-values away from their pI-values should not affect fractionation. However the possibility of complex formation cannot be excluded.

On the other hand, there is a distinct possibility that molecules of a protein might take part in aggregation reactions at the pI under the conditions of low ionic strength associated with electrofocusing. Evidence of aggregate formation as a result of isoelectric focusing has been published for aryl sulfatases.[59]

8.3   Gel Electrofocusing

Even though use of the density gradient apparatus gives preparative separation and also allows the accurate determination of isoelectric point, it is expensive in reagents and sample if analysis only is required; Furthermore, up to three days are needed for the fractionation of a single sample. The technique of gel electrofocusing was developed as an analytical procedure based on the same principle of fractionation involved in preparative electrofocusing. But it requires the use of a much simpler apparatus, smaller samples, and smaller amounts of expensive carrier ampholytes, and it is capable of analyzing a number of samples simultaneously in 3 hours or less. Gel electrofocusing provides the means of fractionating proteins according to isoelectric point in such applications as the rapid assessment of purity during a multistage isolation procedure, the survey of protein composition in a large number of samples, or establishment of the feasibility and best conditions for performing preparative isoelectric focusing.

8.3.1   *Principle*

Figure 8 shows the diagram of a disc electrophoresis apparatus set up for gel electrofocusing. The polyacrylamide gel, set in the running tubes, contains carrier ampholytes throughout the sample either throughout or layered on top. To protect the carrier ampholyte from anodic oxidation and cathodic reduction, the electrodes are surrounded by acid and base. During electrolysis the acid remains in the anode vessel due to the attraction of sulfate ions to the anode. The acid above the gel gives a positive charge to the carrier ampholytes, which are repelled from the anode and thus are confined to the tube. Conversely, at the cathode the ampholytes in contact with the base are negatively charged and are therefore repelled from the cathode. Continued passage of current arranges the carrier ampholytes down the tube in order of their isoelectric points and focuses sample ampholytes to the region of their isoelectric pH-values. The gels removed from the run-

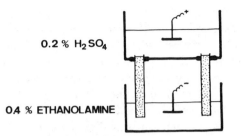

0.2 % H₂SO₄

0.4 % ETHANOLAMINE

Figure 8. Diagram of apparatus for column gel electrofocusing.

ning tubes can then be analyzed using the wide range of procedures developed for handling electrophoresis gels provided care is taken to avoid interference from carrier ampholytes.

Table II
Proteins Fractionated by Gel Electrofocusing

Protein	Gel composition	Reference
Rabbit immunoglobulins	5% acrylamide slab, 230 × 150 × 1 mm	Awdeh *et al.*, 79, 127, Awdeh and Williamson, p. 433 in 4
Soybean whey proteins	5% acrylamide column, 65 × 5 mm	Catsimpoolas, 80
Rabbit serum proteins	5% acrylamide column, 65 × 5 mm	Catsimpoolas, 81–83
Rabbit serum proteins	1% agarose, 2 mm on microscope slide	Catsimpoolas, 84
Human serum proteins	6.5% acrylamide column, 70 × 2–4 mm	Dale and Latner, 85, 86, p. 427 in 4
Hemoglobins, myoglobins	8% acrylamide column, 150 × 3–6 mm	Fawcett, 87, p. 409 in 4
Human serum proteins	4.5% acrylamide column, 70 × 3.5 mm	Kenrick and Margolis, 88
Human serum proteins	3–5% acrylamide slab, 140 × 80 × 2 mm	Leaback and Rutter, 89 Leaback *et al.*, p. 423 in 4
Hemoglobin	7% acrylamide column, 65 × 4 mm	Riley and Coleman, 78
Serum proteins	1.5% agarose on microscope slide	Riley and Coleman, 78
Wheat grain proteins	7.5% acrylamide column, 65 × 5 mm	Wrigley, 90–94 and p. 417 in 4 Wrigley and Moss, 95

During 1968, reports appeared describing gel electrofocusing in various forms as developed independently and almost simultaneously by a number of researchers. The various accounts of the technique differed mainly in the shape of the gel and electrode solutions. The contents of articles on gel electrofocusing are summarized in Table II. In most articles the use of columns of gel is described. The alternative is a flat slab of gel. In general, the apparatus for gel columns is simpler in makeup and easier to use; however the use of slabs offers the advantage that samples can be run side by side for close comparison. Furthermore, the whole length of the gel is accessible for the incorporation of sample and carrier ampholytes after gel polymerization. The procedure for gel electrofocusing in 65 × 5-mm columns of polyacrylamide is described below. The formulas and most experimental details are also applicable to the use of gel slabs.

8.3.2 Apparatus

8.3.2.1 Gel electrofocusing apparatus

Column gel electrofocusing can be performed in apparatus similar to that described by Davis[99] for disc electrophoresis. Many versions of this apparatus are available commercially[78, 80, 92, 93] (see Figure 9). For certain applications the capacity of the gel apparatus to take double-length tubes (110 mm) is an advantage. An apparatus similar to that used by Fawcett[87] is also manufactured by Buchler Instruments, Inc., Fort Lee, New Jersey.

8.3.2.2 Power supply

A power supply having a range of 0–400 V, 0–30 mA is adequate in most cases for gel electrofocusing. For removal of carrier ampholytes by electrophoresis, much higher currents are required: longitudinal electrophoresis, 0–100 V, 0–100 mA; transverse electrophoresis, 0–20 V, 0–500 mA.

8.3.2.3 Gel sectioning

Various devices for cutting transverse sections from gel columns have been described.[96, 97, 98] A convenient device is a set of razor blades held on two threaded rods and separated from one another by groups of 1-mm thick washers.

8.3.2.4 pH Measurement

The pH can be measured in aqueous extracts of gel sections, or directly by applying a miniature glass electrode to the gel surface with the reference electrode nearby.

Figure 9. Commercial apparatus for gel electrofocusing, illustrating spacer ring for raising upper vessel to accommodate double-length tubes (see Section 8.3.4.8). (Reproduced by courtesy of Shandon Scientific Company, Willesden, London.)

8.3.3 Procedure

Many details of the procedure are similar to those employed in disc electrophoresis, described fully by Davis.[99] However preparation for gel electrofocusing is quicker and less exacting. Multiple gel layers are not involved and a perfectly flat top to the gel is not essential for good resolution.

8.3.3.1 Gel formulas

The volumes below are sufficient for one tube (65×5 mm). Immediately before use, mix enough for the number of tubes required.

For Chemical Polymerization

Water	0.8 ml
Acrylamide solution	0.3 ml
Carrier ampholytes (40%)	0.03 ml

Potassium persulfate
    (10 mg/ml, fresh solution)   0.07 ml
Protein sample   30–300 μg

Acrylamide Solution (used above)

N,N'-methylene bisacrylamide   1   g
Acrylamide   30   g
Water   to 100   ml

For photopolymerization, replace the potassium persulfate solution in the above gel formula by an equal volume of riboflavin solution (0.15 mg/ml).

If there is difficulty[78, 89, 91] in setting photopolymerized gels containing ampholytes with an average pH above 7, add sufficient dilute sulfuric acid to reduce the pH to about 7. (The sample should be applied to gels set in this manner either by incorporation in the gel or by application to the alkaline end, not the acid end.)

8.3.3.2 Polymerization

Close off the lower end of each gel tube and fill with gel mixture to within about 5 mm of the top. Layer about 3 mm water on top using a fine-tipped pipet. Chemically polymerized gels become set 20–30 minutes after mixing at 20–25°C. Exposure close to a strong light for about 30 minutes is needed for photopolymerization. In either case, polymerization can be delayed by cooling the gel mixture. Before filling the tubes dissolved air should be removed from the gel solution by applying a vacuum to the gel solution flask with gentle shaking. This precaution is to prevent the appearance of bubbles in the gel during electrofocusing.

8.3.3.3 Sample application

The sample can be incorporated throughout the gel or applied to the top in the manner used for disc electrophoresis. When the former procedure is used, allowance should be made for the volume of the sample by reducing the amount of water used to make the gel. The alternative procedure of zone layering on top of the gel is performed after setting up the apparatus or even after establishing the pH gradient by preliminary electrolysis. The top of each gel tube is filled with 1% carrier ampholyte in 5% sucrose. Electrode solution is carefully poured into the apparatus containing the gel tubes and the sample (in 10% sucrose) is introduced onto the top of the gel, under the protecting 5% sucrose solution, using a microsyringe or fine-tipped pipet.

8.3.3.4 Electrolysis

Mount gel tubes in the apparatus. Fill the anode vessel (generally on top) with 0.2% sulfuric (or phosphoric) acid and the cathode vessel with 0.4% ethanolamine (or ethylenediamine). Turn on the current, adjusted to 2 mA per gel. As the conductivity drops, the voltage can be increased maintaining the same current to a maximum of 400 V.

At the end of the run (generally 1–3 hours), turn off the current and remove gels from their tubes by rimming between the gel and inner wall using a fine hypodermic needle with water running through it.

8.3.3.5 Determination of pH gradient

Cut up one of the gels from the analysis into 10 or 20 equal lengths. Soak each in 1–2 ml water and measure the pH of the extracts after several hours.

8.3.3.6 Fixation of protein in the gel

For total protein staining, fix protein zones by immersing gels in 5% trichloroacetic acid (TCA). For many proteins this is sufficient for detection since zones show very quickly as white precipitation bands. However if protein staining is desired, carrier ampholytes must first be removed because they stain strongly with most protein stains.

8.3.3.7 Removal of carrier ampholytes

Wash ampholytes from gels by agitating occasionally in at least five changes of 5% TCA (10 ml per gel) at 1–2-hour intervals. More efficient removal is obtained by electrophoresis of the gels in 2.5% TCA either longitudinally or transversely. In the former case, gels are placed in slightly oversize glass tubes constricted at the lower end to retain the gel. A current of 10 mA per tube is passed for at least 20 hours. At the low pH of electrophoresis the carrier ampholytes are all positively charged and migrate toward the cathode (lower vessel). Faster removal at high currents (70 mA per gel) can be obtained by transverse electrophoresis in an apparatus of the type described by Schwabe[101] and Prusik[102] or in commercially available equipment.[80, 92] For easily fixed proteins, acetic acid–ethanol–water (1:5:4)[78] or 7% acetic acid[80] are suitable for ampholyte removal, but the possibility of removing proteins together with carrier ampholytes should be checked.

8.3.3.8 Protein staining

Gels, with carrier ampholytes removed, are stained for 1 hour in Amido Black (1% in 7% acetic acid) and destained by electrophoresis or by washing in 7% acetic acid. Direct staining without removal of carrier ampholytes can be obtained[78] using the dyes Fast Green FCF (C.I. 42053) or Light Green SF (C.I. 670), but the color yield has been found to be about 1/20 of that derived from using Amido Black. The requirement of preliminary ampholyte removal is less stringent with Coomassie Blue[78, 79, 87, 103] (0.05% in 10% TCA) than with Amido Black. However Coomassie Blue gives only about 1/5 of the color yield obtained with Amido Black.

8.3.3.9 Recording of gel electrofocusing patterns

Gels fixed in TCA are photographed against a black background using side lighting. Stained gels should be illuminated by transmitted light. A technique described recently[104] for UV photography of patterns may be useful in overcoming the problem of ampholyte interference. If gels are scanned directly by ultraviolet light (280 nm) in a densitometer,[105] a blank scan should be performed on a control gel. For densitometry of stained gels using visible light a blank scan need not be performed.

8.3.4 *Design of the Experiment*

Certain aspects of experimental design for density gradient electrofocusing also apply to gel electrofocusing, *e.g.*, choice of temperature and the concentration and range of carrier ampholytes. Other points specific to gel electrofocusing must also be considered.

8.3.4.1 Application of sample

The amount of sample to be loaded depends on the number of separable components and on the sensitivity of the detection procedures used. For a sample containing about ten components, a loading of 100–200 μg should be satisfactory for detection by TCA fixation or staining. Unlike sucrose gradients, the gel medium does not restrict the upper limit of loading. Urea may be incorporated in the gel to maintain the solubility of otherwise insoluble samples (see Figures 10 and 14 below).

Very low concentrations of protein (down to 0.003%) can be accommodated by setting the protein in the gel. In some cases resolution is slightly better if the sample is applied on top of the gel but in this case preliminary concentration of the sample may be necessary.

8.3.4.2 Polymerization of acrylamide

Chemical polymerization is a more reliable procedure than photopolymerization, but it has been identified as a source of artifact formation in certain applications of gel electrophoresis.[106–109] In most cases reported, the anomalous behavior could be avoided by using gels polymerized with riboflavin instead of persulfate. Alternatively, residual persulfate can be removed by electrophoresis before sample application. At 25–50 V,[110] this takes about 30 minutes for 40 × 5-mm gels at pH 4 or 8.

The formation of a large number of spurious bands during the gel electrofocusing of hemoglobin applied to a chemically polymerized gel has been reported by Wrigley.[4] In this case the anomalous behavior was not prevented by preliminary electrolysis, even for periods of up to 80 minutes. This observation suggests that preelectrolysis might not be an adequate precaution in the case of chemically polymerized gels used for electrofocusing. For this reason, photopolymerization is recommended whenever possible.

8.3.4.3 Nature of gel

Polyacrylamide gel has been used in nearly all articles describing gel electrofocusing (Table II). It is preferred to other media because it has very few charged groups over the pH range 3–10 and because of its low electroendosmosis (1–2 cm in 16 hours[89] for gel slab electrofocusing). Despite the slow shift of the pH gradient toward the cathode[87,92] its stability is adequate for the time usually required for gel electrofocusing.

The pore size should be chosen so that the focusing of zones is not unduly restricted by the gel. An acrylamide concentration of 7.5% is a satisfactory compromise between providing adequate gel strength for handling, on the one hand, and allowing the movement of proteins with molecular weight of up to about 100,000, on the other. For larger proteins, lower gel concentrations should be used, strengthened if necessary by the incorporation of agarose.[111,112] Agarose by itself has proved useful for gel electrofocusing preceding immunodiffusion.[78,84]

8.3.4.4 Electrode solutions and polarity

Figure 10 shows that for various combinations of acids and bases as electrode solutions quite similar patterns are obtained for a heterogeneous protein mixture. Reversal of polarity is also shown to have little effect apart from simply inverting the pattern.

8.3.4.5 Time

Electrical resistance in the gels increases during development of the pH gradient and becomes constant when the gradient is established. This takes about 15 minutes for the range pH 3–10 and a little longer for narrow pH ranges. Gradients of pH after 10 and 30 minutes of electrolysis are shown in Figure 11.

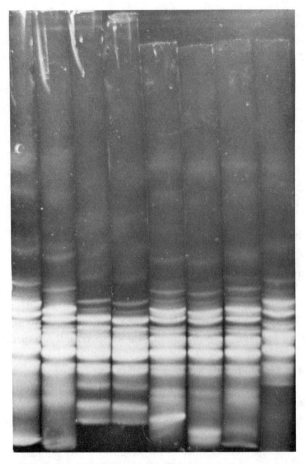

Figure 10. Use of various electrode solutions for gel
electrofocusing of gluten proteins extracted from wheat
flour. Sample (300 μg) was set in the gel. Ampholyte
range, pH 3 (top) to 10. Counting from the left, the following solutions were used for upper and lower electrode
vessels, respectively: gels 1 and 2, 0.2% sulfuric acid and
0.4% ethanolamine; 3 and 4, 0.4% ethanolamine and
0.2% sulfuric acid, gels inverted; 5 and 6, 0.2% ortho-
phosphoric acid and 0.4% ethylenediamine; 7 and 8,
0.04N sulfuric acid and 0.08N sodium hydroxide (from
Wrigley[92]).

Focusing of proteins takes considerably longer in general than the
establishment of the pH gradient. Factors involved include pH range
and length of the gel, method of application, voltage gradient, charge
of components away from their pI-values, and molecular size of the
components. The time required for the sample under consideration is
best determined by experiment.

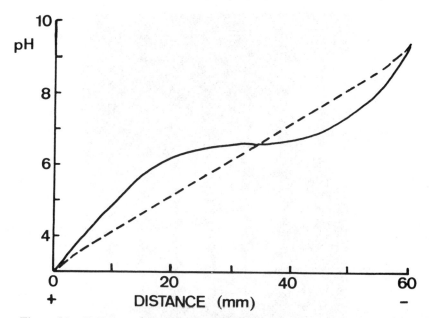

Figure 11. Variation of pH of gel with distance from top (anode) end after electrofocusing for 10 minutes (solid line) and 30 minutes (broken line). The gel contained pH 3–10 ampholytes (from Wrigley[92].).

Electrofocusing of a sample applied as a zone is faster than focusing of the protein set in the whole gel (20 minutes compared with about 45 minutes for ovalbumin or hemoglobin).

Figure 12 illustrates the progressive focusing of components of a heterogeneous protein mixture applied to the top of a gel containing prefocused pH 3–10 carrier ampholytes. Electrofocusing is complete after about 50 minutes but the pattern is virtually unchanged after the electrolysis has been continued for considerably longer. By way of comparison, about 75 minutes are required for focusing the same mixture using a narrower pH range (5–8).

It is important to ensure that electrofocusing is continued to completion. In the case shown in Figure 11, a distorted view of heterogeneity would be obtained if gels were routinely fixed after only about 30 minutes. Furthermore, there would be difficulty in obtaining reproducible patterns. A similar point can be illustrated in the case of serum albumin.[91] The monomer focuses in a pH 3–10 gel in 15–20 minutes but the polymers need about 40 minutes to focus from the top of a 7.5% polyacrylamide gel to the same region as the monomer. Fixation of such a gel after less than 40 minutes would reveal apparent heterogeneity of pI due in fact to size differences. It is therefore important to establish experimentally that zones are properly focused.

Figure 12. Gel electrofocusing of wheat flour proteins soluble in neutral sodium pyrophosphate. Ampholytes (pH 3–10) were prefocused for 30 minutes before sample was applied to the top of the gel (anode end). After electrofocusing for 15, 30, 45, 60, 75, 90, 120 and 180 minutes, gels were fixed in trichloroacetic acid (from Wrigley[92]).

8.3.4.6   Detection of components

The popularity of gel electrophoresis has led to the development of techniques for specific detection and quantitation in the gel of proteins, glycoproteins, lipoproteins, nucleoproteins, and a large number of enzymes.[113, 114] The procedures can be applied to gel electrofocusing provided possible interference from carrier ampholytes is avoided. In detecting certain enzymes, addition of excess metal ion cofactor may be necessary due to the tendency of carrier ampholytes to form metal ion complexes. Immunological detection of proteins following gel electrofocusing (immunoelectrofocusing) has been described by Riley and Coleman[78] and by Catsimpoolas.[81-84] Electrofocusing for this purpose is usually performed in agarose gels. The equipment and many details of the procedure are similar to immunoelectrophoresis.

8.3.4.7 Comparison of patterns

The difficulty of precisely aligning corresponding zones in patterns for different samples on separate gel columns can be largely overcome by using the duplex tube illustrated in Figure 13. It consists of a single tube having a double compartment at the top for zone application of samples to either half of the tube. In the tube illustrated, the dividing wall is of glass fused into the tube. However it may be stiff plastic

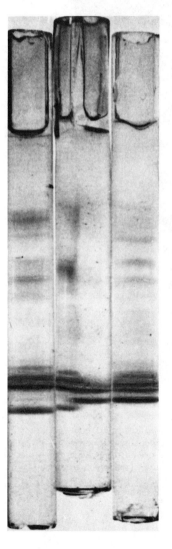

Figure 13. Patterns obtained for hemoglobin preparations (Sigma Chemical Company) from sheep (left) and ox immediately following gel electrofocusing for 40 minutes. Samples (150 μg) were applied to the tops of gels in single tubes or in the respective compartments of the duplex tube (center) before commencing electrolysis. Gradient, pH 3 (top) to 10 (from Wrigley[4]).

cemented or wedged in place. Figures 13 and 14 illustrate, respectively, the use of duplex tubes in aligning hemoglobin components of similar isoelectric points and in comparing the gliadin protein composition of different wheat varieties. The technique is also useful in comparing the pattern of a somewhat purified protein sample with that of the original crude extract.

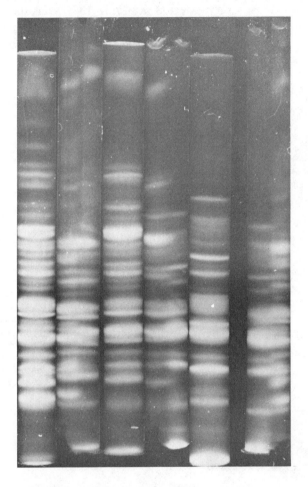

Figure 14. Use of single and duplex tubes to compare patterns obtained by gel electrofocusing of gluten proteins (400 μg) extracted with 2M urea from wheats of the varieties Sonora (gel 1), Glaive (gel 3), and Dural (gel 5). Gels 2, 4, and 6 are duplex patterns representing Sonora-Glaive, Glaive-Dural, and Dural-Sonora, respectively. Gels were fixed in 5% TCA. Range of pH 5 (top) to 9. Gels contained 2M urea (from Wrigley[4]).

8.3.4.8 Long gel columns

Tubes 65 mm long have proved a convenient size for routine gel electrophoresis and electrofocusing. There is a little advantage in using longer tubes for gel electrophoresis because of the increased diffusion of zones during extended electrophoresis. However this factor is not a problem with gel electrofocusing, especially if a similar voltage gradient is used for longer gels. Double-length gels (110 mm) have advantages when the results of gel electrofocusing are to be assessed by densitometry. In this case, good separation of zones is needed for satisfactory quantitation. Resolution of zones in a double-length tube is illustrated by the low absorbance minima between peaks in the densitometric scan in Figure 15.

8.3.5 *Results and Applications*

Table II shows that colored proteins such as hemoglobin and myoglobin have been popular for demonstrating gel electrofocusing. Proteins of this type or amphoteric dyes are useful in preliminary experiments with the technique. Fractionation of some suitable dyes is shown in Figure 16.

Fractionation of serum proteins by gel electrofocusing has received considerable attention. Awdeh *et al.*[79] and Leaback and Rutter[89] have resolved immunoglobulin G into a complex pattern of discrete components with pI-values in the range 5–8. Gel electrophoresis of this mixture, on the other hand, reveals a continuous spectrum of mobility with no band formation. The sequence of events during gel electrofocusing of serum proteins has been examined[88] by combining gel electrofocusing and gel electrophoresis. Examples of isoenzyme analysis by gel electrofocusing are given, using specific staining for serum lactic dehydrogenases[85] and erythrocyte acid phosphatases.[89] The potential value of gel electrofocusing for biochemical genetics and taxonomy is suggested by the results of protein analysis of wheat genotypes[95] (see Figure 14). The possibility that heterogeneity revealed by gel electrofocusing may be due to complex formation between carrier ampholytes and a homogeneous protein has been examined by cutting individual zones from an unfixed gel and rerunning each section in a separate electrofocusing gel. Components examined in this manner refocused to the pH region from which each was originally taken (see References 87 and 89, and Wrigley in Reference 4). The only exception is the formation of multiple zones from hemoglobin under certain conditions, but this anomaly appears to be an artifact of gel polymerization (see Section 8.3.4.2). Complex formation thus appears unlikely but is not excluded.

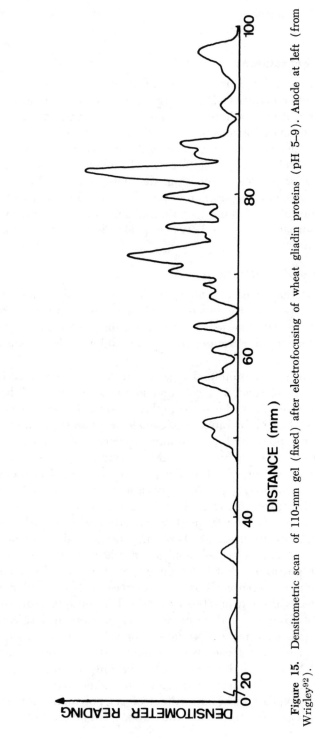

DISTANCE (mm)

DENSITOMETER READING

Figure 15. Densitometric scan of 110-mm gel (fixed) after electrofocusing of wheat gliadin proteins (pH 5–9). Anode at left (from Wrigley[92]).

330

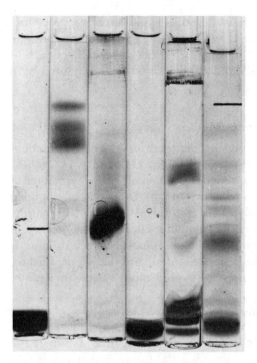

Figure 16. Gel electrofocusing of amphoteric dyes, 15 minutes after applying 0.1 ml of 0.2% dye solution to the top (cathode end) of the gel. Range, pH 10 (top) to 3. Left to right: Acid Fuchsin (BDH), Hematein (BDH), Erie Garnet (E. Gurr), Light Green (G. Gurr), Carmine (BDH), and Orcein (BDH).

Gel electrofocusing provides a rapid estimate of isoelectric point but cannot be expected to show the accuracy of density gradient electrofocusing. Gel electrofocusing has mainly been used with carrier ampholytes of wide pH range. Use of narrow pH ranges provides increased resolution in a selected region of the isoelectric spectrum in the way already discussed for density gradient electrofocusing (Section 8.2.5.3). Gel electrofocusing on an analytical scale is thus useful in assessing the feasibility of density gradient electrofocusing and in choosing suitable conditions for the use of preparative columns. However it can be expected that the relative separation of zones will be more satisfactory using gel electrofocusing than will separation following elution of the preparative column, due to the loss of definition generally encountered with elution analysis (see Section 8.2.5.10).

The high resolving power of gel electrofocusing can be used for preparative fractionation by eluting protein from gel sections. Components are located in the gel by comparison with colored markers, by staining a strip cut from the gel (for the slab technique), or by measuring the pH on the gel surface with a microglass electrode (see Section 8.3.3.5). However, recovery of protein from gel is generally rather difficult and only limited quantities can be handled in this way.

Leaback and Rutter[89] have briefly described an alternative procedure for preparative gel electrofocusing which involves about 30 mg protein. Sample is applied in a slot to a gel slab. After focusing, each component becomes carried by endosmosis to another slot where it is eluted by a cross flow of liquid.

8.3.6 Electrofocusing on Paper Media

Paper has been used as supporting material for isoelectric fractionation in artificial pH gradients[115,116] but the procedure has not gained wide acceptance, due partly to difficulties inherent in the formation of artificial pH gradients. More satisfactory pH gradients can be obtained using synthetic carrier ampholytes for natural pH gradient formation. Compared with gel, paper offers the advantages of more convenience in preparation and use. In addition, ampholyte removal before staining is quicker because protein can be heat-denatured onto the paper; furthermore, fractionated proteins can be recovered more easily from paper than from gel.

Satisfactory staining of proteins on paper in the presence of carrier ampholyte has been obtained (St. Clair and Wrigley ,unpublished) with an aqueous solution containing 0.05% Bromophenol Blue, 1% mercuric chloride, and 2% acetic acid[117] provided the oven-dried paper is first immersed in hot 10% TCA. Color can be intensified by exposing the dry paper to ammonia after staining.

Endosmosis due to the charged groups present on most paper supports tends to make the pH gradient unstable especially at alkaline pH-values. This difficulty has been largely overcome for Whatman No. 1 paper by including 0.01M sodium chloride with the 1.2% carrier ampholyte solution used to wet the paper and by making direct contact between paper and electrodes, using carbon rods soaked in either 0.2% sulfuric acid or 0.4% ethanolamine. Incorporation of sodium chloride is unnecessary with Rhovyl PVC paper (No. 1001, Schleicher and Schüll) since this paper does not produce endosmosis.[118]

The pH gradient was reasonably stable for about 2 hours using 15-cm strips of either of the above papers and a voltage gradient of about 20 V/cm. Hemoglobin focused to the region of its isoelectric point on paper but was not separated into individual components. In these studies (St. Clair and Wrigley, unpublished), paper electrofocusing gives poorer resolution of proteins and amphoteric dyes than gel electrofocusing. However, the procedure might prove useful for fractionation of some compounds of medium or low molecular weight, or as one of the two dimensions used for peptide mapping.

8.4 Electrofocusing and Other Fractionation Procedures

8.4.1 *The Basis of Fractionation by Electrofocusing*

Fractionation by electrofocusing appears to depend solely on the isoelectric point. Most other procedures, on the other hand, fractionate on the basis of one or several of other properties such as solubility, size, shape, charge. (Exceptions sometimes include pH gradient elution of ion exchangers[119] and electrodecantation.[120]) Since a different property of proteins is involved in separation by electrofocusing, it might be expected that the sequence of separating components from a given mixture would differ from that obtained with other procedures. This has been so in some cases[30, 88, 94] but not for a mixture of similar proteins such as isoenzymes.[25, 26, 32]

8.4.2 *Combined Gel Electrofocusing and Gel Electrophoresis*

The combination of gel electrophoresis and gel electrofocusing has provided very good two-dimensional analyses of protein composition in several cases. The large number of components present in blood serum makes analysis by a single procedure inadequate for proper identification of all of them. These proteins have been resolved in great detail by first electrofocusing in a rod of polyacrylamide and then by electrophoresis from the electrofocusing gel into a slab of polyacrylamide gel.[86, 88, 91] The resulting two-dimensional array of components (Figure 17) is useful in diagnosing disease states. This technique, also referred to as "protein mapping," has been applied to the fractionation of proteins from potato tuber.[122]

Figure 18 illustrates another example of this type of two-dimensional analysis. Gliadin proteins are fractionated into about 20 components by either gel electrophoresis or gel electrofocusing. Nearly twice as many components are resolved by combining the two procedures.[94]

8.4.3 *Advantages of Electrofocusing*

The efficiency of most procedures for fractionation of a solute mixture depends on the sharpness of the starting zone because as fractionation proceeds, definition of zones is reduced by diffusional spreading. However, electrofocusing offers the advantage that zones become sharpened during fractionation since the forces producing separation act against spreading due to diffusion.

In many instances electrofocusing, in a density gradient or in a gel, has revealed a greater degree of heterogeneity than has been shown

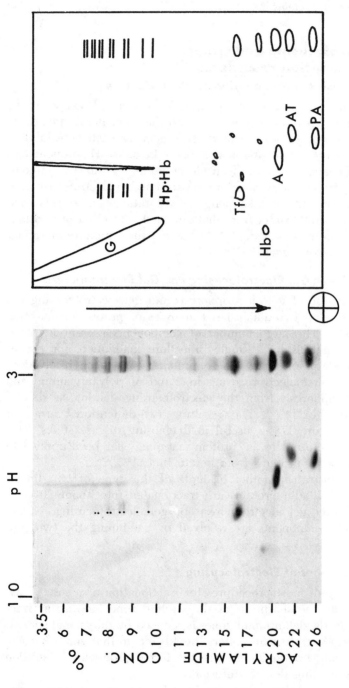

Figure 17. Two-dimensional fractionation of a mixture of human plasma proteins. First dimension, gel electrofocusing for 120 minutes in a pH 3–10 gradient (4.5% polyacrylamide). Second dimension, gradient gel electrophoresis[121] for 20 hours at pH 8.3 in a slab of gel in which the concentration of polyacrylamide increases from 3.5 to 26%. (G, gammaglobulin, 50 μg applied; Hp·Hb, polymer series of haptoglobin-hemoglobin complex, 50 μg; Tf, transferrin, 25 μg; Hb, hemoglobin, 25 μg; AT, α_1-antitrypsin, 25 μg; PA, prealbumin, 25 μg.) The pattern obtained by gel electrophoresis alone appears on the right. (By courtesy of Kenrick and Margolis, Children's Medical Research Foundation, Sydney, Australia.)

334

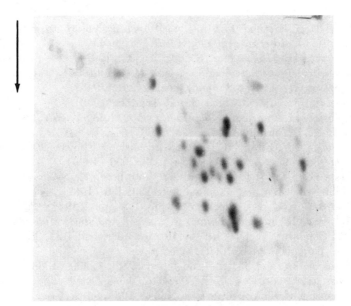

Figure 18. Two-dimensional fractionation of wheat gliadin proteins. First dimension, gel electrofocusing of 1 mg protein in a 5 × 110-mm gel for 4 hours at 600 V, pH range, 5 (left) to 9. Second dimension, starch gel electrophoresis in aluminum lactate, pH 3.2.

previously (*e.g.*, References 26, 39, 45). The resolving power of electrofocusing is derived partly from zone sharpening, as mentioned above, but also from the fact that any part of an isoelectric spectrum can be expanded to cover the full length of the separation column by use of a shallow pH gradient in the pertinent range. This property permits a high degree of purification in a single fractionation step.

REFERENCES

1. Haglund, H.   Sci. Tools **14**, 17 (1967).
2. Vesterberg, O.   Svensk. Kem. Tidskr. **80**, 213 (1968).
3. Peeters, H. (ed.).   *Protides Biol. Fluids, Proc. 16th Colloq., 1968* (London: Pergamon Press, 1969).
4. Peeters, H. (ed.).   *Protides Biol. Fluids, Proc. 17th Colloq., 1969* (London: Pergamon Press, 1970).
5. Edsall, J. T., and J. Wyman.   *Biophysical Chemistry*, Vol. 1, (New York: Academic Press, 1958).
6. Bier, M. (ed.).   *Electrophoresis: Theory, Methods, and Applications* (New York: Academic Press, 1957).
7. Florkin, M., and E. H. Stotz.   *Comprehensive Biochemistry*, Vol. 7, Part I (Amsterdam: Elsevier, 1963), pp 23, 113.

8. Kolin, A., in *Methods of Biochemical Analysis*, Vol. 6, Ed. by D. Glick. (New York: Interscience, 1958), 259.
9. Svensson, H.   Acta Chem. Scand. **15**, 325 (1961).
10. Kolin, A.   Proc. Nat. Acad. Sci. U.S. **41**, 101 (1955).
11. Svensson, H.   Advan. Protein Chem. **4**, 251 (1948).
12. Svensson, H.   Acta Chem. Scand. **16**, 456 (1962).
13. Svensson, H., in *Protides Biol. Fluids, Proc. 15th Colloq., 1967.* Ed. by H. Peeters (Amsterdam: Elsevier, 1968), 515.
    Colloq., 1967. Ed. by H. Peeters (Amsterdam: Elsevier, 1968), p 515.
14. Vesterberg, O.   Acta Chem. Scand. **23**, 2653 (1969).
15. Vesterberg, O., T. Wadstrom, K. Vesterberg, H. Svensson, and B. Malmgren.   Biochim. Biophys. Acta **133**, 435 (1967).
16. Svensson, H.   Arch. Biochem. Biophys., Suppl. 1 **1962**, 132.
17. Svensson, H., in *A Laboratory Manual of Analytical Methods of Protein Chemistry*, Vol. 1, Ed. by P. Alexander, and R. J. Block (London: Pergamon Press, 1960), 193.
18. Svensson, H., and S. Pettersson.   Separ. Sci. **3**, 209 (1968).
19. Jonsson, M., and E. Pettersson.   Acta Chem. Scand. **22**, 712 (1968).
20. Dempster, K. T.   Spectrovision No. 18 **1968**, 5.
21. Valmet, E.   Sci. Tools **15**, 8 (1968).
22. Rust, P.   Anal. Biochem. **17**, 316 (1966).
23. Tappan, D. V.   Anal. Biochem. **18**, 392 (1967).
24. Vesterberg, O., and H. Svensson.   Acta Chem. Scand. **20**, 820 (1966).
25. Flatmark, T., and O. Vesterberg.   Acta Chem. Scand. **20**, 1497 (1966).
26. Carlstrom, A., and O. Vesterberg.   Acta Chem. Scand. **21**, 271 (1966).
27. Pettersson, U., L. Philipson, and S. Hoglund.   Virology **35**, 204 (1968).
28. Wadstrom, T.   Biochim. Biophys. Acta **168**, 228 (1968).
29. Jeppsson, J-O.   Biochim. Biophys. Acta **140**, 468 (1967).
30. Brown, D. G., C. Baron, and A. Abrams.   Biochim. Biophys. Acta **168**, 386 (1968).
31. Bjork, I.   Acta Chem. Scand. **22**, 1355 (1968).
32. Bloemendal, H., and J. G. G. Shoenmakers.   Sci. Tools **15**, 6 (1968).
33. Schoemakers, J. G. G., and H. Bloemendal.   Nature **220**, 790 (1968).
34. Grasbeck, R.   Acta Chem. Scand. **22**, 1041 (1968).
35. Vesterberg, O., and B. Berggren.   Arkiv. Kemi. **27**, 119 (1967).
36. Ahlgren, E., K-E. Eriksson, and O. Vesterberg.   Acta Chem. Scand. **21**, 937 (1967).
37. Mach, O., and L. Lacko.   Anal. Biochem. **22**, 393 (1968).
38. Kamat, V. B., and D. F. H. Wallach.   Science **148**, 1343 (1965).
39. Quast, R., and O. Vesterberg.   Acta Chem. Scand. **22**, 1499 (1968).
40. Wenn, R. V., and J. Williams.   Biochem. J. **108**, 69 (1968).
41. Lowry, O. H., N. J. Rosebrough, A. L. Farr, and R. J. Randall.   J. Biol. Chem. **193**, 265 (1951).
42. Gornall, A. G., C. J. Bardawill, and M. M. David.   J. Biol. Chem. **177**, 751 (1949).

43. Goa, J.   Scand. J. Clin. Lab. Invest. **5**, 218 (1953).
44. Earland, C., and D. B. Ramsden.   J. Chromatog. **35**, 575 (1968).
45. Vesterberg, O.   Acta Chem. Scand. **21**, 206 (1967).
46. Paleus, S., and O. Vesterberg.   *Intern. Symp. Comparative Hemoglobin Structure* (Thessaloniki, Greece: M. Triantofylo Sons Publishers, 1966), p 149.
47. Grasbeck, R., in *Protides Biol. Fluids, Proc. 16th Colloq., 1968,* Ed. by H. Peeters (London: Pergamon Press, 1969), p 401.
48. Grasbeck, R., and K. Visuri.   Scand. J. Clin. Lab. Invest. **101**, 13 (1968).
49. Flodh, H., B. Bergrahm, and B. Oden.   Life Sci. **7**, 155 (1968).
50. Aares, E., K. Closs, and V. Lehmann.   Sci. Tools **15**, 36 (1968).
51. Wadstrom, T.   Biochim. Biophys. Acta **147**, 441 (1967).
52. Vesterberg, O.   Biochim. Biophys. Acta **168**, 218 (1968).
53. Jonsson, M., E. Pettersson, and B. Reinhammer.   Acta Chem. Scand. **22**, 2135 (1968).
54. Ahlgren, E., and K-E. Eriksson.   Acta Chem. Scand. **21**, 1193 (1967).
55. Bucht, B., and K-E. Eriksson.   Arch. Biochem. Biophys. **124**, 135 (1968).
56. Li, Y-T., and S-C. Li.   J. Biol. Chem. **243**, 3994 (1968).
57. Sandhoff, K.   Hoppe-Seylers Z. Physiol. Chem. **349**, 1095 (1968).
58. Bernheimer, A. W., P. Grushoff, and L. S. Avigad.   J. Bacteriol. **95**, 2439 (1968).
59. Allen, E., and A. B. Roy.   Biochim. Biophys. Acta **168**, 243 (1968).
60. Weller, D. L., A. Heaney, and R. E. Sjogren.   Biochim. Biophys. Acta **168**, 576 (1968).
61. Mitchell, W. M.   Biochim. Biophys. Acta **178**, 194 (1969).
62. Rebeyrot, P., and J. P. Labbe.   Compt. Rend., Ser. D **268**, 1125 (1969).
63. Berndt, W., U. Hoffman, and K. Muller-Wieland.   Z. Gastroenterol. **6**, 185 (1968).
64. Catsimpoolas, N.   Arch. Biochem. Biophys. **131**, 185 (1969).
65. Van Baelen, H., W. Heyns, and P. DeMoor.   Ann. Endocrinol. (Paris) **30**, 199 (1969).
66. Kaplan, L. J., and J. F. Foster.   Federation Proc. **28**, 865 (1969).
67. Reis, H. E., and O. Wetter.   Klin. Wochschr. **47**, 426 (1969).
68. Pechet, L., and J. A. Smith.   Federation Proc. **28**, 828 (1969).
69. Smith, M. B., and J. F. Back.   *Proc. Australian Biochem. Soc.,* 13th Meeting **1969**, 36.
70. Toom, P. M., P. G. Squire, and A. T. Tu.   Federation Proc. **28**, 903 (1969).
71. Weber, T., and R. Grasbeck.   Scand. J. Clin. Lab. Invest. **101**, 14 (1968).
72. Herschman, H. R., and D. R. Helinski.   J. Biol. Chem. **242**, 5360 (1967).
73. Jonsson, M., and E. Pettersson.   Sci. Tools **15**, 2 (1968).

74. Schultze, H. E., and J. F. Heremans. *Molecular Biology of Human Proteins*, Vol. 1 (Amsterdam: Elsevier, 1966), p 176.
75. Catsimpoolas, N., C. Ekenstam, and E. W. Meyer. Cereal Chem. **46**, 357 (1969).
76. Heins, J. N., J. R. Suriano, H. Taniuchi, and C. B. Anfinsen. J. Biol. Chem. **242**, 1016 (1967).
77. Catsimpoolas, N., C. Ekenstam, and E. W. Meyer. Biochim. Biophys. Acta **175**, 76 (1969).
78. Riley, R. F., and M. K. Coleman. J. Lab. Clin. Med. **72**, 714 (1968).
79. Awdeh, Z. L., A. R. Williamson, and B. A. Askonas. Nature **219**, 66 (1968).
80. Catsimpoolas, N. Anal. Biochem. **26**, 480 (1968).
81. Catsimpoolas, N. Biochim. Biophys. Acta **175**, 214 (1969).
82. Catsimpoolas, N. Immunochemistry **6**, 501 (1969).
83. Catsimpoolas, N. Sci. Tools **16**, 1 (1969).
84. Catsimpoolas, N. Clin. Chim. Acta **23**, 237 (1969).
85. Dale, G., and A. L. Latner. Lancet **1**, 847 (1968).
86. Dale, G., and A. L. Latner. Clin. Chim. Acta **24**, 61 (1969).
87. Fawcett, J. S. FEBS Letters **1**, 81 (1968).
88. Kenrick, K. G., and J. Margolis. Anal. Biochem. **33**, 204 (1970).
89. Leaback, D. H., and A. C. Rutter. Biochem. Biophys. Res. Commun. **32**, 447 (1968).
90. Wrigley, C. W. J. Chromatog. **36**, 362 (1968).
91. Wrigley, C. W. Sci. Tools **15**, 17 (1968).
92. Wrigley, C. W. Shandon Instrument Applications **29**, 1 (1969).
93. Wrigley, C. W. *Instructions for Performing Gel Electrofocusing* (Fort Lee, N.J.: Buchler Instruments, Inc., 1970).
94. Wrigley, C. W. Biochem. Genet. **4**, 509 (1970).
95. Wrigley, C. W., and H. J. Moss, in *Proc. Third Intern. Wheat Genetics Symp.* Ed. by K. W. Finlay and K. W. Shepherd (Sydney, Australia: Butterworth, 1968), p 439.
96. Chrambach, A. Anal. Biochem. **15**, 544 (1966).
97. Aronson, J. N., and D. P. Boris. Anal. Biochem. **18**, 27 (1967).
98. Gressel, J., and J. Wolowelsky. Anal. Biochem. **22**, 352 (1968).
99. Davis, B. J. Ann. N.Y. Acad. Sci. **121**, 404 (1964).
100. Huebner, F. R., J. A. Rothfus, and J. S. Wall. Cereal Chem. **44**, 221 (1967).
101. Schwabe, C. Anal. Biochem. **17**, 201 (1966).
102. Prusik, Z. J. Chromatog. **32**, 191 (1968).
103. Chrambach, A., R. A. Reisfeld, M. Wycoff, and J. Zacchari. Anal. Biochem. **20**, 150 (1967).
104. Luner, S. J. Anal. Biochem. **23**, 357 (1968).
105. Wieme, R. J. *Agar Gel Electrophoresis* (Amsterdam: Elsevier 1965), pp 80, 135.
106. Brewer, J. M. Science **156**, 256 (1967).
107. Mitchell, W. M. Biochim. Biophys. Acta **147**, 171 (1967).
108. Fantes, K. H., I. G. S. Furminger. Nature **215**, 750 (1967).

109. Schyns, R.   J. Chromatog. **36**, 549 (1968).
110. Bennick, A.   Anal. Biochem. **26**, 453 (1968).
111. Ringborg, U., E. Egyhasi, B. Daneholt, and B. Lambert.   Nature **220**, 1037 (1968).
112. Peacock, A. C., and C. W. Dingman.   Biochemistry **7**, 668 (1968).
113. Whitaker, J.   *Electrophoresis in Stabilizing Media* (New York: Academic Press, 1967).
114. Latner, A. L., and A. W. Skillen.   *Isoenzymes in Biology and Medicine* (London: Academic Press, 1968).
115. Hoch, H., and G. H. Barr.   Science **122**, 243 (1955).
116. McDonald, H. J., and M. B. Williamson.   Naturwiss. **42**, 461 (1955).
117. Kunkel, H. G., and A. Tiselius.   J. Gen. Physiol. **35**, 89 (1951).
118. Stegmann, H.   Naturwiss. **43**, 518 (1956).
119. Lampson, G. P., and A. A. Tytell.   Anal. Biochem. **11**, 374 (1965).
120. Polson, A., and J. F. Largier, in *A Laboratory Manual of Analytical Methods of Protein Chemistry*, Vol. 1, Ed. by P. Alexander and R. J. Block (London: Pergamon Press, 1960).
121. Margolis, J., and K. G. Kenrick.   Nature **221**, 1056 (1969).
122. Macko, V., and H. Stegeman.   Hoppe-Seylers Z. Physiol. Chem. **350**, 917 (1969).
123. Vesterberg, O.   Sci. Tools **16**, 24 (1969).
124. Blatt, W. F., and B. G. Hudson.   Anal. Biochem. **26**, 329 (1968).
125. Nilsson, P., T. Wadstrom, and O. Vesterberg.   Anal. Biochem. (in press).
126. Valmet, E.   Sci. Tools **16**, 8 (1969).
127. Awdeh, Z. L., A. R. Williamson, and B. A. Askonas.   Biochem. J. **116**, 241 (1970).
128. Haglund, H., in *Methods of Biochemical Analysis*, Ed. by D. Glick (New York: Interscience Publishers, Inc., in press).
129. Vesterberg, O., in *Methods in Microbiology*, Ed. by J. R. Norris and D. W. Ribbons (New York: Academic Press, in press).

Chapter 9

Chemical Accessibility and Environment of Amino Acid Residues in Native Proteins

by Kazuo Shibata

9.1 Amino Acid Residues in Native Proteins

Amino acid residues in proteins exist in a variety of states depending on their location and environment in the native protein structures. Some are buried in the interior of molecules and some are exposed on the surface; other local environmental conditions further differentiate their states. For example, hydrogen bonding between amino acid side chains changes their chemical and physicochemical properties, and the catalytic property of enzymes is generated as a result of concerted action of certain amino acid residues arranged in a specific manner in the active site. These environmental conditions naturally affect the chemical accessibility of amino acid residues.

Chemical accessibility and environment of amino acid residues as a subject has drawn much attention in recent years because of the following points of significance. (a) Knowledge concerning reactivity of each residue in the native structure in relation to environmental conditions will lead us to an understanding of various chemical properties of proteins and enzymes. For example, the mechanism of enzyme action may be described in terms of affinity and reactivity with the substrate of the residues in the active site. (b) Chemical accessibilities of amino acid residues are sensitive to conformational changes of proteins effected by pH, temperature, ionic strength, substrate, and so on. One may deduce the native protein structures from the accessibilities and their changes under various conditions. (c) The molar fractions of different states of residues are determined by physicochemical measurements of circular dichroism (optical rotatory dispersion), ionization characteristics, and spectral shifts by hydrogen bonding or some other changes in the environment. These measurements, however, do not afford information about the location of each state of residues in the primary structure—which residue in the structure belongs to what state. By labeling with a state-discrimination reagent followed by a common procedure of sequence determination, one can locate the modified residues in the primary structure. (d) One can study the environment of a specific amino acid residue in proteins by use of various types of bifunctional reagents.

9.2 Types of Reagents

The reagents for reactivity examination must have at least one functional group reactive with a specific amino acid. Such "monofunctional" reagents are expressed in this context as R—O, in which R is the reactive group and O represents the remaining moiety of reagent; the protein to be modified is expressed as X—P or X—P—Y, in which X or Y is an amino acid residue and P stands for the protein molecule lacking the X and/or Y residue.

Three types of "bifunctional" reagents also are reviewed in this chapter. One type, which may be expressed as R—O—R, has two reactive groups. Intermolecular or intramolecular linkages are formed when proteins are treated with this type of reagent. The second type has a single reactive group, R, and another group, A, having a specific affinity with a certain amino acid side chain. The affinity group is adsorbed on a specific site, such as the adsorption, specificity-determining and regulatory site of an enzyme, or on sites possessing certain other characteristics such as hydrophobic, basic or acidic, and negatively or positively charged sites or regions on proteins. The third type is expressed by S—O—A or S—O—R where S stands for a state-sensitive group. When this group is adsorbed on or linked chemically to a site of protein, it conveys information (through color or fluorescence change or ESR signal) about the environmental conditions around itself.

Strictly speaking, only very simple molecules are monofunctional. The moiety expressed by O may affect the reactivity of R or the affinity of A and may have some interaction with residues in protein, and group R or S in bifunctional reagents may act as an affinity group. This argument does not, however, kill the importance of the above simplification in classifying reagents.

9.3 Reactivity Examination with Monofunctional Reagent, Type R-O

For reactivity examination of amino acid residues in native proteins the reagent should satisfy the following conditions: (a) it reacts with the side chains of few amino acids, preferably of a single amino acid; (b) the reaction proceeds under the conditions where most proteins are native, *e.g.*, at room temperature and neutral pH; (c) the reactivity of the reagent is moderate in that it reacts with free or exposed residues but does not react with buried or bound residues; (d) the degree of reaction is determined by a simple technique such as absorption spectrophotometry, ESR spectrophotometry, fluorometry, radioactivity counting, etc.; and (e) the method of labeling the residues by the modification enables one to easily identify the labeled residues in the peptides obtained by digestion of the modified protein. When the residues are labeled or stained with radioactivity or fluorescence, this analysis is easily achieved. None of these conditions is completely satisfied through use of a single reagent, which may be superior to others in one respect but inferior in other respects. The purpose of the experiment and the particular sample of protein will determine the reagent to be used.

9.3.1 Tyrosine

9.3.1.1 Ionization

The absorption band of the nonionized form of tyrosine is shifted and intensified by ionization to a band at 295 nm. The degree of ionization can therefore be determined by measuring the absorbance increment, ΔE, at 295 nm with the aid of $\Delta\epsilon_{295} = 2305$ cm^{-1}M^{-1}, the change in the molar extinction coefficient due to ionization.[1] The difference spectrum between neutral and alkaline solutions of tyrosine or protein shows a maximum at 242–244 nm[2,3] in addition to this maximum at 295 nm. The ionization curves measured at these different wavelengths for stem bromelin[4] were identical.

Since the first measurement by Crammer and Neuberger,[5] spectrophotometric titration has been applied for discrimination of tyrosine residues in different states. Free tyrosine in neutral solution ionizes instantaneously on addition of alkali, and the pK-value of the sigmoidal ionization curve of the first order (designated as $m = 1$) is 10.0 ± 0.1. These characteristics are not affected by the presence of guanidine, a denaturation reagent for protein.[1] Tyrosine residues in proteins exhibit a variety of characteristics: a pK-value higher than 10.0, an m-value greater than unity and a slow rate of ionization. An m-value greater than unity indicates that more than one mole of alkali is required for ionization of that tyrosine residue. When $m = 2$, for example, the additional mole of alkali is used to expose or to free the buried or bound ionizable tyrosine residue. The rate of ionization is another aspect showing the difference in state of tyrosine residues. For example, it takes a few hours for one of the 4 tyrosine residues in insulin[1] to ionize at pH 12.0, while the other 3 residues ionize instantaneously.

The curves in Figure 1 show two types of tyrosine residues in trypsin.[6] Curves B and C measured immediately and 3 hours, respectively, after addition of alkali indicate instantaneous ionization with pK = 10.0

($m = 1$) of 6 of the total 10 tyrosine residues, and slow ionization of the remaining 4 residues below pH 11.4. Curve F for these slowly ionizing residues obtained by subtraction of the data on curve C from those on curve B indicates pK $= 10.8$ ($m = 2$). After denaturation of trypsin with 3.2M guanidine, all of the 10 residues ionize instantaneously with a normal pK-value of 10.0 ($m = 1$).

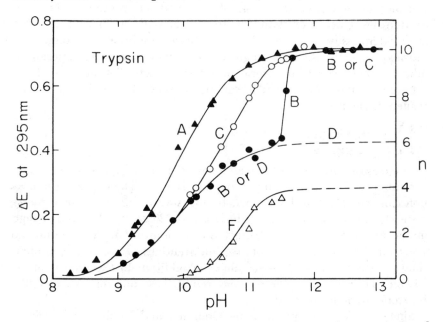

Figure 1. Ionization curves of trypsin: curve A, with guanidine; curves B and C, without guanidine, immediately after preparation of alkaline trypsin solution and after equilibration of the ionization, respectively; curves D and F, theoretical curves for six residues with pK $= 10.0$ ($m = 1$) and for four residues with pK $= 10.8$ ($m = 2$), respectively. Open triangles on curve F were obtained by subtracting the values for the six residues from the observed values on curve C. Reproduced from Reference 6 by permission of Elsevier Publishing Co., Amsterdam.

Various ionization characteristics of tyrosine residues in proteins have been reported in a number of papers.[1–4, 6–19, 57, 62, 64–66] These data are summarized in Table I (see below) where n_t stands for the total number of tyrosine residues in the protein molecule, and n_i for the number of each state of residues with different reactivity or ionization characteristics.

9.3.1.2 Cyanuric fluoride (CyF)

Cyanuric fluoride (CyF)[20] reacts with tyrosine and some other side chains such as amino and sulfhydryl groups but does not react with other aromatic side chains. The UV absorption band of tyrosine disappears on the reaction with CyF. This reagent is stable in dioxane but is hydrolyzed slowly in water to yield cyanuric acid. Neither cyanuric acid nor the reaction products with amino acids show any absorption above 290 nm. In the reaction mixture between protein and CyF, the hydrolysis of CyF and the reaction with amino acid residues, including tyrosine residues, proceed in parallel. The optimum pH for the modification of tyrosine residues is 9.0–12.6. The reaction product with amino groups is hydrolyzed back to amino groups in the procedure of amino acid analysis. The degree of reaction can be determined by difference spectrophotometric determination of intact tyrosine not modified by CyF; the reaction mixture is divided into two parts, one adjusted to neutral pH and the other to pH 12.6, and the absorbance difference between these solutions gives the concentration of tyrosine. Addition of guanidine is helpful in ionizing some buried intact tyrosine residues completely.

Application of CyF to insulin[20, 21] at pH 9.0 revealed that Tyr A19 and Tyr B16 of the 4 tyrosine residues are reactive but Tyr A14 and Tyr B26 are not reactive (see Figure 2). Tyr B26 is transformed rapidly with alkali into the reactive type while a few hours of alkali treatment are required for Tyr A14 to become reactive. These nonreactive residues seem to be the same residues as those found from the spectral shifts by tryptic digestion of insulin and by acidification of the tryptic digest.[22–25] The presence of two abnormal residues was also derived by evaluation of the solvent effect on the absorption spectrum.[26]

9.3.1.3 Acetylimidazole (AcI)

Acetylimidazole, explored by Vallee *et al.*,[12, 13, 27, 28] acetylates the hydroxyl group of tyrosine residues in proteins.[27–29] The 275-nm band of N-acetyltyrosine with $\epsilon_{275} = 1370$ and $\epsilon_{262} = 620M^{-1}cm^{-1}$ at pH 7.5 is transformed by the acetylation to N,O-diacetyltyrosine into a band at 262 nm with $\epsilon_{275} = 95$ and $\epsilon_{262} = 280M^{-1}cm^{-1}$. The O-acetyl group in diacetyltyrosine is deacetylated by treatment with hydroxylamine. The moles of acetylated tyrosine in AcI-treated protein can be determined by means of this back-reaction with the aid of $\Delta\epsilon_{278} = 1160M^{-1}cm^{-1}$.

Simpson *et al.*[12] successfully applied AcI to carboxypeptidase A. Approximately 4 of the total 19 tyrosine residues are acetylated with AcI, and of the 4 acetylated residues 2 are deacetylated more rapidly than the others. The presence of β-phenylpropionate suppresses acetylation of 2 of the 4 residues. Acetic anhydride acetylates 6–7 of these 19 residues, and iodine modifies 6 residues, so that AcI is the least reactive reagent in this example. In the case of trypsin,[28] 6.7 of the total 10 tyrosine residues are modified with AcI. This is close to 6.0, the number of CyF-reactive residues[30] and the number of rapidly ionizing residues.[6]

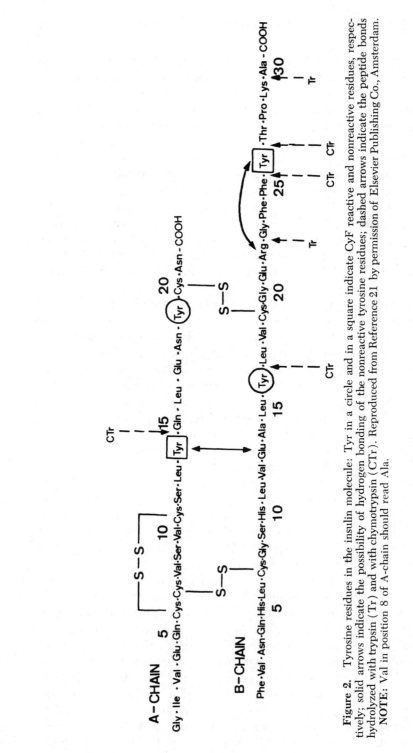

Figure 2. Tyrosine residues in the insulin molecule: Tyr in a circle and in a square indicate CyF reactive and nonreactive residues, respectively; solid arrows indicate the possibility of hydrogen bonding of the nonreactive tyrosine residues; dashed arrows indicate the peptide bonds hydrolyzed with trypsin (Tr) and with chymotrypsin (CTr). Reproduced from Reference 21 by permission of Elsevier Publishing Co., Amsterdam. NOTE: Val in position 8 of A-chain should read Ala.

The OH group of serine residues and the SH group of cysteine residues are partially modified with AcI.

9.3.1.4 Iodination

Iodination of tyrosine affords monoiodotyrosine (MIT) and diiodotyrosine (DIT). The rate of iodination from MIT to DIT is higher than the rate from tyrosine to MIT so that the major product is DIT. The degree of iodination and the distribution of iodinated tyrosine in the peptides obtained by digestion have been determined in two ways: radioactivity counting and difference spectrophotometry.

De Zoeten[31,32] applied radioactive iodine, ^{131}I, for modification of tyrosine residues in insulin and measured the distribution of radioactivity in the 4 tyrosine residues. The rates of iodination are in the order of Tyr A19 > Tyr A14 > Tyr B16 = Tyr B26; Tyr A19 is iodinated to MIT but not to DIT, whereas the other three residues are iodinated to DIT. Springell[33,34] obtained similar results except that Tyr B16 was more reactive than Tyr B26. Hashizume and Imahori[35] iodinated chymotrypsinogen under the conditions used by Hughes and Straessle[36] and determined the degree of iodination by difference spectrophotometry between neutral and alkaline solutions of iodinated protein. Results showed that 2 of the total 4 tyrosine residues are iodinated to DIT while the other 2 residues are not iodinated. This agrees with the data that indicate 2 of the 4 residues ionizing rapidly[6] and reacting with CyF.[37] The absorption bands of tyrosine, MIT and DIT in their ionized forms are appreciably different: λ_{max}(absorption maximum) = 295, 303 and 311 nm; $\Delta\epsilon$ for the ionization at these maxima = 2305, 3600 and $5500 M^{-1}cm^{-1}$ (Reference 38); and pK = 10.0, 8.2 and 6.4 (Reference 39), respectively. The pK-values found for iodinated tyrosine in proteins are 7.9 for MIT and 6.0 for DIT in lysozyme,[39] 7.9 for DIT in insulin,[40] and 8.2 for MIT and 7.5 for DIT in thyroglobulin.[41] Iodination of ribonuclease A was most extensively carried out by the group of Scheraga[42–46] and three pairs of hydrogen bonding, Tyr 92–Asp 38, Tyr 25–Asp 14 and Tyr 97–Asp 83, were postulated from the results.

9.3.1.5 Tetranitromethane (TNM)

TNM, noticed by Herriott[47] as a mild nitration reagent was used by the group of Vallee[48–50] for state discrimination of tyrosine residues. This reagent reacts with the sulfhydryl group of cysteine in addition to tyrosine, but not with other amino acid residues. Much caution must be used in the experiment since TNM is poisonous and violently explosive. Hydrogen ion is liberated on the reaction so that the reaction is followed by a pH-stat. The degree of nitration can be determined by spectrophotometry or by amino acid analysis. The 360-nm band

($\epsilon = 2790$ cm^{-1}M^{-1}) of the nonionized form of N-acetyl-3-nitrotyrosine in weakly acidic solution is shifted and intensified by ionization (pK $= 7.0$) to a band at 427 nm ($\epsilon = 4100$ cm^{-1}M^{-1}); isobestic point at 381 nm ($\epsilon = 2200$M^{-1}cm^{-1}). The treatment of carboxypeptidase with 64-fold molar excess TNM nitrates 6.7–7.1 of the total 19 tyrosine residues. The treatment with fourfold molar excess, however, nitrates only one of the residues and with this modification the peptidase activity decreases to 10% and the esterase activity increases to 170%.[49] No activity change takes place with treatment in the presence of β-phenylpropionate, an inhibitor of this enzyme. Participation of a tyrosine residue has been proposed from these data.

9.3.1.6 Enzymatic oxidation

Tyrosinase[51–54] or polyphenoloxidase[55] has been applied for partial oxidation of tyrosine residues in native proteins. For instance, in zinc-free insulin 1 of the 4 tyrosine residues, probably Tyr B26, is more rapidly oxidized with tyrosinase. The degrees of oxidation found with these different enzymes are not identical. For example, $n = 5$ with tyrosinase[53,54] and $n = 3$ with polyphenoloxidase[55] for α-lactalbumin ($n_t = 5$); $n = 1$ with tyrosinase and $n = 0$ with polyphenoloxidase for ribonuclease ($n_t = 6$). For abbreviations, see Table I.

9.3.1.7 Other reagents and methods

In addition to the above reagents, diisopropyl phosphorofluoridate (DFP) modifies tyrosine residues,[56] and ultraviolet irradiation destroys tyrosine as well as tryptophan and histidine. Small spectral shifts (see Section 9.6.1) of tyrosine and tryptophan bands have been applied for state discrimination. The data obtained by the above different methods are summarized in Table I.

9.3.2 Tryptophan
9.3.2.1 N-Bromosuccinimide (NBS)

Explored by the Witkop group, NBS has been used for various purposes:[68–72] (a) to cleave the peptide bond next to a tryptophan residue for determination of the primary protein structures; (b) to determine the tryptophan content; and (c) to classify different states of tryptophan residues. This reagent when used under appropriate conditions[73] partially modifies tryptophan residues in proteins. For example, about four of the total eight tryptophan residues in the α-chymotrypsin molecule are oxidized with NBS at pH 5.5–6.0, while almost all of them are oxidized at pH 4.0–4.5.[73] Similar results have been reported for trypsin.[62,73] The treatment of chymotrypsin and trypsin under such moderate conditions modifies some of the tyrosine residues but does

Table I

Reactivity and Ionization Characteristics of Tyrosine Residues in Proteins

Sample	n_t	Reagent or method (reference)	$n_1 > n_2 > \ldots \quad n_i$* (pK, rate, etc.)
Insulin	4	Ionz[1]	3(10.4, r) > 1(11.4, s)
		Spectral shift[22,23]	2 > 1(digestion) > 1(digestion and denaturation)
		Spectral shift[26]	2 > 2(solvent effect)
		I_2[31,32]	1 > 1 > 2(Tyr A19 > Tyr A14 > Tyr B16, Tyr B26)
		I_2[33,34]	2 > 1 > 1(Tyr A14, Tyr A19 > Tyr B16 ≫ Tyr B26)
		CyF[20,21]	2 > 1* > 1*(Tyr A19, Tyr B16 > Tyr B26* > Tyr A14*)
		AcI[28]	4 > 0*
		TNM[50]	2 > 2*
		Tyrosinase[51]	1(Tyr B26) > 3
Glucagon	2	Ionz[7]	2(10.0, r)
		CyF[7]	1 > 1
Oxytocin	1	Ionz[7]	1(10.0, r)
		CyF[7]	1(reactive)
Stem Bromelin	19	Ionz[4]	9 > 10
		CyF[4]	(7–8) > (11–12)*
Lysozyme (Hen)	3	Ionz[1,8]	1(9.95, r) > 1(11.6, r) > 1(12.7, s)
		I_2[39,58]	2 > 1*
		CyF[20]	2 > 1*
		TNM[50]	2.7 > 0.3*
		Tyrosinase[59,60]	0 > 3*
Lysozyme (Duck)	5	Ionz[8]	3(10.15, r) > 2(12.6, s)
α-Lactalbumin	5	Ionz[19]	4(10.4, r) > 1(irre, s)
		CyF[19]	3 > 1 > 1*
		Tyrosinase[53,54]	5 > 0*
		Polyphenol oxidase[55]	3 > 2*

*This number of residues is nonionizable or not reactive with the reagent.

Table I, continued

Sample	n_t	Reagent or method (reference)	$n_1 > n_2 > \ldots \quad n_i$* (pK, rate, etc.)
β-Lactoglobulin	4	Ionz[19]	3(10.9, r) >1 (irre, s)
		CyF[19]	2>1>1*
		Tyrosinase[53, 54]	0>4*
Myoglobin	3	Ionz[2, 3]	1(10.3) >1(11.5) >1(12.8)
		AcI[28]	1>2*
		Tyrosinase[52]	0>3* (MetMb)
Hemoglobin A	12	Ionz[17, 28]	CO–Hb; 8(10.6, r) >4(11.4, s) Deoxy–Hb; 6(10.8, r) >6(11.4, s)
		AcI[28]	8>4*
Hemoglobin A + Carboxypeptidase	10	Ionz[17]	CO–Hb; 6(10.6) >4(11.4) Deoxy–Hb; 6(10.6) >4(11.4)
Hemoglobin F	10	Ionz[17]	CO–Hb; 6(10.5) >4(11.4) Deoxy–Hb; 4(10.7) >6(11.4)
Hemoglobin H + Iodoacetate	12	Ionz[17]	8(10.6) >4(11.4)
Ribonuclease A	6	Ionz[9]	3(rev, r) >3(irre, s)
		Spectral shift[26]	4(exposed with 8M urea) >2
		I_2[38, 42, 43, 44, 45]	2>1>3* (Tyr 73, Tyr 76 >Tyr 115>Tyr 25*, Tyr 92*, Tyr 97*)
		CyF[61]	2>2(denaturation) >2*
		CyF[19]	1>1>1>3*
		AcI[28]	3>3*
		Tyrosinase[53, 54]	1>5*
		Tyrosinase[59]	0>6*
		Polyphenol oxidase[55]	0>6*

*This number of residues is nonionizable or not reactive with the reagent.

Table I, continued

Sample	n_t	Reagent or method (reference)	$n_1 > n_2 \ldots > n_i$* (pK, rate, etc.)
Carboxypeptidase A	19	7(9.5)>12	Ionz[12,13]
		I_2[13]	6>13*
		AcI[12,28]	6(2>2>2)>13*
		Acetic anhydride[13]	(6–7)>(12~13)*
		TNM[50]	7>12*
Trypsinogen	10	Ionz[11]	4(*rev, r*)>4(>11.5)>2(irre)
Trypsin (Bovine)	10	Ionz[6]	6(10.0, *r*)>4(10.8, *s*)
		I_2[62]	(8 ± 1)>(2 ± 1)*
		CyF[30]	6>4*
		AcI[28]	7>3*
Trypsin (Porcine)	8	Ionz[57]	3(10.2, *r*)>3(10.5, *s*) >2(12.2, *r*)
		CyF[57]	2>1>5*
DIP–Trypsin	10	CyF[30]	4>6*
Chymotrypsinogen	4	Ionz[6]	1(10.7, *r*)>1(12.4, *r*) >1(partial ionz, *s*)>1*
		I_2[35]	2>2*
α-Chymotrypsin	4	Ionz[6]	1(10.2, *r*)>1(11.3, *r*) 1(12.5, *s*)>1*
		I_2[63]	1>1>1≫1 (Tyr 146>Tyr 94 >Tyr 171≫Tyr 229)
		CyF[37]	2>2*
		AcI[28]	2>2*
π-Chymotrypsin	4	I_2[35]	2>2*
DIP–Chymotrypsin	4	CyF[37]	1>1>2*
Pepsinogen	17	TNM[50]	10>7*
Alcohol dehydrogenase (Liver)	10	AcI[28]	6>4*
Alcohol dehydrogenase (Yeast)	50	Tyrosinase[52]	(10–13)>(29–32)>8*
Lactate dehydrogenase	40	Tyrosinase[52]	4>36*

*This number of residues is nonionizable or not reactive with the reagent.

Table I, continued

Sample	n_t	Reagent or method (reference)	$n_1 > n_2 > \ldots \quad n_i$ * (pK, rate, etc.)
Glyceraldehyde-3-phosphate dehydrogenase	25	Tyrosinase[52]	0>25*
Aldolase (Rabbit muscle)	42	Ionz[64]	13>29*
		Tyrosinase[52]	4>38*
Trypsin inhibitor (Soybean)	4	Tyrosinase[52]	2>2*
Iron conalbumin	18	Ionz[66]	5>13
		AcI[28]	5>13*
Ovomucoid	5	AcI[28]	5>0*
Ovalbumin	9	AcI[28]	1.5>(7.5–8.5)*
	10	TNM[50]	6>(3–4)*
Ferricytochrome c (Horse heart)	4	Ionz[15]	4(12.1)
		Ionz[16]	1(10.1)>1(11.0)>1(12.4) >1(13.1)
		Ionz[65]	2>2
		AcI[29]	2>2* (n_2*>3.5; reduced form)
Ferricytochrome c (Bovine heart)	4	Ionz[14]	2(11.0, r)>2(12.7, s)
Ferricytochrome c (*Candida krusei*)	5	Ionz[14]	3(10.9, r)>2(11.9, s)
Ferricytochrome c (*Saccharomyces oviformis*)	5	Ionz[14]	5(10.8, r)
Serum albumin (Bovine)	20	AcI[28]	4>16*
		TNM[50]	4.6>15.4*
		Spectral shift[67]	6>4>10
Myosin A	18	Ionz[18]	KCl; 3.7(10)>14.3(11.6) KCl + PP; 6.6(10)>11.4(11.6) KCl + ATP; 5.5(10) >12.5(11.6)

Abbreviations: n, the moles of an amino acid modified per mole of protein; n_t, total number of residues of the amino acid (here, tyrosine) in the protein molecule; n_i, number of each state of residues with different reactivity and ionization (ionz) characteristics (pK; *rev*, reversible; *irre*, irreversible; r, rapid or instantaneous rate of ionization; and s, slow rate of ionization, all in parentheses).

*This number of residues is nonionizable or not reactive with the reagent.

not modify histidine, methionine and cystine residues.[74,75] Trp 62 in lysozyme[76] is more reactive with this reagent than the remaining 5 tryptophan residues. Partial modification data have also been reported for bacterial α-amylase[77] and horse cytochrome c.[28] The degree of modification can be determined from the drop of absorbance at 280–282 nm.[68,69]

9.3.2.2 H₂O₂-dioxane

A mixture of H_2O_2 (varied concentration) and dioxane (10%) when applied to a protein in bicarbonate buffer oxidizes its tryptophan residues;[78] optimum pH $= 8.1$–9.4. The degree of reaction is estimated from the drop of absorbance at 282 nm. No further process of oxidation takes place with this reagent at high H_2O_2 concentrations so that clear isobestic points are found throughout the experiment. This is one of the advantages of using this reagent. Dioxane to be used must be purified to remove more reactive organic peroxides. Tryptophan residues in some proteins are oxidized stepwise with this reagent. For example, five of the total seven tryptophan residues in chymotrypsinogen are oxidized below 2 mM of H_2O_2, one of the remaining 2 residues between 3 and 4 mM, and the other between 4 and 5 mM of H_2O_2.[78] This seems to indicate stepwise conformational changes of the protein molecule by oxidation. This reagent has been applied to lysozyme, chymotrypsinogen, α-chymotrypsin, DIP-chymotrypsin, trypsinogen, and trypsin.[37,78]

9.3.2.3 2-Hydroxy-5-nitrobenzyl bromide (HNBB)

The Koshland reagent,[79,80] HNBB, reacts with tryptophan residues to give a colored derivative. This reagent has been used for determination of the tryptophan content and for state discrimination of tryptophan residues.[81,82] The modification is carried out between pH 2.0 and 7.5 where only cysteine residues are modified in addition to tryptophan. At alkaline pH, tyrosine is modified besides these amino acids. The reaction is followed by recording the amount of HBr liberated on a pH-stat.[82] The degree of modification is determined by photometry at 410 nm ($\epsilon = 18,000 M^{-1} cm^{-1}$) for a solution of the modified protein separated by gel filtration. Low degrees of modification can be estimated more simply from the ratio of the absorbance at 280 nm to that at 410 nm. The eight tryptophan residues in the α-chymotrypsin molecule are modified to different extents at different pH-values;[82] $n = 7.5$ at pH 2.0, 0.7–1.3 at pH 3.5–5.6 and 2.1 at pH 3.0. Bewly and Li[81] applied this reagent to lysozyme.

9.3.2.4 Other reagents

Tryptophan is a labile amino acid, so that several other methods have been employed for partial modification: oxidation by ozone,[83]

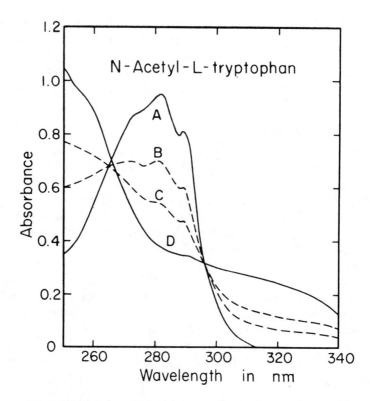

Figure 3. The spectral change on the oxidation of N-acetyl-tryptophan by H_2O_2 with 10% dioxane in 0.5M bicarbonate buffer (pH 8.4): curve A, the spectrum before oxidation; curves B, C and D, the spectra obtained by treatment with 1.50, 2.50 and 20.0 mM H_2O_2 for 30 minutes, respectively. Reproduced from Reference 78 by permission of Elsevier Publishing Co., Amsterdam.

photo-oxidation,[84,85] iodination,[86] and modification with 2-nitrophenyl-sulfenyl chloride (NPS-Cl) and 2,4-dinitrophenylsulfenyl chloride (DNPS-Cl).[87]

9.3.3   Histidine

9.3.3.1   Diazobenzenesulfonic acid

This reagent modifies histidine and tyrosine residues[88] to afford colored bisazo derivatives, both having a strong absorption band at 480 nm. Cytochrome c is inactivated by treatment with this reagent.[89]

9.3.3.2   Diazonium-1-*H*-tetrazole (DHT)

DHT modifies histidine and tyrosine residues in proteins to yield monoazo and bisazo derivatives having different absorption bands[90,91] (see Figure 4). Bisazotyrosine residue (B-Tyr) possesses a band near

550 nm while bisazohistidine residue (B-His) possesses a band at 480 nm. Monoazotyrosine residue (M-Tyr) at pH 10.0 shows a maximum at 480 nm, while M-Tyr at pH 8.0 and M-His at pH 8.0 and pH 10.0 show only end-absorption in the visible region. The band heights[90]

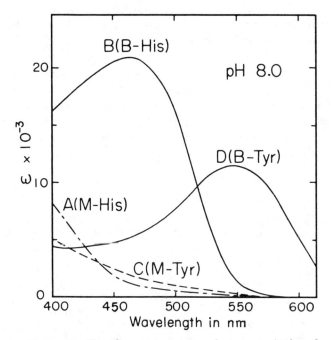

Figure 4. The absorption spectra of monoazo (M) and bisazo (B) derivatives of Val-Tyr-Val and Cbz-His-Ala obtained by treatment with DHT. Reproduced from Reference 91 by permission of Elsevier Publishing Co., Amsterdam.

estimated for these derivatives of free amino acids, tyrosine and histidine are considerably different from those for these derivatives in peptides and proteins.[91] Different methods must be employed, therefore, for determination of the histidine content in an amino acid mixture[90] and for reactivity examination of histidine and tyrosine residues in proteins.[91] The ϵ-values in peptides at pH 8.0 and 10.0 are listed in Table II. Contribution of the ultraviolet band of M-His to the absorption at 480 nm or longer wavelengths is negligible. We may set up the following simultaneous equations from the ϵ-values in Table II:

$$E(8; 480) = 20200\,C(\text{B-His}) + 1500\,C(\text{M-Tyr})$$
$$+ 5900\,C(\text{B-Tyr}) \tag{1}$$
$$E(8; 570) = 400\,C(\text{B-His}) + 10100\,C(\text{B-Tyr}) \tag{2}$$
$$E(10; 480) = 25700\,C(\text{B-His}) + 8400\,C(\text{M-Tyr})$$
$$+ 6400\,C(\text{B-Tyr}) \tag{3}$$

where C is the molar concentration of the azo derivative and E represents the absorbance values at the pH and wavelength indicated in the parentheses immediately following.

The reaction mixture of DHT and protein in borate buffer of pH 8.8 ± 0.2 after a certain period of incubation is divided into two parts. One part is adjusted to pH 8.0 by addition of 0.7M borate buffer, pH 6.8, for the photometry at 480 nm and 570 nm, and the other is adjusted to pH 10.0 by addition of 0.7M borate buffer, pH 10.8, for the photometry at 480 nm. The reaction may be stopped by addition of sodium azide. *DHT in a dry state is violently explosive; much caution must be used in experiments.* Unused DHT solutions should be discarded. The error caused by a deviation of ± 0.1 in the pH adjustment of the solutions for photometry is $\pm 7\%$, $\pm 18\%$ and $\pm 2\%$ for $C(\text{B-His})$, $C(\text{M-Tyr})$ and $C(\text{B-Tyr})$, respectively, when these concentrations are identical.

Table II

Molar Extinction Coefficients (ϵ) of Monoazo (M) and Bisazo (B) Derivatives of Cbz-His-Ala and Val-Tyr-Val at pH 8.0 and 10.0.[91]

pH	Wavelength (nm)	M-Tyr	B-Tyr	B-His
8.0	480	1,500	5,900	20,200
	548	300	11,400	1,300
	570	0	10,100	400
	600	0	5,150	0
10.0	480	8,400	6,400	25,000
	548	1,900	13,700	2,000
	570	600	12,200	400
	600	0	6,300	0

[Printed, by permission, from A. Takenaka *et al.*, Biochimica Biophysica Acta **194**, 293 (1969).]

The application of DHT to bovine zinc insulin revealed that one of the 2 histidine residues and 3 of the 4 tyrosine residues are biscoupled, and other histidine and tyrosine residues remain as the monoazo derivative. In addition to histidine and tyrosine DHT modifies tryptophan, the SH group of cysteine, and the amino groups of lysine and at peptide terminals, but the products obtained by these modifications do not interfere with the photometry at 480 nm and longer wavelengths. The reactivities of the amino groups of insulin determined by amino acid analysis are in the order of Gly A1 > Phe B1 ≫ Lys B29.[91,92] Kasai and Ando[93] modified ribonuclease T₁ with DHT and found that a specific tyrosine residue (Tyr 4 or Tyr 11) and a lysine and alanine residue are modified with DHT, accompanied by a remarkable drop of activity. The color of bisazotyrosine or bisazohistidine residues on a peptide map in an acidic medium is faint (violet), but a brilliant orange color is developed when sprayed with a neutral buffer.

9.3.3.3 Photo-oxidation

Photo-oxidation in the presence of Methylene Blue has been extensively employed for modification of histidine residues. Participation of a histidine residue in the activities of chymotrypsin,[94] ribonuclease A[95] and cytochrome c[96] was postulated from the data obtained by this method. Weil *et al.*[97] examined the effect of pH on the degree of photo-oxidation of lysozyme with the Warburg manometer; the rate is maximum at pH 9.0; $n = 1(n_t = 1)$ for histidine (abbreviations, see Table I), $n = 1.2(n_t = 8)$ for tryptophan and $n = 0$ for tyrosine, cysteine and methionine when 2 moles of oxygen is absorbed; $n = 1$ for histidine, $n = 4$ for tryptophan, $n = 0.57(n_t = 3)$ for tyrosine when 6 moles of oxygen is absorbed. The photo-oxidation is not specific to histidine, but is a useful tool for the study of histidine when applied in a low degree of oxidation.

9.3.4 Amino Groups

A number of reagents modify amino groups of lysine or at peptide terminals. Some of these reagents are, however, too reactive for state discrimination. Only the reagents with moderate reactivities are reviewed below.

9.3.4.1 Naphthoquinone sulfonic acid (NQS) and disulfonic acid (NQDS)

β-Naphthoquinone-4-sulfonic acid employed by Folin[98, 99] for assay of amino acids in urine and blood has a moderate reactivity.[100] The yellow color of this reagent purified by the Folin method[98] changes to strong yellowish brown upon reaction with the amino group. The reagent is linked by the reaction to the amino group, with liberation of the sulfonic group. The procedure[100] includes (a) a certain period of incubation in complete darkness of a reaction mixture made up of a protein solution, bicarbonate buffer of pH 8.8, dioxane and an NQS solution in 50% methanol; and (b) the addition of an acetic acid solution and dioxane to this reaction mixture. The difference in absorbance between this sample mixture and a blank without protein is measured at 480 nm; $\Delta\epsilon_{480} = 3800\text{M}^{-1}\text{cm}^{-1}$. The blank solution is rather strongly stained during incubation but this color is almost completely faded out by the acidification with acetic acid. The color developed by the reaction with amino groups is not altered by the acidification. The use of dioxane before and after the acidification is helpful in precluding precipitation of modified protein. Table III gives the moles of NQS-modified amino groups in several proteins in comparison with the data obtained with other reagents.

β-Naphthoquinone-4,6(or -4,7)-disulfonic acid (NQDS-4,6 or NQDS-4,7) modifies amino groups of proteins to the same extent as that found with NQS, and gives more stable modified proteins;[101] the $\Delta\epsilon_{480}$-values obtained for lysine with NQDS-4,6 and NQDS-4,7 are 7180 and 6500M^{-1}cm^{-1}, respectively. One may use one-half these $\Delta\epsilon$-values for calculation of n. Naphthoquinone disulfonic acid-4,5 does not react with amino groups.

Table III

The Moles of Amino Groups Modified with NQS(NQDS), CFQ and TNBS per Mole of Protein[100, 102, 103]

Sample protein	n_t*	NQS(NQDS)	CFQ	TNBS
Glucagon	2	—	1	—
Insulin	3	2	1	—
Lysozyme	7	4	3	—
Ribonuclease	11	7	7	—
Chymotrypsinogen	15	12	8	—
α-Chymotrypsin	17	13–14	10	12
DIP–Chymotrypsin	17	12	8	—
Trypsin	15	13	4	14

*n_t Stands for the total content of amino group.

9.3.4.2 Monochlorotrifluoro-*p*-benzoquinone (CFQ)

Monochlorotrifluoro-*p*-benzoquinone,[102] a yellow solid, is stable in dioxane but slowly hydrolyzed in water. The difference spectra for its reaction with amino acids show a maximum at 350 nm ($\Delta\epsilon_{350} = 16,600$ M^{-1}cm^{-1} for glycine and 35,200M^{-1}cm^{-1} for lysine). The difference spectra obtained with proteins show a maximum at 353 nm, and the $\Delta\epsilon_{353}$-values are higher. The $\Delta\epsilon_{353}$-value determined for bacitracin containing a single amino group was 21,600 cm^{-1}M^{-1}, which can be used for calculation of n-values of proteins. The reagent CFQ is less reactive than NQS, NQDS and TNBS (Table III).[100, 102, 103]

9.3.4.3 2,4,6-Trinitrobenzene sulfonic acid (TNBS)

This reagent used by Okuyama and Satake[104] is linked to an amino group of protein with liberation of the sulfonic group. After a certain period of incubation at pH 9, dioxane in one-third the volume of the reaction mixture is added to the mixture[103] and photometry is made at 425 nm between sample and blank mixtures. The addition of dioxane stops the gradual and almost endless rise of absorbance which is probably due to adsorption of TNBS by protein and/or aggregation of modified proteins. The $\Delta\epsilon$-value should be determined under the same conditions. The value obtainable with alanine as the relative standard ranges from 15,000 to 20,000$M^{-1}cm^{-1}$.

The treatment of trypsin with NQDS, CFQ and TNBS modified 13, 4 and 14 of the total 15 amino groups, and respectively inactivated this enzyme to 50%, 20% and 5% of the original activity.[103] In the case of α-chymotrypsin (Table III), the activity was lowered to 60%, 5% and 0% by modification of 13–14, 10, and 12 amino groups with NQDS, CFQ and TNBS, respectively. Tonomura *et al.*[105, 106] determined the amino acid sequence of the TNBS-modified lysine residue in G-actin. They also found[107, 108] the reaction of 2 moles of TNBS with myosin and determined the amino acid sequences of the TNBS-modified peptides. Lys 72 and Lys 73 in cytochrome c are modified with TNBS according to Takemori *et al.*[109] The inhibitory activities of trypsin inhibitors are abolished by modification of one or two amino groups with TNBS.[110]

9.3.4.4 Methylisoureas

The ϵ-amino group of lysine[111, 112] is converted to a guanidyl group by treatment with O-methylisourea or S-methylisothiourea:

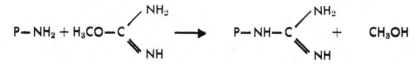

The α-amino group of amino acids is not modified with these reagents. This reaction was first applied for modification of serum albumin by Hughes *et al.*,[113] who found that 54–57 of the total 68 amino groups are modified by treatment with O-methylisourea for 3 days at 0°C and pH 10.5. This reagent has since been applied to various proteins: casein, hemoglobin, lysozyme and albumin,[114] chymotrypsinogen,[115, 116] ribonuclease,[117, 118] peptide hormones,[117] insulin,[119] cytochrome c,[120] and trypsin inhibitor.[121] In the case of insulin, for example, the ϵ-amino group of Lys B29 is completely modified with O-methylisourea, and the α-amino groups of Gly A1 and Phe B1 at peptide terminal are modified to the extents of about ½ and 1/10, respectively.[119] Modification of nine out of the total ten ϵ-amino groups in the ribonuclease molecule does not affect the activity, but modification of the least reactive ϵ-amino group stops the activity.[119] S-Methylisothiourea is less reactive than O-methylisourea.[114] Maekawa and Liener[122, 123] treated trypsin with S-methylglycosylisothiourea for 3 days at pH 8.8 and 0°C, and verified that three lysine residues and three histidine residues in the trypsin molecule are modified with this reagent.

9.3.4.5 Succinic anhydride

Succinic anhydride introduces a succinyl group onto amino groups of proteins, so that the proteins are more negatively charged. Suzuki et al.[124–126] found activation of bacterial amylase by treatment with this reagent. They also applied S-acetylmercaptosuccinic anhydride to Taka-amylase in order to introduce carboxyl and sulfhydryl groups by modification with this reagent, followed by treatment with hydroxyl-amine.

9.3.4.6 Isocyanate and isothiocyanate

Phenyl isocyanate and phenyl isothiocyanate have been used for modification of amino groups.[127, 128] Recently, fluorescein isothiocyanate (FTC) became preferred because of easier identification of modified amino groups. The application of FTC to insulin revealed the order of reactivity, Phe B1 > Gly A1 ≫ Lys B29,[129, 130] which is in agreement with the order found with NQS, NQDS and CFQ as reagents.

9.3.4.7 Other reagents

The Sanger reagent,[131–133] 1-fluoro-2,4-dinitrobenzene, and several other alkylation reagents such as monoiodoacetic acid and its methyl ester or amide may be used for modification of amino groups. These reagents modify cysteine, methionine and histidine residues in addition to amino groups. Their reactivities in commonly used conditions are too high, however, to be employed for reactivity examination. Partial modification under milder conditions was tried with bromoacetic acid for ribonuclease,[134–136] and the carboxymethyl group was introduced to a histidine residue rather than to an amino group. Acid chlorides and acid anhydrides modify amino groups, the OH group of serine, the phenol group of tyrosine, and carboxyl groups. These reagents are generally too reactive. It was reported[136, 137] that acetylation in organic solvents selectively modifies amino groups of proteins. Ethyl acetimidate was used for amidination of amino groups.[121, 138]

9.3.5 Arginine

The guanidyl group of arginine is stable and inert, to be modified with common alkylation reagents under mild conditions. In 1946, Fraenkel-Conrat and Olcott[139] modified arginine residues by treat-ment with formaldehyde at 70°C, and recently glyoxal[140] was applied as a more suitable reagent. The reaction between arginine residues and glyoxal proceeds at room temperature and pH 9.2, and reaches a stationary level after 3 hours of incubation. The degree of modification can be determined quantitatively by amino acid analysis or semi-quantitatively by the Sakaguchi reaction.[141, 142] The B22 arginine residue of insulin is modified 79% after 3 hours of treatment with 58 mM glyoxal at pH 9.2 while it is modified 7% with formaldehyde. Only 10% of the ε-amino group of Lys B29 is modified with glyoxal as examined by amino acid analysis. The lower degree of modification found for the more reactive amino group is likely due to occurrence of a back-reaction during the amino acid analysis. The bifunctional nature of glyoxal seems to be essential in obtaining a stable derivative of arginine not easily hydrolyzed. Trypsin inhibitor is not inactivated by glyoxal modification of five to six residues of the total ten arginine residues but is completely inactivated by modification of eight arginine residues.[143] The presence of trypsin in a trypsin inhibitor solution suppresses the modification of four residues out of the eight reactive residues. Recently, Takahashi[144] found phenylglyoxal to be more suitable for selective modification of arginine residues in the active site of ribonuclease A and T_1; in addition, he postulated the participation of Arg 39 in the activity of ribonuclease A and of Arg 77 in the activity of ribonuclease T_1. Malonaldehyde[145] in 10N HCl at 25°C modifies 83–100% of the arginine residues in serum albumin, ribonuclease and lysozyme.

9.3.6 Carboxyl Group

9.3.6.1 Water-soluble carbodiimide

Water-soluble carbodiimides mediate the formation of a peptide bond between a carboxyl group of protein and the amino group of amine or amino acid ester added to the reaction mixture:

$$Q-NH_2 + HOOC-P \longrightarrow Q-NH-CO-P + H_2O$$

This reaction was applied for partial modification of carboxyl groups of proteins for reactivity examination.[46, 146, 147] Hoare and Koshland[146] used 1-benzyl-3(3-dimethylamino-(N)-propyl)-carbodiimide metho-p-toluenesulfonate (BDC) with glycine methyl ester, and Horinishi *et al.*[147] used 1-ethyl-3(3-morpholinyl-(4)-propyl)-carbodiimide (EMPC) with glycine ethyl ester. The degree of modification can be determined by titration of unmodified carboxyl groups or by amino acid analysis. In the case of lysozyme, 8 of the total 11 carboxyl groups are modified with BDC while 6 of them are modified with EMPC. EMPC modifies one of the 2 carboxyl groups in bacitracin and three of the six carboxyl groups of insulin.

9.3.6.2 Diazo compounds

Carboxyl groups are esterified with diazo compounds. Diazomethane[148] in 85% ethanol partially methylates the carboxyl groups of βlactoglobin, and diazoacetamide and methyl diazoacetate[149] esterify the carboxyl groups of human serum albumin. Diazoacetoglycinamide was recently applied for the modification.[150-152] Hydrolysis of the carboxyl group modified with this reagent gives free glycine so that one can estimate the degree of modification from the excess glycine content in the hydrolysate. Asp 14, Asp 38 and Asp 83 in the ribonuclease molecule are not reactive with this reagent, probably due to hydrogen bonding with tyrosine residues.

9.3.7 Cysteine

9.3.7.1 Mercurials

Of various organic mercurials, p-chloromercuribenzoic acid (PCMB)[153] has been used most extensively for the assay of the sulfhydryl groups in proteins. The strong absorption band of PCMB at 230–240 nm shifts toward longer wavelengths on reaction with the SH group: $\Delta\epsilon_{250} = 7600$ cm^{-1}M^{-1}. Photometric titration is more easily conducted by use of colored organic mercurials such as sodium-4-(4-acetoxymercuriphenylazo)-1,7-Cleve's acid (MPAC)[154] and 1-(4-chloromercuriphenylazo)-2-naphthol (Mercury Orange).[155-157] The sulfhydryl content in Mercury Orange-modified protein is measured by photometry at 592 nm: $\epsilon = 2 \times 10^4$ cm^{-1}M^{-1}.

9.3.7.2 N-Ethylmaleimide (NEM)

NEM has weak absorption band ($\epsilon = 620$ cm^{-1}M^{-1}) at 300 nm,[158] and this band disappears on the reaction with the SH group. The degree of reaction is estimated by difference photometry at 300 nm of a reaction mixture containing NEM and protein against a blank protein solution without NEM. NEM is less reactive than PCMB, being more suitable for state discrimination. Sekine[159-161] succeeded in modifying a single SH group in the subunit of myosin. Other SH groups in the subunit are reactive with PCMB. Modified SH groups are located in the primary structure by use of ^{14}C-labeled NEM or chromophore-containing derivatives of NEM. Yellow color is developed on the reaction of N-(4-dimethylamino-3,5-dinitrophenyl)-maleimide (DDPM)[162] with the SH group. The ϵ-value at 390 nm of DDPM-modified cysteine (S-DDPM-cysteine) in glacial acetic acid is 4.9×10^3 cm^{-1}M^{-1}. Yamashita et al.[163] applied DDPM to myosine A and ATPase and determined the amino acid sequence near the DDPM-modified SH groups. Stem bromelein was successfully modified with DDPM after pretreatment with cysteine.[164, 165] Replacement of the ethyl group of NEM by some other groups has been tried in an effort to develop colored maleimides.

9.3.7.3 Ellmann's reagent

Ellman[166] used bis(3-carboxy-4-nitrophenyl)-disulfide for the assay of SH groups in tissues. The disulfide bond of this reagent is reduced on the reaction with an SH group in protein, and the thiol compound thus derived shows a strong absorption band at 412 nm.

9.3.7.4 Other reagents

Monoiodoacetic acid or its amide alkylates the SH groups. The degree of modification is determined by amino acid analysis of S-carboxylmethylcysteine, by measuring HI with a pH-stat or by radio-activity-counting of ^{14}C in the modified protein. In general, the high reactivity of such halides is not suitable for reactivity examination of SH groups.

9.4 Modification of the Residues in the Active Site of an Enzyme

9.4.1 Modification with Monofunctional Reagent O-R by Means of Activity

In enzymatic catalysis, the substrate is first linked to a residue in the catalytic site of enzyme to form an enzyme-substrate complex. This residue in the catalytic site, therefore, possesses a higher reactivity than do other residues of the same amino acid with respect to the reaction to form the complex. Thus, a specific serine residue in esterases, phosphorylases and some proteolytic enzymes is selectively acylated or phosphorylated. Such high reactivity of a residue in the catalytic site can therefore be used for selective modification of this residue. The complex with a good substrate, however, is unstable and converts rapidly to yield a product and free enzyme. For identification of the residue in the catalytic site by selective modification, the complex should be stable for subjection to further examinations. Poor or pseudo substrates which will form a complex not rapidly yielding the product are preferable for such studies.

Selective modification of the specific serine residue in "serine proteinases" such as chymotrypsin, trypsin and thrombin was first achieved by Jansen *et al.*[167-169] who used diisopropylfluorophosphate (DFP) for its phosphorylation. Urea treatment of these enzymes[170, 171] and photo-oxidation with Methylene Blue of a histidine residue near this serine residue[85, 95] destroy its high reactivity, so that the serine residue in the denatured or photo-oxidized protein is not modified with DFP. DFP does not inactivate stem bromelin, papain or Taka-amylase A.[172] This is because these enzymes are not a serine enzyme and contain an

SH group in place of the specific serine residue; DFP modifies some tyrosine residues of these enzymes but does not modify the active SH group.[56, 173] By contrast, p-nitrophenylacetate (NPA) acetylates the active SH group of glyceraldehyde-3-phosphate dehydrogenase and inactivates this enzyme.[174, 175]

The specific serine residue in serine proteinases is phosphorylated or alkylated with other, similar reagents: tetraethylpyrophosphate (TEPP), diethyl-4-nitrophenylphosphate, isopropylmethylphosphonofluoridate, dimethylamidoethoxyphosphoryl cyanide, pinacolyloxymethylphosphoryl fluoride, methylfluorophosphorylcholine iodide, ethoxymethylphosphorylthiocholine iodide. In addition to these reagents with intricate structures, methanesulfonyl fluoride (MSF), phenylmethane sulfonyl fluoride (PMSF), m-nitrobenzene sulfonyl fluoride (NBSF),[176, 177] p-chloromercuribenzene sulfonyl fluoride (CMBF),[178] and diphenylcarbanyl chloride (DPCC)[179–181] are known to modify the serine residue:

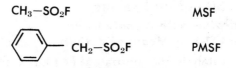

$$CH_3-SO_2F \qquad\qquad\qquad \text{MSF}$$

$$\text{—} CH_2-SO_2F \qquad\qquad \text{PMSF}$$

The effect of the aromatic ring(s) must be taken into consideration in the inactivation of chymotrypsin by PMSF, NBSF, CMBF and DPCC, as stressed in the next section. Inactivation of chymotrypsin by DPCC is protected by the presence of indole, an inhibitor of this enzyme. DPCC inactivates trypsin in addition to chymotrypsin but does not react with chymotrypsinogen, diethylphosphorylchymotrypsin, and pepsin. Addition of hydroxylamine or isonitroacetone reactivates DPC-chymotrypsin and DPC-trypsin. Proteins are stained yellow by use of 4-phenylazodiphenylcarbanyl chloride or fluoride (PADPCC or PADPCF).[182]

The acetylated serine residue of chymotrypsin treated with p-nitrophenylacetate (NPA) is stabilized by acidifying the reaction mixture.[183–185] NPA was thus found to modify the same serine residue as that phosphorylated with DFP. Treatment of chymotrypsin with p-nitrophenyldiazoacetate[186, 187] affords diazoacetylchymotrypsin. The modified group when irradiated with 320 nm light forms reactive carbene, which then reacts with water to yield O-carboxymethyl derivative of the serine residue. The complex, in which the active serine residue is phosphorylated with the substrate, is rather stable.[188] Two active serine residues in phosphoglucomutase, which are alternatively phosphorylated with the substrate in the catalysis of mutation,[189–191] were identified by use of a ^{32}P-labeled substrate. A stable phosphorylated complex is also known for phosphoglyceromutase.[192] The Schiff base formed between the ϵ-amino group of an active lysine residue and a carbonyl group of a substrate molecule is stabilized by treatment with NaBH$_4$ which affords a stable secondary amine. [193,194] This reaction was applied to aldolase, *trans*-aldolase and acetoacetic dehydrogenase.[195–200] The linkage between pyridoxal phosphate and an active lysine residue of pyridoxal enzymes is converted by urea treatment to the Schiff base which is stabilized by reduction with NaBH$_4$.

There are several reagents known to modify active residues of other amino acids. Diazoacetylnorleucine methyl ester[201] and 1-diazo-4-phenylbutanone-2 (DPB)[202] with cupric ion inactivate pepsin but do not inactivate pepsinogen. An aspartic acid residue in the active site of pepsin is esterified by treatment with *p*-bromophenacylbromide.[180,181] In the case of carboxypeptidase B, 4-bromoacetamidobutylguanidine alkylates a specific tyrosine residue, leading to inactivation.[203] A glutamic acid residue in the active site of myosin is modified with *p*-nitrothiophenol in the presence of Mg^{++} and ATP.[204–207]

9.4.2 Affinity Modification with Bifunctional Reagent, Type A-O-R

The high substrate specificity of enzymes may be used for selective modification of a residue in the active site. Bifunctional reagents, type A-O-R, have been explored intentionally for this purpose. Affinity group A is selectively adsorbed on the specificity-determining (regulatory or adsorption) site Y of the enzyme molecule $-P-X-Y-$, and the reactive group, R, modifies the residue, X:

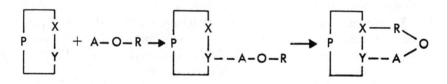

The local concentration of R near the active site is increased by the effect of adsorption, so that the modification proceeds more rapidly compared with the rate obtainable with a monofunctional reagent. This is reflected in the data shown in Table IV.[176] The second order rate constant for inactivation of chymotrypsin by PMSF with a phenyl group is about 55 times the rate for inactivation of trypsin by the same reagent. This contrasts with the fact that MSF containing no aromatic group inactivates chymotrysin and trypsin at almost identical rates. The ratio obtained with DPCC containing two phenyl groups as the reagent is more than 70.[197-181]

Table IV

**Second Order Rate Constants for Inactivation of Chymotrypsin (pH 7.0)
and Trypsin (pH 7.2) with Methanesulfonyl Fluoride (MSF)
and Phenylmethanesulfonyl Fluoride (PMSF)[176]**

Reagent	Chymotrypsin	Trypsin
PMSF	14,900.0	271.00
MSF	1.3	0.75

Successful achievement on this line was made by the group of Schoellman and Shaw,[208-211] who developed N-tosyl-L-phenylalanyl-chloromethane (TPCK) for chymotrypsin and N-tosyl-L-lysylchloro-methane (TLCK) for trypsin:

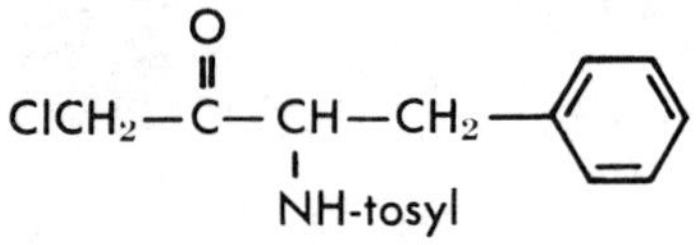

TPCK with a phenyl group modifies a specific histidine residue in chymotrypsin, but not the corresponding histidine residue in trypsin. With TLCK containing an ε-amino group of lysine, on the other hand, the histidine residue in trypsin is modified, while the histidine residue in chymotrypsin is not modified. His 57 in chymotrypsin and His 46 in trypsin are the residues modified with these reagents.[212-217]

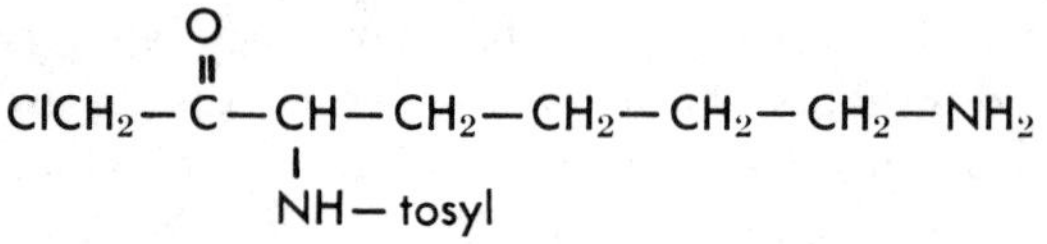

The residue X to be modified is determined not only by the reactive group R but also by another part of the reagent molecule. For example, reagent (II) modifies Met 192 in chymotrypsin[218-220] while TPCK (I) modifies His 57. The L-form of reagent (III), similar to TPCK, does not inactivate chymotrypsin while its D-form when incubated with chymotrypsin for 14 days inactivates this enzyme by 35%.[221] Reagent (IV)[222] modifies Ser 195 in chymotrypsin as does DFP, while this reagent does not react with trypsin. Similarly, reagent (V) modifies Met 192 while DPCC (VI) modifies Ser 195.[179,223]

$$BrCH_2-\overset{\overset{\displaystyle O}{\|}}{C}-NH-CH_2-C_6H_5 \qquad \text{(II)}$$

$$ICH_2-\overset{\overset{\displaystyle O}{\|}}{C}-NH-\underset{\underset{\displaystyle O=C-OCH_3}{|}}{CH}-CH_2-C_6H_5 \qquad \text{(III)}$$

$$Cl-SO_2-C_6H_4-CH_3 \qquad \text{(IV)}$$

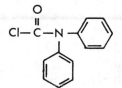

$$CH_2-CH-CH_2-O-C_6H_5 \qquad \text{(V)}$$

$$Cl-\overset{\overset{\displaystyle O}{\|}}{C}-N(C_6H_5)_2 \qquad \text{DPCC} \qquad \text{(VI)}$$

In addition to the nature of R, its direction and distance from the affinity group A adsorbed on the enzyme molecule, and some other steric factors may be essential.

Inagami[224, 225] made a new approach, using two reagents of O-A and Q-R, one carrying an affinity group and the other with a reactive group. Methylguanidine and monoiodoacetic amide were chosen as O-A and Q-R, respectively, and these reagents when added together to a trypsin solution were found to inactivate this enzyme; either of these alone, however, is not effective for the inactivation. Inagami found that butylguanidine as O-A does not cooperate with monoiodoacetic amide to inactivate trypsin, so that the length of the O group is also an essential factor.

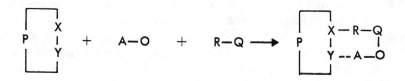

9.4.3 Negative Labeling

If a specific residue is masked or protected against modification, one may label other residues. This is termed "negative labeling." Cohen and Warringa[226] treated cholinesterase with DFP in the presence of butyrylcholine which masks the active site. They then treated the modified protein with ^{32}P-labeled DFP in the absence of butyrylcholine. A similar technique was applied by Koshland *et al.*[227] who first iodinated antibody protein toward *p*-azobenzenearsonate haptane in the presence of *p*-azobenzenearsonate and, then, further iodinated the modified protein with radioactive ^{131}I.

The residues in the active site of an enzyme are masked by an inhibitor or a substrate molecule. Modification data in the presence and in the absence of an inhibitor or substrate provide information about the active site. Five of the total seven tryptophan residues in α-chymotrypsin are oxidized with H_2O_2-dioxane at low concentrations and one of the remaining two residues is oxidized at high reagent concentrations. The presence of benzoylglycine ethyl ester transforms one of the more reactive residues into the less reactive state.[37] One of the two histidine residues in α-chymotrypsin becomes less reactive with DHT by diisopropylphosphorylation of the specific serine residue or by modification of the specific histidine residue with TPCK.[37, 103] The presence of benzoylglycine ethyl ester or diisopropylphosphorylation with DFP makes one of the two CyF-reactive tyrosine residues not reactive with CyF.[37] These effects of inhibitor (substrate) and prior modification of the active site are interpreted in two ways: (a) direct effect of the inhibitor (substrate) or modification on the residues in the vicinity of the active site, or (b) indirect effect on the residues remote from the active site through a conformational change induced by the inhibitor (substrate) or modification.

The molecule of carboxypeptidase A contains one zinc (Zn) atom, which can be completely removed by dialysis against a chelating agent solution.[228–231] One can therefore examine this apoenzyme in the presence and in the absence of Zn^{++} ion. In its absence, the apoenzyme reacts with Ag, PCMB and ferricyanide in 1:1 molar ratio, and the Zn ion added after these reactions is not linked to the apoenzyme. Formation of zinc mercaptide is essential for generation of activity and makes the α-amino group of the N-terminal aspartic amide residue not susceptible to modification by fluorodinitrobenzene (FDNB) or phenylisocyanate. Furthermore, addition of Zn ion to an apoenzyme solution releases two hydrogen ions, and the presence of β-phenylpropionic acid in a Zn–enzyme solution protects the enzyme from the action of mercaptoethanol on one of the SH groups. It was deduced from these data that the Zn atom on the active site is chelated with two groups, an SH group and an α-amino group of aspartic amide. The presence of two serine residues near the active cysteine residues has been found by a similar technique.[232]

9.5 Dimensional Analysis with Bifunctional Reagent, Type R-O-R

Bifunctional reagents with two reactive groups were developed initially to increase the chemical resistivity and mechanical strength of wool by cross-linkages.[233, 234] Recently, linkages made by these reagents in an enzyme molecule or between an enzyme and a synthetic polymer were applied to stabilize or to solidify enzymes and proteins.[235–237] Measurement of the distance between two amino acid residues in a native protein structure by use of such a reagent, R-O-R or R-O-R', is a subject of greater interest from the standpoint of protein chemistry. One of the reagents first applied for this purpose is 1,5-difluoro-2,4-dinitrobenzene (FFD):[238, 239]

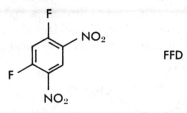

FFD

A 2,4-dinitrophenylene-1,5(DPE) bridge is formed between two amino groups of protein by treatment with this reagent. The formation of a DPE bridge between the α-amino group of Gly A1 at the A chain terminus of insulin and the ϵ-amino group of Lys B29 suggests flexibility of these chains in solution. In the case of lysozyme[240] a DPE bridge is formed between Lys 7 and Lys 41 which are not so far apart in crystalline lysozyme.[241] Fasold synthesized 2,2'-dicarboxy-4,4'-azophenyldiisocyanate and applied it to sperm whale myoglobin.[242–245] Four pairs of intramolecular linkages, Lys 16–Lys 34, Lys 34–Lys 47, Lys 56–Lys 62 and Lys 145–Lys 147 were identified in the myoglobin molecule.

About two moles of lysine in ribonuclease are modified with a similar reagent of hexamethylenediisocyanate.[246] Another reagent, p,p'-difluoro-m,m'-dinitrodiphenylsulfone (FNPS)[247] modifies 20 moles of lysine residues and some of the tyrosine residues in bovine serum albumin.[247–250] Herzig *et al.*[251] synthesized the water-soluble bifunctional reagents, phenol-2,4-disulfonyl chloride and α-naphthol-2,4-disulfonyl chloride, and found intramolecular linkages between lysine residues in lysozyme. The application of α,α'-dibromoxylene sulfonic acid

(DBX)[252] labeled with ^{35}S or ^{3}H to lysozyme revealed two pairs of intramolecular linkages, Lys 33–Lys 116 and Lys 96–Lys 97. This is com-
patible with the fact[251] that Lys 13, Lys 33 and Lys 116 are modified
with phenol-2,4-disulfonyl chloride. The positive charges of amino
groups on protein molecules in neutral solution are neutralized by
these modifications. This makes the protein molecule unstable in solution or induces a conformational change. From this point of view
Singer *et al.*[138, 253] synthesized diethylmalonimide dihydrochloride
(DEM)

$$\begin{array}{ccc} Cl^-NH_2^+ & & NH_2^+Cl^- \\ \| & & \| \\ C_2H_5-O-C-CH_2-C-O-C_2H_5 \end{array}$$

which bridges two amino groups with liberation of ethanol. The loss
of the positive charges of amino groups are compensated for by the
positive charges carried on the DEM molecule. They applied this
reagent for immunochemical studies.

A number of reagents such as ferricyanide, iodine, sodium iodo-
benzoic acid, porphyrindin, and hydrogen peroxide are known to oxi-
dize SH groups to afford a disulfide bridge. Edsall *et al.*[254] used dioxane derivatives for making intermolecular bridges between two SH
groups

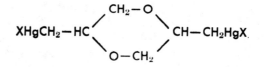

where X is Cl, acetate or nitrate. A bridge, which may be expressed by
P-S-Hg-R-Hg-S-P, is formed between two protein molecules. The Edsall group found that the formation of this type of dimer is 2000–3000
times faster than the formation of the type P-S-Hg-S-P.

Bifunctional reagents, type R-O-R', with different reactive groups
have been applied to search for the residues near the active site of an
enzyme. When chymotrypsin is treated with p-nitrophenylbromo-
acetyl-α-aminoisobutyrate,[219, 255] the active serine residue is first modified by reaction with one of the two reactive groups; the remaining
reactive group then attacks a methionine residue at the third position
from the modified serine residue in the primary structure. Hydrolysis
of the product yields a protein in which only the methionine residue is
modified. In the case of trypsin in which the methionine residue is replaced by a glutamine residue, the active serine residue is modified
with this reagent but no further reaction takes place.

9.6 Environment Analysis with Bifunctional Reagent, Type A-O-S or R-O-S

The state-sensitive group S in the bifunctional reagents A-O-S and R-O-S, when linked chemically to a protein molecule through the R group or adsorbed on a site of protein through the A group, provides information about the environment around this S group. The kinds of information are spectral shifts, color or fluorescence changes, and ESR signals.

9.6.1 Spectral Shifts of Aromatic Amino Acid Residues

The ultraviolet absorption bands of tryptophan, tyrosine and phenylalanine residues, by themselves, could provide the information around these residues. The ultraviolet bands are different in maximum wavelength and height, depending on their environmental conditions. Wetlaufer et al.[256] and Donovan et al.[257] studied the spectra of aromatic amino acid derivatives to discover the charge effect due to ionization or protonation of their carboxyl and amino groups and other factors affecting their spectra. Bigelow et al.[258,259] concentrated their study on the effects of solvents and other chemicals added to the medium.

An aromatic residue may be exposed on the surface of the protein molecule because the residue is surrounded by the medium, or it may be embedded in the interior of the protein molecule because in this case various amino acid side chains surround it. The spectrum is slightly different, depending on the medium or side chains surrounding the residue, and these small differences are measurable by difference spectrophotometry between native and denatured or digested proteins, or between proteins in different media. For example, the $\Delta\epsilon$-value at the difference maximum of 298 nm for a state change of tryptophan residues in pepsin is 365 $cm^{-1}M^{-1}$ according to Inada et al.[10] and is 240 $cm^{-1}M^{-1}$ according to Donovan et al.[257] The absorbance at 298 nm of a pepsin solution plotted against pH show two steps in state change of the 6 tryptophan residues;[10, 260, 261] $n = 2$ with pK = 4.0 and $n = 4$ with pK = 7.2 (irreversible denaturation). Two steps of absorbance change are also observed for 2 of the 6 tryptophan residues of lysozyme: $n = 1$ with pK = 3.15 ($\Delta\epsilon = 375$ $cm^{-1}M^{-1}$) and $n = 1$ with pK = 6.20 ($\Delta\epsilon = 438$ $cm^{-1}M^{-1}$).

One can measure the number of exposed residues if the spectrum of these residues is changed by addition of a chemical or a different solvent.[262] Hamaguchi et al.[263,264] made an interesting observation on lysozyme by addition of denaturation reagents such as guanidine, urea and LiCl. With increasing guanidine concentration, for example, three steps of absorbance change are observable: (1) the spectral change of exposed or solvent-accessible tryptophan residues ($n = 3$–4) by the perturbation with guanidine; (2) the change due to exposure of buried tryptophan residues by the denaturation with guanidine; and (3) the effect of guanidine on all the tryptophan residues thus exposed. Heat denaturation is also applicable for such a study.[265] One or two tryptophan residues in chymotrypsin undergo a similar spectral change upon diisopropylphosphorylation,[266–269] and activation of chymotrypsinogen also changes its spectrum.[270, 271] A change of the tyrosine band is more difficult to estimate because of overlapping of the tryptophan band. It was found in insulin containing no tryptophan that one of the 4 tyrosine residues undergoes a spectral change on digestion and another one undergoes a change on acidifying the digest.[22, 23]

9.6.2 Koshland's "Reporter"

A yellow "reporter" group of 2-acetamido-4-nitrophenol or 4-acetamido-3-nitrophenol is introduced onto a protein molecule when modified with 2-bromoacetamido-4-nitrophenol or 4-bromoacetamido-3-nitrophenol, and the spectrum of this group is dependent on its environmental conditions.[272, 273] In the case of triosephosphate dehydrogenase, 3.7 moles of 2-bromoacetamido-4-nitrophenol react with 1 mole of this protein made up of 4 subunits, each containing a single NAD-binding site and an active SH group. The ϵ-value of this reporter group on the protein is 7100 cm^{-1}M^{-1} at $\lambda_{max} = 390$ nm between pH 7.0 and 7.6. By modification with this reagent, about the same moles ($n = 3.8$) of SH groups are lost and the enzyme is completely inactivated.

When NAD is added to a solution of the modified protein from which NAD is removed by charcoal treatment after the modification, the band of the reporter group at 390 nm shifts toward longer wavelengths. The difference spectrum of this shift shows a maximum at 420 nm and a minimum at 370 nm. Nearly the same moles (4.1 moles) of 4-bromoacetamido-3-nitrophenol react with one mole of dehydrogenase, and its band at 426 nm shifts to 436 nm ($\epsilon = 3000$ cm^{-1}M^{-1} at 435 nm and pH 6.9).[274] The band of this reporter, however, shifts toward shorter wavelengths on addition of NAD to the NAD-free enzyme. These reagents react with a methionine residue of chymotrypsin, and addition of benzoylphenylalanine shifts their bands toward opposite directions. TNM possesses similar state-sensitive properties.[49]

9.6.3 The Spin-Label Method

The reagents for the spin-label method originated by Ohnishi and McConnell[275] contain a radical whose ESR spectrum is sensitive to its environment. Such radicals, which have been applied to proteins, are chloropromazine and nitroxides, and more stable nitroxides have been extensively used

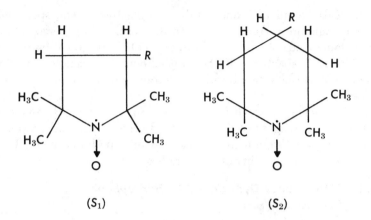

(S_1) (S_2)

where R stands for the reactive group. Several reactive groups for different amino acid side chains have been used in combination with S_1 and S_2. They are

R_a	$-NCO$	Ref. 276
R_b		Ref. 277–279
R_c	$-NHCOCH_2Br$ or $-NHCOCH_2I$	Ref. 280
R_d	$-HNCO-\!\!\langle\ \rangle\!\!-HgCl$	Ref. 278
R_e	$-CO-O-\!\!\langle\ \rangle\!\!-NO_2$	Ref. 281

Reagent S_1-R_e was applied to bovine serum albumin, and two states of the S_1 radicals on this protein were identified from analysis of the fine structures of spectrum: a freely rotatable state and a state in which the radical is immobile or some restriction is imposed on its rotation. A more stable reagent, S_1-R_b, was applied to 15 kinds of proteins and enzymes, and the immobile state of the S_1 radical linked to the SH group was found for creatine kinase and hemoglobin. The radicals linked to other proteins were freely rotatable as was the radical linked to polylysine. Hemoglobin was investigated with S_1-R_b, S_1-R_c, S_1-R_d and S_2-R_b.[278–280, 282] Spin-labeled oxyhemoglobin shows an ESR spectrum composed of wide and narrow bands superimposed, which arise from a radical bound to the SH group of Cys 93 in the β-subunit and radicals bound to lysine residues, respectively. The spectrum of deoxyhemoglobin in comparison with this result on oxyhemoglobin indicates some restriction imposed on the SH-bound radical by oxygenation. The ESR spectrum of the radical bound to isolated β-subunit changes when the subunit forms a tetramer of $\alpha_2\beta_2$ by addition of α-subunit.

Another means of application is found in reagent S_1-R_e for chymotrypsin.[281] This reagent of type A-O-S adsorbed on the adsorption site of chymotrypsin through the aromatic R_e group shows an ESR spectrum indicating immobility of the S_1 radical.

9.6.4 Fluorescence Dyes Bound to Hydrophobic Regions of Proteins

The presence of hydrophobic regions in protein structures has been demonstrated experimentally from solubilities of hydrocarbons in protein solutions[283-286] and from fluorescence intensities of dyes, type R-O-S or A-O-S, bound to or adsorbed on such hydrophobic regions. The use of fluorescence for such a study was first pointed out by Oster and Nishijima[278, 288] who stressed that a molecule of basic dye with an internally rotatable chromophore group becomes strongly fluorescent when the rotation of the group is restricted by adsorption. Quenching of fluorescence by thermal dissipation of excitation energy through the internal rotation may be suppressed by holding the planar structure on a biopolymer molecule. Recently, such enhancement of fluorescence was studied in connection with the hydrophobic regions of proteins. For example, 1-anilino-8-naphthalene sulfonate (ANS) undergoes a fluorescence change when adsorbed on apomyoglobin or apohemoglobin free from the heme group, and the addition of heme reverses this fluorescence change.[289] On the adsorption, the fluorescence band at 515 nm shifts to 454 nm and the quantum yield increases about 200 times from 0.004 to 0.98. In general the π–π excited states of a π-electron system, compared with the ground state, are stabilized to a relatively greater extent by perturbation with solvent molecules so that the release from such perturbation will intensify the fluorescence and shift the band toward longer wavelengths. A model experiment with various solvents having different dipole moments supported this interpretation. The presence of hydrophobic regions in bovine serum albumin was demonstrated by application of ANS to this protein. With 2-p-toluidinylnaphthalene-6-sulfonate (TNS), McClure and Edelman[290] found fluorescence enhancement for bovine serum albumin and β-lactoglobulin. Deranleau and Neurath[291] used a reagent with the following general structure;

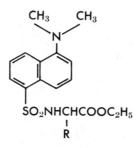

which is synthesized by the reaction of 1-dimethylaminonaphthalene-5-sulfonyl chloride with an amino acid ethyl ester. By changing the amino acid ester, one can synthesize various reagents with different specificities. The presence of an adsorption site on chymotrypsinogen as well as on chymotrypsin was demonstrated by use of this reagent synthesized from tryptophan ethyl ester. Stilbene derivatives with hydrophobic side chains—for example, Na-4,4-bis[2-chloro-4-diethanolamino-1,3,5-triazyl-(6)]-diaminostilbene-2,2′-disulfonate (TAS)[292] —and coumarin derivatives[293] are also applicable for such investigations. Coumarin derivatives seem to be superior to other fluorescent dyes. For example, 4-methyl-7-diethylamino coumarin (MDC) carries no charge on the side chains, being soluble in a variety of solvent, and is applicable to aqueous protein solutions in a wide range of pH above 5. Application of MDC to proteins showed the presence of a hydrophobic binding site(s) on the molecules of insulin, bovine β-lactoglobulin and bovine serium albumin. Recently, Sekine *et al.*[294,295] used a fluorescent derivative of NEM as a reagent of type *R-O-S*.

REFERENCES

1. Inada, Y.   J. Biochem. (Tokyo) **49**, 217 (1961).
2. Hermans, J., Jr.   Biochemistry **2**, 193 (1962).
3. Hermans, J., Jr.   Biochemistry **2**, 453 (1962).
4. Tachibana, A., and T. Murachi.   Biochemistry **5**, 2756 (1966).
5. Crammer, J. L., and A. Neuberger.   Biochem. J. **37**, 302 (1943).
6. Inada, Y., M. Kamata, A. Matsushima, and K. Shibata.   Biochim. Biophys. Acta **81**, 323 (1964).
7. Matsushima, A., Y. Inada, and K. Shibata.   Biochim. Biophys. Acta **121**, 338 (1966).
8. Tojo, T., K. Hamaguchi, M. Imanishi, and T. Amano.   J. Biochem. (Tokyo) **60**, 538 (1966).
9. Tanford, C., J. D. Hauenstein, and D. G. Rands.   J. Am. Chem. Soc. **77**, 64()9 (1956).

10. Inada, Y., A. Matsushima, M. Kamata, and K. Shibata.   Arch. Biochem. Biophys. **106**, 326 (1964).
11. Smillie, L. B., and C. M. Kay.   J. Biol. Chem. **236**, 112 (1961).
12. Simpson, R. T., J. F. Riordan, and B. L. Vallee.   Biochemistry **2**, 616 (1963).
13. Simpson, R. T., and B. L. Vallee.   Federation Proc. **22**, 493 (1963).
14. Hamaguchi, K., K. Ikeda, H. Sakai, K. Sugeno, and K. Narita.   J. Biochem. (Tokyo) **62**, 99 (1967).
15. Stellwagen, E.   Biochemistry **3**, 919 (1964).
16. Rupley, J. A.   Biochemistry **3**, 1648 (1964).
17. Nagel, R. L., H. M. Ranney, and L. K. Kucinkis.   Biochemistry **5**, 1934 (1966).
18. Tonomura, Y., K. Sekiya, K. Imamura, and T. Tokiwa.   Biochim. Biophys. Acta **69**, 305 (1963).
19. Gorbunoff, M. J.   Biochemistry **6**, 1606 (1967).
20. Kurihara, K., H. Horinishi, and K. Shibata.   Biochim. Biophys. Acta **74**, 678 (1963).
21. Aoyama, M., K. Kurihara, and K. Shibata.   Biochim. Biophys. Acta **107**, 257 (1965).
22. Laskowski, M., Jr., J. M. Widom, M. L. McFadden, and H. A. Scheraga.   Biochim. Biophys. Acta **19**, 581 (1956).
23. Laskowski, M., Jr., S. J. Leach, and H. A. Scheraga.   J. Am. Chem. Soc. **82**, 571 (1960).
24. Laskowski, M., Jr., S. J. Leach, and H. A. Scheraga.   J. Am. Chem. Soc. **82**, 2154 (1960).
25. Leach, S. J., and H. A. Scheraga.   J. Am. Chem. Soc. **82**, 4790 (1960).
26. Herskovits, T. T.   J. Biol. Chem. **240**, 628 (1965).
27. Riordan, J. F., W. E. C. Wacker, and B. L. Vallee.   Nature **208**, 1209 (1965).
28. Riordan, J. F., W. E. C. Wacker, and B. L. Vallee.   Biochemistry **4**, 1758 (1965).
29. Ulmer, D. D.   Biochemistry **5**, 1886 (1966).
30. Hachimori, Y., A. Matsushima, M. Suzuki, Y. Inada, and K. Shibata.   Biochim. Biophys. Acta **124**, 395 (1966).
31. De Zoeten, L. W., and O. A. De Bruin.   Rec. Trav. Chim. **80**, 907 (1961).
32. De Zoeten, L. W., and E. Havinga: Rec. Trav. Chim. **80**, 917 (1961).
33. Springell, P. H.   Biochim. Biophys. Acta **63**, 136 (1962).
34. Springell, P. H.   Biochem. J. **83**, 7P (1962).
35. Hashizume, H., and K. Imahori.   J. Biochem. (Tokyo) **58**, 60 (1965).
36. Hughes, W. L., Jr., and R. Straessle.   J. Am. Chem. Soc. **72**, 452 (1950).
37. Hachimori, Y., K. Kurihara, H. Horinishi, A. Matsushima, and K. Shibata.   Biochim. Biophys. Acta **105**, 167 (1965).
38. Woody, R. W., M. E. Friedman, and H. A. Scheraga.   Biochemistry **5**, 2034 (1966).
39. Wolff, J., and I. Covelli.   Biochemistry **5**, 867 (1966).

40. Gruen, L., M. Laskowski, and H. A. Scheraga. J. Biol. Chem. **234**, 2050 (1959).
41. Edelhoch, H. J. Biol. Chem. **237**, 2778 (1962).
42. Cha, C.-Y., and H. A. Scheraga. J. Biol. Chem. **238**, 2958, 2965 (1963).
43. Fujioka, H., and H. A. Scheraga. Biochemistry **4**, 2206 (1965).
44. Li, L.-K., J. P. Riehm, and H. A. Scheraga. Biochemistry **5**, 2043 (1966).
45. Friedman, M. E., H. A. Scheraga, and R. Goldberger. Biochemistry **5**, 3770 (1966).
46. Riehm, J. P., and H. A. Scheraga. Biochemistry **5**, 99 (1966).
47. Herriott, R. M. Advan. Protein Chem. **3**, 170 (1947).
48. Riordan, J. F., M. Sokolovsky, and B. L. Vallee. J. Am. Chem. Soc. **88**, 4104 (1966).
49. Riordan, J. F., M. Sokolovsky, and B. L. Vallee. Biochemistry **6**, 358 (1967).
50. Sokolovsky, M., J. F. Riordan, and B. L. Vallee. Biochemistry **5**, 3582 (1966).
51. Cory, J. G., and E. Frieden. Biochemistry **6**, 116 (1967).
52. Cory, J. G., and E. Frieden. Biochemistry **6**, 121 (1967).
53. Yasunobu, K. T., and W. B. Dandliker. J. Biol. Chem. **224**, 1065 (1957).
54. Yasunobu, K. T., E. W. Peterson, and H. S. Mason. J. Biol. Chem. **234**, 3291 (1959).
55. Lissitzky, S., M. Rolland, and S. Lasry. Biochim. Biophys. Acta **39**, 379 (1960).
56. Murachi, T., T. Inagami, and M. Yasui. Biochemistry **4**, 2815 (1965).
57. Nakamura, K., and A. Matsushima. J. Biochem. (Tokyo) **65**, 785 (1969).
58. Covelli, I., and J. Wolff. Biochemistry **5**, 860 (1966).
59. Frieden, E., and M. Ottesen. Biochim. Biophys. Acta **34**, 438 (1959).
60. Yasunobu, K. T., and P. E. Wilcox. J. Biol. Chem. **223**, 309 (1948).
61. Takenaka, O., H. Horinishi, and K. Shibata. J. Biochem. (Tokyo) **62**, 501 (1967).
62. Steiner, R. F. Biochemistry **5**, 1964 (1966).
63. Dube, S. K., O. A. Roholt, and D. Pressman. J. Biol. Chem. **241**, 4665 (1966).
64. Hass, L. F., and M. F. Lewis. Biochemistry **2**, 1368 (1963).
65. Flatmark, T. Acta Chem. Scand. **18**, 1796 (1964).
66. Wishnia, A., I. Weber, and R. C. Warner. J. Am. Chem. Soc. **83**, 2071 (1961).
67. Herskovits, T. T., and M. Laskowski, Jr. J. Biol. Chem. **237**, 2481 (1962).
68. Patchornik, A., W. B. Lawson, and B. Witkop. J. Am. Chem. Soc. **80**, 4747 (1958).
69. Patchornik, A., W. B. Lawson, and B. Witkop. J. Am. Chem. Soc. **80**, 4748 (1958).

70. Schmir, G. L., L. A. Cohen, and B. Witkop. J. Am. Chem. Soc. **81**, 2225 (1959).
71. Patchornik, A., W. B. Lawson, E. Gross, and B. Witkop. J. Am. Chem. Soc. **82**, 5923 (1960).
72. Ramachandran, L. K., and B. Witkop. J. Am. Chem. Soc. **81**, 4028 (1959).
73. Spande, T. F., N. M. Green, and B. Witkop. Biochemistry **5**, 1926 (1966).
74. Viswanatha, T., W. B. Lawson, and B. Witkop. Biochim. Biophys. Acta **40**, 216 (1960).
75. Viswanatha, T., and W. B. Lawson. Arch. Biochem. Biophys. **93**, 128 (1961).
76. Hayashi, K., T. Imoto, G. Funatsu, and M. Funatsu. J. Biochem. (Tokyo) **58**, 227 (1965).
77. Okada, Y., K. Onoue, S. Nakashima, and Y. Yamamura. J. Biochem. (Tokyo) **54**, 477 (1963).
78. Hachimori, Y., H. Horinishi, K. Kurihara, and K. Shibata. Biochim. Biophys. Acta **93**, 346 (1964).
79. Koshland, D. E., Jr., Y. D. Karkhanis, and H. G. Latham. J. Am. Chem. Soc. **86**, 1448 (1964).
80. Horton, H. R., and D. E. Koshland, Jr. J. Am. Chem. Soc. **87**, 1126 (1965).
81. Bewly, T. A., and C. H. Li. Nature **206**, 624 (1965).
82. Oza, N. B., and C. J. Martin. Biochem. Biophys. Res. Commun. **26**, 7 (1967).
83. Previero, A., M. A. Coletti, and L. Galzigna. Biochem. Biophys. Res. Commun. **16**, 195 (1964).
84. Weil, L., S. Jones, and A. R. Buchert. Arch. Biochem. Biophys. **46**, 266 (1953).
85. Piras, R., and B. L. Vallee. Biochemistry **5**, 849 (1966).
86. Ramachandran, L. K. Chem. Rev. **56**, 199 (1956).
87. Scoffone, E., A. Fontana, and R. Rocchi. Biochem. Biophys. Res. Commun. **25**, 170 (1966).
88. Fraenkel-Conrat, H. *Methods in Enzymology*, Vol. 4 (New York: Academic Press, 1957), p 247.
89. Ishikura, H., K. Takahashi, K. Titani, and S. Minakami. J. Biochem. (Tokyo) **46**, 719 (1959).
90. Horinishi, H., Y. Hachimori, K. Kurihara, and K. Shibata. Biochim. Biophys. Acta **86**, 477 (1964).
91. Takenaka, A., T. Suzuki, O. Takenaka, H. Horinishi, and K. Shibata. Biochim. Biophys. Acta **194**, 293 (1969).
92. Suzuki, T., O. Takenaka, and K. Shibata. J. Biochem. (Tokyo) **66**, 815 (1969).
93. Kasai, H., and T. Ando. Seikagaku **39**, 570 (1967).
94. Jandorf, B. J., H. O. Michel, N. K. Schaffer, R. Egan, and W. H. Summerson. Discussions Faraday Soc. **20**, 134 (1955).
95. Weil, L., and T. S. Seibles. Arch. Biochem. Biophys. **54**, 368 (1955).

96. Nakatani, M. J. Biochem. (Tokyo) **48,** 633 (1960).
97. Weil, L., A. R. Buchert, and J. Maher. Arch. Biochem. Biophys. **40,** 245 (1952).
98. Folin, O. J. Biol. Chem. **51,** 377 (1922).
99. Folin, O. J. Biol. Chem. **51,** 393 (1922).
100. Matsushima, A., Y. Hachimori, Y. Inada, and K. Shibata. J. Biochem. (Tokyo) **61,** 328 (1967).
101. Matsushima, A., K. Sakurai, M. Nomoto, Y. Inada, and K. Shibata. J. Biochem. (Tokyo) **64,** 507 (1968).
102. Nakaya, K., H. Horinishi, and K. Shibata. J. Biochem. (Tokyo) **61,** 337 (1967).
103. Shibata, K. Abstract, 7th Intern. Cong. Biochem. Tokyo, Col. I-2 (1967).
104. Okuyama, T., and K. Satake. J. Biochem. (Tokyo) **47,** 454 (1960).
105. Tonomura, Y., S. Tokura, and K. Sekiya. J. Biol. Chem. **237,** 1074 (1962).
106. Tokura, S., and Y. Tonomura. J. Biochem. (Tokyo) **53,** 422 (1963).
107. Kubo, S., S. Tokura, and Y. Tonomura. J. Biol. Chem. **235,** 2835 (1960).
108. Tokuyama, A., S. Kubo, and Y. Tonomura. J. Biochem. (Tokyo) **60,** 701 (1966).
109. Takemori, S., K. Wada, K. Ando, M. Hosokawa, I. Sekuzu, and K. Okunuki. J. Biochem. (Tokyo) **52,** 28 (1962).
110. Haynes, R., D. T. Osuga, and R. E. Feeney. Biochemistry **6,** 541 (1967).
111. Kampfhammer, J., and H. Muller. Z. Physiol. Chem. **225,** 1 (1934).
112. Greenstein, J. P. J. Biol. Chem. **109,** 541 (1935).
113. Hughes, W. L., Jr., H. A. Saroff, and A. L. Carney. J. Am. Chem. Soc. **71,** 2476 (1949).
114. Roche, J., M. Mourgue, and R. Baret. Bull. Soc. Chim. Biol. **36,** 85 (1954).
115. Chervenka, C. H., and P. E. Wilcox. J. Biol. Chem. **222,** 621 (1956).
116. Chervenka, C. H., and P. E. Wilcox. J. Biol. Chem. **222,** 635 (1956).
117. Geschwind, I. I., and C. H. Li. Biochim. Biophys. Acta **25,** 171 (1957).
118. Klee, W. A., and F. M. Richards. J. Biol. Chem. **229,** 489 (1957).
119. Evans, R. L., and H. A. Saroff. J. Biol. Chem. **228,** 295 (1957).
120. Hettinger, T. P., and H. A. Harbury. Proc. Nat. Acad. Sci. U.S. **52,** 1469 (1964).
121. Kassell, B., and R. B. Chow. Biochemistry **5,** 3449 (1966).
122. Maekawa, K., and I. E. Liener. Arch. Biochem. Biophys. **91,** 101 (1960).
123. Maekawa, K., and I. E. Liener. Arch. Biochem. Biophys. **91,** 108 (1960).
124. Suzuki, S., and Y. Hachimori. Nippon Kagaku Zasshi **89,** 614 (1968).
125. Suzuki, S., Y. Hachimori, and R. Matoba. Biochim. Biophys. Acta **167,** 641 (1968).

126. Suzuki, S. Private communication.
127. Anderson, W. Acta Chem. Scand. 8, 1723 (1954).
128. Christensen, H. N. Compt. Rend. Trav. Lab. Carlsberg, 28, 265 (1951).
129. Tietze, F., G. E. Mortimore, and N. R. Lomax. Biochim. Biophys. Acta 59, 336 (1962).
130. Bromer, W. W., S. K. Sheehan, A. W. Berns, and E. R. Arquilla. Biochemistry 6, 2378 (1967).
131. Sanger, F. Biochem. J. 39, 507 (1945).
132. Sanger, F. Biochem. J. 40, 261 (1946).
133. Porter, R. R., and F. Sanger. Biochem. J. 42, 287 (1948).
134. Bernard, E. A., and W. D. Stein. J. Mol. Biol. 1, 339 (1959).
135. Stein, W. D., and E. A. Barnard. J. Mol. Biol. 1, 350 (1959).
136. Vratsanos, S. M. Arch. Biochem. Biophys. 90, 132 (1960).
137. Vratsanos, S. M., M. Bier, and F. F. Nord. Arch. Biochem. Biophys. 77, 216 (1958).
138. Wofsy, L., and S. J. Singer. Biochemistry 2, 104 (1963).
139. Fraenkel-Conrat, H., and H. S. Olcott. J. Am. Chem. Soc. 68, 34 (1946).
140. Nakaya, K., H. Horinishi, and K. Shibata. J. Biochem. (Tokyo) 61, 337 (1967).
141. Sakaguchi, S. J. Biochem. (Tokyo) 5, 25 (1925).
142. Sakaguchi, S. J. Biochem. (Tokyo) 37, 231 (1950).
143. Ozawa, K., R. Ishida, T. Kaino, and S. Tanaka. Seikagaku 39, 568 (1967).
144. Takahashi, K. J. Biol. Chem. 243, 6171 (1968).
145. King, T. P. Biochemistry 5, 3454 (1966).
146. Hoare, D. G., and D. E. Koshland, Jr. J. Am. Chem. Soc. 88, 2057 (1966).
147. Horinishi, H., K. Nakaya, A. Tani, and K. Shibata. J. Biochem. (Tokyo) 63, 41 (1968).
148. Chibnall, A. C., J. L. Mangan, and M. W. Rees. Biochem. J. 68, 114 (1958).
149. Wilcox, P. E. Abstracts 12th Intern. Cong. Pure Appl. Chem. New York 1951, 61, 62.
150. Broomfield, C. A., J. P. Riehm, and H. A. Scheraga. Biochemistry 4, 751 (1965).
151. Riehm, J. P., C. A. Broomfield, and H. A. Scheraga. Biochemistry 4, 760 (1965).
152. Riehm, J. P., and H. A. Scheraga. Biochemistry 4, 772 (1965).
153. Boyer, P. D. J. Am. Chem. Soc. 76, 4331 (1954).
154. Nosoh, Y. J. Biochem. (Tokyo) 50, 450 (1961).
155. Bennett, H. S., and D. A. Yphantis. J. Am. Chem. Soc. 70, 3522 (1948).
156. Murachi, T. Nippon Kagaku Zasshi 88, 899 (1967).
157. Murachi, T., and M. Yasui. Seikagaku 33, 586 (1961).
158. Alexander, N. M. Anal. Chem. 30, 1292 (1958).

159. Sekine, T., L. M. Barnett, and W. W. Kielley. J. Biol. Chem. **237**, 2769 (1962).
160. Sekine, T., and W. W. Kielley. Biochim. Biophys. Acta **81**, 336 (1964).
161. Sekine, T., and M. Yamaguchi. J. Biochem. (Tokyo) **54**, 196 (1963).
162. Witter, A., and H. Tuppy. Biochim. Biophys. Acta **45**, 429 (1960).
163. Yamashita, T., Y. Soma, S. Kobayashi, T. Sekine, K. Titani, and K. Narita. J. Biochem. (Tokyo) **55**, 576 (1964).
164. Miyake, T., M. Kanda, and T. Murachi. Seikagaku **37**, 535 (1965).
165. Murachi, T., T. Miyake, and M. Mizuno. Abstracts 7th Intern. Cong. Biochem. Tokyo **1967**, 765.
166. Ellman, G. L. Arch. Biochem. Biophys. **82**, 70 (1959).
167. Jansen, E. F., M. D. F. Nutting, R. Jang, and A. K. Balls. J. Biol. Chem. **179**, 189 (1949).
168. Jansen, E. F., M. D. F. Nutting, R. Jang, and A. K. Balls. J. Biol. Chem. **179**, 201 (1949).
169. Jansen, E. F., R. Jang, and A. K. Balls. J. Biol. Chem. **196**, 247 (1952).
170. Dixon, G. H., and H. Neurath. Biochim. Biophys. Acta **20**, 572 (1956).
171. Dixon, G. H., W. J. Dreyer, and H. Neurath. J. Am. Chem. Soc. **78**, 4810 (1956).
172. Murachi, T. Biochim. Biophys. Acta **71**, 239 (1963).
173. Murachi, T., and M. Yasui. Biochemistry **4**, 2275 (1965).
174. Taylor, E. L., B. P. Meriwether, and J. H. Park. J. Biol. Chem. **238**, 734 (1963).
175. Harris, J. I. *Structure and Activity of Enzymes*. Ed. by T. W. Goodwin, J. I. Harris, and B. S. Hartley (New York: Academic Press, 1965), p 97.
176. Fahrney, D. E., and A. M. Gold. J. Am. Chem. Soc. **85**, 907 (1963).
177. Gold, A. M. Biochemistry **4**, 897 (1965).
178. Rizok, D., and J. Kallos. Biochem. Biophys. Res. Commun. **18**, 478 (1965).
179. Erlanger, B. F., and W. Cohen. J. Am. Chem. Soc. **85**, 348 (1963).
180. Erlanger, B. F., A. G. Cooper, and W. Cohen. Biochemistry **6**, 190 (1966).
181. Erlanger, B. F., S. M. Vratsanos, N. Wasserman, and A. G. Cooper. Biochem. Biophys. Res. Commun. **23**, 243 (1966).
182. Kaufman, H., and B. F. Erlanger. Biochemistry **6**, 1579 (1967).
183. Oosterbaan, R. A., and M. E. van Androchem. Biochim. Biophys. Acta **27**, 423 (1958).
184. Oosterbaan, R. A., P. Kunst, J. van Rotterdam, and J. A. Cohen. Biochim. Biophys. Acta **27**, 556 (1958).
185. Cohen, J. A., R. A. Oosterbaan, H. S. Jansz, and F. Berends. J. Cellular Comp. Physiol. **54**, Suppl. No. 1, 231 (1959).
186. Singh, A., E. R. Thornton, and F. H. Westheimer. J. Biol. Chem. **237**, PC 3006 (1962).

187. Shafer, J., P. Baronowsky, R. Laursen, F. Finn, and F. H. Westheimer. J. Biol. Chem. **241**, 421 (1966).
188. Jagannathan, V., and J. M. Luck. J. Biol. Chem. **179**, 569 (1949).
189. Koshland, D. E., and M. J. Erwin. J. Am. Chem. Soc. **79**, 2657 (1957).
190. Milstein, C., and F. Sanger. Biochem. J. **79**, 456 (1961).
191. Harshman, S., and V. A. Najjar. Biochemistry **4**, 2526 (1965).
192. Pizer, L. I. J. Am. Chem. Soc. **80**, 4431 (1958).
193. Fischer, E. H. *Structure and Activity of Enzymes.* Ed. by T. W. Goodwin, J. I. Harris, and B. S. Hartley (New York: Academic Press, 1965), p 111.
194. Fischer, E. H., A. B. Kent, E. R. Snyder, and E. G. Krebs. J. Am. Chem. Soc. **80**, 2906 (1958).
195. Horecker, B. L., S. Pontremoli, C. Ricci, and T. Cheng. Proc. Nat. Acad. Sci. U.S. **47**, 1949 (1961).
196. Grazi, E., T. Cheng, and B. L. Horecker. Biochem. Biophys. Res. Commun. **7**, 250 (1962).
197. Grazi, E., P. T. Rowley, T. Cheng, O. Tchola, and B. L. Horecker. Biochem. Biophys. Res. Commun. **9**, 38 (1962).
198. Grazi, E., E. Meloche, G. Martinez, W. A. Wood, and B. L. Horecker. Biochem. Biophys. Res. Commun. **10**, 4 (1963).
199. Fridovich, I., and F. H. Westheimer. J. Am. Chem. Soc. **84**, 3208 (1962).
200. Warren, S., B. Zerner, and F. H. Westheimer. Biochemistry **5**, 817 (1966).
201. Rajagopalan, T. G., W. H. Stein, and S. Moore. J. Biol. Chem. **241**, 4295 (1966).
202. Hamilton, G. A., J. Spona, and L. D. Crowell. Biochem. Biophys. Res. Commun. **26**, 193 (1957).
203. Plummer, T. H., Jr., and W. B. Lawson. J. Biol. Chem. **241**, 1648 (1966).
204. Imamura, K., T. Kanazawa, M. Tada, and Y. Tonomura. J. Biochem. (Tokyo) **57**, 627 (1965).
205. Tonomura, Y., and T. Kanazawa. J. Biol. Chem. **240**, PC 4110 (1965).
206. Kanazawa, T., and Y. Tonomura. J. Biochem. (Tokyo) **57**, 604 (1965).
207. Kubo, S., N. Kinoshita, and Y. Tonomura. J. Biochem. (Tokyo) **60**, 476 (1966).
208. Schoellmann, G., and E. Shaw. Biochem. Biophys. Res. Commun. **7**, 36 (1962).
209. Schoellmann, G., and E. Shaw. Biochemistry **2**, 252 (1963).
210. Shaw, E., M. Mares-Guia, and W. Cohen. Biochemistry **4**, 2219 (1965).
211. Petra, P. H., W. Cohen, and E. Shaw. Biochem. Biophys. Res. Commun. **21**, 612 (1965).
212. Meloun, B., and D. Pospisilova. Biochim. Biophys. Acta **92**, 152 (1964).

213. Smillie, L. B., and B. S. Hartley. Abstract Federation Eur. Biochem. Soc. **1964**, A-30.
214. Smillie, L. B., and B. S. Hartley. Biochem. J. **101**, 232 (1966).
215. Ong, E. B., E. Shaw, and G. Schoellmann. J. Biol. Chem. **240**, 694 (1965).
216. Tomasek, V., E. S. Severin, and F. Sorm. Biochem. Biophys. Res. Commun. **20**, 545 (1965).
217. Shaw, E., and S. Springhorn. Biochem. Biophys. Res. Commun. **27**, 391 (1967).
218. Schramm, H. J., and W. B. Lawson. Z. Physiol. Chem. **332**, 97 (1963).
219. Lawson, W. B., and H. J. Schramm. Biochemistry **4**, 377 (1965).
220. Hartley, B. S. *Structure and Activity of Enzymes.* Ed. by T. W. Goodwin, J. I. Harris, and B. S. Hartley. (New York: Academic Press, 1965), p 47.
221. Gundlach, G., and F. Turba. Biochem. Z. **335**, 573 (1962).
222. Kallos, J., and D. Rizok. J. Mol. Biol. **7**, 599 (1963).
223. Brown, J. R., and B. S. Hartley. Abstract Federation Eur. Biochem. Soc. **1964**, A-29.
224. Inagami, T. J. Biol. Chem. **239**, 787 (1964).
225. Inagami, T. J. Biol. Chem. **240**, PC 3453 (1965).
226. Cohen, J. A., and M. G. P. J. Warringa. Biochim. Biophys. Acta **11**, 52 (1953).
227. Koshland, M. E., F. M. Englberger, and D. E. Koshland, Jr. Proc. Nat. Acad. Sci. U.S. **45**, 1470 (1959).
228. Vallee, B. L., T. L. Coombs, and F. L. Hoch. J. Biol. Chem. **235**, PC 45 (1960).
229. Vallee, B. L., J. F. Riordan, and J. E. Coleman. Proc. Nat. Acad. Sci. U.S. **49**, 109 (1963).
230. Coombs, T. L., and Y. Omote. Federation Proc. **21**, 234 (1962).
231. Coombs, T. L., Y. Omote, and B. L. Vallee. Biochemistry **3**, 653 (1964).
232. Walsh, K. A., K. S. V. Sampath Kumar, J. P. Bargetzi, and H. Neurath. Proc. Nat. Acad. Sci. U.S. **48**, 1443 (1962).
233. Alexander, P., M. Fox, K. A. Stacey, and L. F. Smith. Biochem. J. **52**, 177 (1952).
234. Grubhofer, N., and L. Schleith. Hoppe-Seylers Z. Physiol. Chem. **297**, 108 (1954).
235. Bar-Eli, A., and E. Katchalski. Nature **188**, 856 (1960).
236. Bar-Eli, A., and E. Katchalski. J. Biol. Chem. **238**, 1690 (1963).
237. Rimon, A., B. Alexander, and E. Katchalski. Biochemistry **5**, 792 (1966).
238. Zahn, H., and J. Meienhofer. Makromol. Chem. **26**, 153 (1958).
239. Zahn, H. Sixth Intern. Cong. Biochem. New York **1964**.
240. Marfey, P. S., M. Uziel, and J. Little. J. Biol. Chem. **240**, 3270 (1965).
241. Phillip, D. C. Sci. Am. **215**, No. 5 **1966**, 78.
242. Fasold, H., and F. Turba. Biochem. Z. **337**, 80 (1963).

243. Fasold, H.   Biochem. Z. **339**, 482 (1964).
244. Fasold, H.   Biochem. Z. **342**, 288 (1965).
245. Fasold, H.   Biochem. Z. **342**, 295 (1965).
246. Ozawa, H.   J. Biochem. (Tokyo) **62**, 419 (1967).
247. Zahn, H., and H. Zuber.   Ber. **86**, 172 (1953).
248. Tawde, S. S., J. Sri Ram, and M. R. Iyengar.   Arch. Biochem. Biophys. **100**, 270 (1963).
249. Wold, F.   J. Biol. Chem. **236**, 106 (1961).
250. Wold, F.   Biochim. Biophys. Acta **54**, 604 (1961).
251. Herzig, D. J., A. W. Rees, and R. A. Day.   Biopolymers **2**, 349 (1964).
252. Hiremath, C. B., and R. A. Day.   J. Am. Chem. Soc. **86**, 5027 (1964).
253. Dutton, A., M. Adams, and S. J. Singer.   Biochem. Biophys. Res. Commun. **23**, 730 (1966).
254. Edsall, J. T., R. H. Maybury, R. B. Simpson, and R. Straessle.   J. Am. Chem. Soc. **76**, 3131 (1954).
255. Lawson, W. B., and H. J. Schramm.   J. Am. Chem. Soc. **84**, 2017 (1962).
256. Wetlaufer, D. B., J. T. Edsall, and B. R. Hollingworth.   J. Biol. Chem. **233**, 1421 (1958).
257. Donovan, J. W., M. Laskowski, Jr., and H. A. Scheraga.   J. Am. Chem. Soc. **83**, 2686 (1961).
258. Bigelow, C. C., and I. I. Geschwind.   Compt. Rend. Trav. Lab. Carlsberg **31**, 283 (1960).
259. Bigelow, C. C., and M. Sonenberg.   Biochemistry **1**, 197 (1962).
260. Blumenfeld, O. O., and G. E. Perlmann.   J. Gen. Physiol. **42**, 563 (1959).
261. Edelhoch, H.   J. Am. Chem. Soc. **80**, 6640 (1958).
262. Herskovits, T. T., and M. Laskowski, Jr.   J. Biol. Chem. **235**, PC 56 (1960).
263. Hamaguchi, K., and A. Kurono.   J. Biochem. (Tokyo) **54**, 111 (1963).
264. Hamaguchi, K., A Kurono, and S. Goto.   J. Biochem. (Tokyo) **54**, 259 (1963).
265. Foss, J. G.: Biochim. Biophys. Acta **47**, 569 (1961).
266. Wootton, J. F., and G. P. Hess.   Nature, **188**, 726 (1960).
267. Wooton, J. F., and G. P. Hess.   J. Am. Chem. Soc. **82**, 3789 (1960).
268. Wootton, J. F., and G. P. Hess.   J. Am. Chem. Soc. **83**, 4234 (1961).
269. Wootton, J. F., and G. P. Hess.   J. Am. Chem. Soc. **84**, 440 (1962).
270. Chervenka, C. H.   Biochim. Biophys. Acta **26**, 222 (1957).
271. Chervenka, C. H.   Biochim. Biophys. Acta **31**, 85 (1959).
272. Burr, M., and D. E. Koshland, Jr.   Proc. Nat. Acad. Sci. U.S. **52**, 1017 (1964).
273. Kirtley, M. E., and D. E. Koshland, Jr.   Biochim. Biophys. Res. Commun. **23**, 810 (1966).
274. Conway, A., and D. E. Koshland, Jr.   Biochim. Biophys. Acta **133**, 593 (1967).
275. Ohnishi, S., and H. M. McConnell.   J. Am. Chem. Soc. **87**, 2293 (1965).

276. Stone, T. J., T. Buckman, P. L. Nordio, and H. M. McConnell. Proc. Nat. Acad. Sci. U.S. **54**, 1010 (1965).
277. Griffith, O. H., and H. M. McConnell. Proc. Nat. Acad. Sci. U.S. **55**, 8 (1966).
278. Boeyens, J. C. A., and H. M. McConnell. Proc. Nat. Acad. Sci. U.S. **56**, 22 (1966).
279. Ohnishi, S., J. C. A. Boeyens, and H. M. McConnell. Proc. Nat. Acad. Sci., U.S. **56**, 809 (1966).
280. Ogawa, S., and H. M. McConnell. Proc. Nat. Acad. Sci. U.S. **58**, 19 (1967).
281. Berlinger, L. J., and H. M. McConnell. Proc. Nat. Acad. Sci. U.S. **55**, 708 (1966).
282. Ohnishi, S., T. Maeda, T. Itoh, Oh-Kenshu, and I. Tyuma. Abstract Symp. Protein Structure **1967**, 85.
283. Wishnia, A. Proc. Nat. Acad. Sci. U.S. **48**, 2200 (1962).
284. Wishnia, A. J. Phys. Chem. **67**, 2079 (1963).
285. Wishnia, A., and T. Pinder. Biochemistry **3**, 1377 (1964).
286. Wishnia, A., and T. W. Pinder, Jr. Biochemistry **5**, 1534 (1966).
287. Oster, G. J. Polymer Sci. **16**, 235 (1955).
288. Oster, G., and Y. Nishijima. J. Am. Chem. Soc. **78**, 1581 (1956).
289. Stryer, L. J. Mol. Biol. **13**, 482 (1965).
290. McClure, W. O., and G. M. Edelman. Biochemistry **5**, 1908 (1966).
291. Deranleau, D. A., and H. Neurath. Biochemistry **5**, 1413 (1966).
292. Takenaka, O., and K. Shibata. Abstract Symp. Protein Structure **1967**, 1.
293. Takenaka, O., Y. Nishimura, A. Tani, and K. Shibata. Biochim. Biophys. Acta **207**, 1 (1970).
294. Ooyashiki, T., T. Sekine, Y. Kanaoka, and M. Machida. Seikagaku **40**, 563 (1968).
295. Sekine, T., K. Ando, T. Ooyashiki, Y. Kanaoka, and M. Machida. Seikagaku **40**, 568 (1968).

Chapter 10

Methods for Investigation of the Quaternary Structure of Proteins

by H. Sund[*]

10.1 Introduction[**]

A great number of protein molecules, particularly those with high molecular weights, do not consist of a single polypeptide chain. Although these proteins in general behave under physiological conditions as monodisperse substances with a defined molecular weight, they are actually complexes of several polypeptide chains not joined to one another by peptide linkages.[†] In an extension of the Linderstrøm-Lang terminology,[12] Bernal introduced the term *quaternary structure* for this phenomenon.[13]

10.1.1 Terminology

For description of particular aspects of the structure of a protein molecule Linderstrøm-Lang had introduced the terms *primary, secondary,* and *tertiary structure* (*cf.* Figure 1a–1c). The primary structure of a polypeptide chain is given by the number and sequence of the amino acid residues that are joined together by peptide linkages. Secondary structures (helical and pleated sheet structures) result from the hydrogen bonds between the CO and NH groups of the peptide linkages. The tertiary structure is the spatial arrangement due to interactions between the side-chains of the amino acid residues. This structure is less regular than the secondary structure and is characterized by coiling or by

---

[*]Dedicated to Kurt Wallenfels on the occasion of his 60th birthday.
[**]For a more general handling of the topic of protein–protein interaction, and for its treatment from the theoretical point of view, see References 1–7.
[†]In some cases one observes an association-dissociation equilibrium between monomers and polymers (*e.g.*, glutamate dehydrogenase from beef liver[8–10] and α-amylase from *Bacillus subtilis*[11]), but in most cases drastic treatment is necessary for dissociation into subunits or polypeptide chains.

folding. The further complexity that occurs—the *quaternary* structure (Figure 1d)—is due to combinations between a defined number of identical or different polypeptide chains in the coiled or folded state.

10.1.2 *Nomenclature*

A protein molecule consisting of more than two polypeptide chains can dissociate stepwise into these components. A dissociation product that still contains at least two polypeptide chains will be referred to as

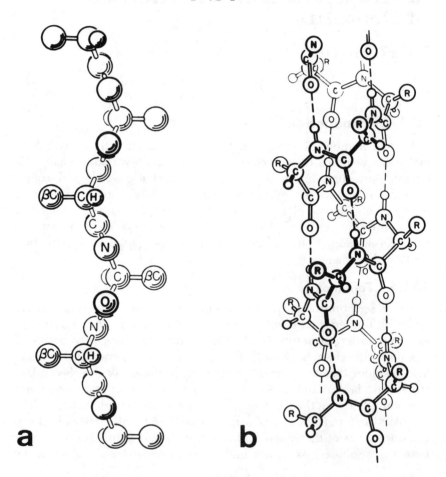

a   **b**

Figure 1. The hierarchy of protein structure, according to Bernal[13]: (a) primary structure (polypeptide chain according to Reference 14); (b) secondary structure (i.e., α-helix after Pauling and Corey[14]); (c) (opposite) tertiary structure (i.e., myoglobin after Kendrew[15]); (d) (opposite) quaternary structure (i.e., hemoglobin after Perutz[16]). White: α-polypeptide chains; gray: β-polypeptide chains; black discs: heme groups; C and N: C— and N—terminal groups of the two α-polypeptide chains.

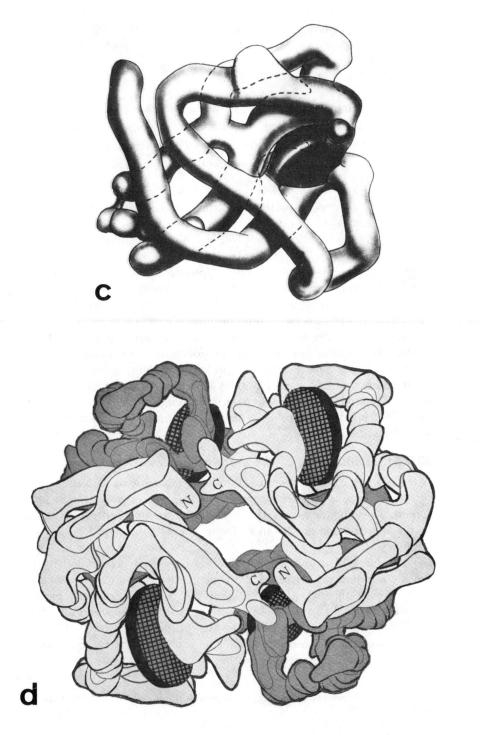

c

d

a subunit. The polypeptide chains themselves, on the other hand, consist entirely of amino acid units joined to one another by peptide linkages. Thus, a protein molecule consisting of several polypeptide chains can dissociate in three ways:

$$\text{protein molecule} \rightleftharpoons a \text{ polypeptide chains} \qquad (1a)$$
$$\text{protein molecule} \rightleftharpoons b \text{ subunits} \rightleftharpoons xb \text{ polypeptide chains} \qquad (1b)$$
$$\text{protein molecule} \rightleftharpoons c \text{ subunits} + d \text{ polypeptide chains} \qquad (1c)$$
$$c \text{ subunits} \rightleftharpoons xc \text{ polypeptide chains}$$

The polypeptide chains of a protein molecule may all be identical (*homogeneous type* of quaternary structure) or they may be different (*heterogeneous type* of quaternary structure).

Under given conditions the protein molecules, subunits, and polypeptide chains (*cf.* Equations 1a–1c) have a characteristic particle weight which can be measured by physicochemical methods. This weight will be referred to as the molecular weight of the particle. This use of the molecular weight has the advantage of considering the functional aspect of a given protein molecule.

For describing the model of allosteric transitions with respect to the quaternary structure of proteins the following terminology was defined:[17] (a) *oligomer*, a polymeric protein containing a finite, relatively small number of identical subunits; (b) *protomer*, the identical subunits associated within an oligomeric protein; and (c) *monomer*, the fully dissociated protomer or, of course, any protein which is not made up of identical subunits.

10.1.3 *Stabilization of the Protein Conformation*

The native conformation of a protein molecule is determined by *covalent bonds* as well as by *noncovalent bonds* or *noncovalent interactions*. These bonds reduce the flexibility of the polypeptide chain and ensure its arrangement in an ordered, compact form instead of as a random coil.*

The folding of a polypeptide chain is necessarily controlled by the specific bonding properties of the constituents. Many experimental results support the hypothesis[22–24] that the conformation of a protein molecule is simply a function of its amino acid sequence and all "structural information" is contained in the primary structure. If in the native state only the associated molecule is observed, it can be assumed that the self-assembly process to the associated molecule (*i.e.*, a spon-

---

*For detailed discussion of the forces involved in protein–protein interactions, see References 18–21.

taneous equilibrium association) leads to the minimum of free energy and therefore that the single polypeptide chain in the native conformation is not stable.

Among the covalent bonds, the only type that appears to be of general occurrence is the disulfide bond which may link different parts of the same polypeptide, as in ribonuclease or lysozyme,[25, 26] or it may join different polypeptide chains, as in insulin,[25] γ-globulin[27] and chymotrypsin.[28, 29] Insulin, γ-globulin, and chymotrypsin contain not only intrachain disulfide bonds but also interchain disulfide bonds.*

No conclusive evidence has yet been found for covalent peptide linkage between different polypeptide chains or between different parts of the same chain (*e.g.*, between the ε-amino group of a lysine residue and the γ-carboxyl groups of a glutamic acid residue). Ester and phosphate links and linkages via carbohydrates or other components are probably peculiar only to a small number of proteins. Such a peculiar covalent bond occurs in collagen.[32–34] It was proposed that the intramolecular interchain cross-link in collagen results from an aldol-type condensation of two lysyl-derived aldehydes (probably δ-semialdehyde of α-aminoadipic acid) on adjacent chains.

Noncovalent bonds include the hydrogen bond, the ionic bond, the hydrophobic bond (or hydrophobic interaction), and the interaction between the protein molecule and the solvent.

The native conformation, with respect also to the quaternary structure, can be determined in addition by metal ions and prosthetic groups.

The association between polypeptide chains is caused by the same bonds that determine and stabilize secondary and tertiary structures. Quaternary structure is only one aspect of the protein structure which cannot be considered isolated from the other aspects especially in viewing protein–solvent interaction. Therefore, when quaternary structure is discussed one should not ask which of all the possible bonds are the most important in holding together the polypeptide chains of a native protein molecule. The native protein molecule is a cooperative phenomenon in which all the bonds mentioned are involved.

10.1.4 *Universal Occurrence*

The molecular weights of proteins vary within wide limits, ranging from a few thousands to several millions. In general, if proteins have

---

*Chymotrypsinogen is converted into chymotrypsin[28, 29] and proinsulin is converted into insulin[30, 31] by specific hydrolysis of peptide linkages. These "activations" involve conversion of the single polypeptide chains of chymotrypsinogen and proinsulin into the three polypeptide chains of chymotrypsin or into the two polypeptide chains of insulin. In both cases some of the intrachain disulfide bonds become interchain disulfide bonds during the activation process which links the polypeptide chains together covalently.

a molecular weight higher than 100,000 they are complexes of several polypeptide chains, the molecular weight of the individual polypeptide chain generally not exceeding 50,000. Table I shows some examples of the quaternary structure of proteins.

Deoxyribonucleic acid polymerase from *E. coli* has a molecular weight of 109,000 and consists probably of only one single polypeptide chain.[59] This conclusion is based on (a) an unchanged molecular weight after unfolding in the presence of 6M guanidine hydrochloride and 0.3M mercaptoethanol; (b) the presence of only one N-terminal methionine residue; and (c) the occurrence of only one single zone on polyacrylamide gel electrophoresis under denaturing conditions. So far as we know, no other enzyme or other protein molecule has been described consisting of one single polypeptide chain and with a molecular weight of more than 100,000. There do exist, however, enzyme molecules consisting of several polypeptide chains of higher molecular weight (*e.g.*, β-galactosidase from *E. coli* and myosin[60]).

Quaternary structure is a general structural feature of proteins since it can be detected in many protein types:[4] in enzymes and respiratory proteins of low and high molecular weight, in seed globulins, serum proteins, and structural proteins such as collagen, or in the flagella of bacteria, as well as in viruses, multienzyme complexes, and ribosomes.

Very likely the genetic phenomenon of interallelic complementation can be explained on the basis of quaternary structure and that it does not occur in enzymes consisting of a single polypeptide chain.[61, 62] The same explanation is valid for the problem of isozymes.[4] So far as it is known, the phenomenon of *allosterism*[17] is connected only with proteins which are built up of more than one polypeptide chain.

The principle of quaternary structure permits the synthesis of macromolecular protein molecules from a relatively small amount of genetic information provided only the complexes formed by association are biologically active. This *economy principle*[4] is particularly obvious for large, ordered biological structures such as the coats of virus particles or structural proteins which consist either of identical polypeptide chains or of only a small number of different types of polypeptide chains. For example, the coat protein of tobacco mosaic virus has a molecular weight of 37×10^6 and consists of 2130 identical polypeptide chains.

10.2 Investigation of the Quaternary Structure

A variety of methods are available for investigation of the quaternary structure of protein molecules. Those methods are commonly applied which use the fact that under denaturing conditions the com-

Table I

Quaternary Structures of Some Proteins

Protein	Molecular Weight	Polypeptide Chain			Notes	Reference
		Molecular Weight	Number	Method[a]		
Hemoglobins from Various Species	65,000	15,000–16,000	4	C, D/M, EA, F, S, X	In general hemoglobin molecules consist of two pairs of identical polypeptide chains.	4, 16, 35
Alkaline Phosphatase from E. coli	80,000	40,000	2	D/M, F	Dissociation into polypeptide chains (probably identical) is reversible with reactivation.	4
Fructose-1, 6-Diphosphate Aldolase from Yeast	80,000	40,000	2	D/M	Dissociation into polypeptide chains is reversible with reactivation. The zinc ion is not necessary for the stabilization of the dimeric structure of the enzyme but is essential for the enzymatic activity.	36
Enolase from Rabbit Skeletal Muscle	82,000	41,000	2	D/M, EA	Dissociation into polypeptide chains is reversible, with reactivation.	4
ATP-Creatine Transphosphorylase from Rabbit Skeletal Muscle	83,000	43,000	2	D/M, EA, F	Fingerprint and end-group analyses indicate that the polypeptide chains are identical.	37
Alcohol Dehydrogenase from Horse Liver	84,000	42,000	2	C, D/M, S, X	Two types of polypeptide chains occur: E, with activity after dimerization only with ethanol and analogs but none on steroids; and S, with somewhat weaker activity on ethanol, but full activity on steroids. With E and S polypeptide chains (and subfractions S' and E') isozymes are formed. Dissociation into polypeptide chains is reversible with reactivation.	4, 38, 39
L-6-Hydroxynicotine Oxidase from A. oxidans	93,000	47,000	2	C, D/M	Storage at −16°C causes dissociation into inactive polypeptide chains.	40
Dihydrolipoic Dehydrogenase from Pig Heart	102,000	50,000	2	D/M, EA, F	End-group and fingerprint analyses indicate that the polypeptide chains are identical.	4
Hemerythrin from Golfingia gouldii	107,000	13,500	8	C, D/M	Dissociation into the polypeptide chains is reversible, with recovery of the ability to bind O_2.	4

393

Table I—continued

Protein	Molecular Weight	Polypeptide Chain Molecular Weight	Number	Method°	Notes	Reference
Lactate Dehydrogenases from Mammalian Tissue	140,000	35,000	4	C, D/M, F, S, X	In general, every animal species contains two basic types of subunit: H (heart type) and M (muscle type). Various combinations of H and M give a maximum of five isozymes. Hybridization between lactate dehydrogenase from various sources can be carried out. Some experimental evidence suggests that the H and M subunits are composed of more—probably two—polypeptide chains.	4, 41–44
Alcohol Dehydrogenase from Yeast	141,000	35,000	4	C, D/M, F, S	Fingerprint and sequence analyses indicate that the polypeptide chains are identical.	45, 46
Glyceraldehyde-3-phosphate Dehydrogenase from Mammalian Muscle and Yeast	145,000	36,000	4	C, D/M, F, S	Within the enzyme molecule the polypeptide chains are identical.	4, 43, 47–50
γ-Globulins from Various Species	150,000	25,000 and 50,000	2 each	D/M, S	The polypeptide chains are held together by disulfide bonds.	4, 27
Aldolase from Rabbit Skeletal Muscle	160,000	40,000	4	D/M		51, 52
Fumarase from Pig Heart Muscle	194,000	48,500	4	D/M, EA, F	Dissociation into polypeptide chains (probably identical) is reversible, with reactivation.	4
L-Histidine Ammonia-Lyase from *Pseudomonas*	212,000	35,000	6	D/M		53
Catalase from Beef Liver	243,000	60,000	4	C, D/M, EM, F, S	Fingerprint and sequence analyses indicate four identical polypeptide chains. At pH 3, catalase dissociates into two subunits having molecular weight 120,000.	54, 55
Acetylcholinesterase from *Electrophorus electricus*	260,000	64,000	4	D/M	Estimation of C-terminal groups suggests a hybrid structure with two α- and two β-polypeptide chains. The molecule contains two active sites.	56

Protein	Molecular Weight	Polypeptide Chain Molecular Weight	Number	Method°	Notes	Reference
Aspartate Transcarbamylase from E. coli	310,000	17,000 and 33,000	6 each	D/M, S, X	The enzyme molecule contains six regulatory (R, molecular weight 17,000) and six catalytic (C, molecular weight 33,000) polypeptide chains: composition R_6C_6.	57, 58
Apoferritin from Horse Spleen	480,000	24,000	20	D/M, EA, EM, F, X		4
Urease from Jackbean Meal	483,000	83,000	6	D/M	Larger associates were observed.	4
Bromegrass Mosaic Virsus	4,600,000	20,000	≈180	F	RNA content 21.4%. Treatment with $1M$ $CaCl_2$ yields subunits with molecular weight 40,000.	4
Turnip Yellow Mosaic Virus	5,000,000	21,300	≈150	D/M, EA, EM, F, X	RNA content 37%. Polypeptide chains are probably identical.	4
Poliomyelitis Virus (Sabin Type II)	5,500,000	27,000	≈130	D/M	RNA content 36%.	4
Alfalfa Mosaic Virus	7,400,000	≈36,000	≈160	D/M	RNA content 21%.	4
Potato X Virus	35,000,000	52,000	≈650	D/M	RNA content 5%. End-group analysis indicates that the polypeptide chains are identical.	4
Tobacco Mosaic Virus	39,400,000	17,530	2,130	D/M, EA, F, S, X	RNA content 5%. Dissociation into identical polypeptide chains and RNA is reversible. Subunits of different sizes occur as dissociation products.	4

°C = determination of "cofactor binding sites"; D/M = molecular weight determination after denaturation; EA = end-group analysis; EM = electron microscopy; F = fingerprint analysis; S = complete or partial sequence analysis; X = X-ray analysis.

pact globular protein molecules are unfolded and dissociate into polypeptide chains. The other methods are too involved, limited in application, not subject to unequivocal answers, or available only in special cases.

10.2.1 *X-Ray Analysis*[63]

With X-ray analysis, even at relatively low resolutions, the individual subunits or polypeptide chains can be distinguished and their arrangement in space recognized (*e.g.*, hemoglobin,[16, 35] aspartate transcarbamylase,[58] alcohol dehydrogenase from horse liver,[39] lactate dehydrogenase from dogfish,[44] and tobacco mosaic virus[64]). However, X-ray analysis is expensive and not used as a routine method.

10.2.2 *Electron Microscopy*[65-68]

The arrangement of subunits or polypeptide chains—the morphology —can be observed directly by means of the electron microscope, particularly when the technique of negative staining is used. Examples

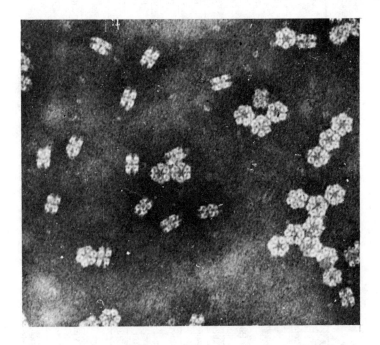

Figure 2. Erythrocruorin from *Eumenia crassa*. When it rests on its flat side, the molecule has the shape of a hexagon. When it stands on edge, the outermost subunits are seen very faintly in comparison with the central ones. Magnification: 200,000 ×. [Reprinted, by permission, from Ö. Levin, Journal of Molecular Biology 6: 95 (1963). © Academic Press.]

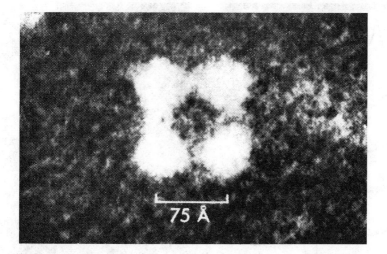

Figure 3. Electron micrograph of pyruvate carboxylase from chicken liver. The superimposed images of five intact molecules show that the four subunits are arranged at the corners of a square. The center-to-center spacing of adjacent subunits is 70–75 Å. Magnification: 2,000,000 ×. [Reprinted, by permission, from R. C. Valentine *et al.,* Biochemistry **5:** 3111 (1966). © American Chemical Society.]

of the application of this method include catalase,[54] erythrocruorin from *Eumenia crassa*[69] (Figure 2), pyruvate carboxylase from chicken liver[70] (Figures 3 and 4), and glutamate dehydrogenase from beef liver[71] (Figure 5).

Pyruvate decarboxylase, molecular weight 660,000, contains four biotine molecules and four manganese ions. The enzyme molecule dissociates into four subunits having a molecular weight of 165,000 on cold inactivation. This process is reversible and can be followed with the electron microscope and by sedimentation behavior, as shown in Figure 4. Each subunit of molecular weight 165,000 is composed of three polypeptide chains (molecular weight 45,000) which are formed in the presence of 1% sodium dodecylsulfate.

Glutamate dehydrogenase exists in an association-dissociation equilibrium with subunits which are enzymically active; the molecular weight of the smallest enzymically active subunit was found to be 310,000.[9, 10] From viscosity and X-ray small-angle measurements it was concluded that the enzyme molecule has a prolate shape in the associated state and that the dissociation involves a transverse cleavage.[8] The results of electron microscopic studies (Figure 5) are consistent with this conclusion. The subunits with molecular weight 310,000 are composed of polypeptide chains (molecular weight 50,000–60,000) which are formed only under denaturing conditions.

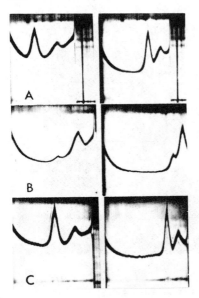

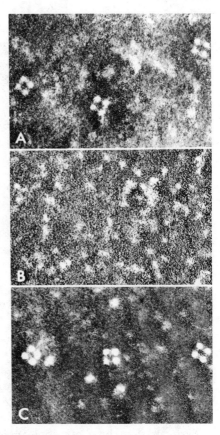

Figure 4. The effect of cold in-activation and subsequent thermal reactivation on the structure and sedimentation behavior of pyruvate carboxylase from chicken liver. Measurements in 0.01*M* phosphate buffer, pH 7.0–7.2. Sedimentation was from right to left at 22.6°C (A), 2.9°C (B) and 20.0°C (C), and 50,740 rpm using 30-mm aluminum cells, protein concentration 1.7 mg/ml. The pictures were taken at 16 and 32 minutes (A and C), and 24 and 56 minutes (B). Magnification of the electron micrographs: 500,000 ×.

The inactivation is accompanied by dissociation of the active enzyme (molecular weight 660,000, $s^0_{20,w} = 14.8S$) into four subunits (molecular weight 165,000, $s^0_{20,w} = 6.75S$).

    (A) Incubation at 23°C for 30 minutes (active enzyme)
    (B) Incubation at 2°C for 30 minutes (inactive enzyme)
    (C) Incubation at 2°C for 30 minutes followed by incubation at 23°C for 30 minutes (reactivated enzyme)

[Reprinted, by permission, from R. C. Valentine *et al.*, Biochemistry **5:** 3111 (1966). © American Chemical Society.]

Based on X-ray diffraction and electron microscopic studies the arrangement of protein subunits in viruses, especially in spherical viruses, has been analyzed with help of shadowgraphs.[72]

10.2.3 *End-group Analysis*[25,73–75]

The "classic" technique in protein chemistry is end-group analysis, in which the N-terminal or C-terminal amino acid of each polypeptide chain is determined with specific reagents or by enzymatic hydrolysis.

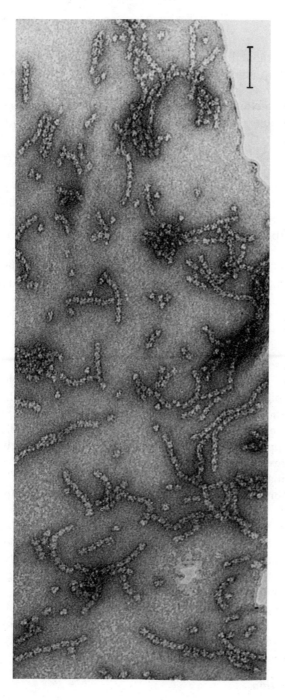

Figure 5. Electron micrograph of beef liver glutamate dehydrogenase. The picture shows that the subunits associated in a linear fashion to form particles of indefinite length. Particles consisting of up to 13 or more subunits can be observed. The spacing between subunits along the length of the rodlike particle is difficult to measure precisely but appears to be 100–120 Å. The particle diameter is slightly less, ranging between 80–100 Å. The calibration bar equals 500 Å. (Reprinted, by permission, from R. Josephs, in *Pyridine Nucleotide-Dependent Dehydrogenases*, ed. H. Sund, © 1970, Springer-Verlag.) These data are in agreement with those obtained from X-ray small-angle measurements: 8 lengths, 124 Å; diameter, 86 Å, assuming a circular cross-section, and 95 Å and 76 Å for the long and short axes respectively, assuming an axial ratio of 0.8:1 for the cross-section.

In many cases it is difficult to calculate the number of polypeptide chains from end-group analysis only because the recovery of the end-group after the chemical or enzymatic reaction is sometimes poor. Another difficulty in detecting polypeptide chains by end-group analysis is that the amino group of the N-terminal amino acid may be blocked by an acetyl group.

10.2.4 Fingerprint Analysis[25,73,75—77]

A technique particularly useful for detecting the presence of identical polypeptide chains is the procedure of peptide mapping or "fingerprinting." In this procedure the protein is broken down with trypsin, which hydrolyzes specifically those peptide bonds which involve the carboxyl groups of the basic amino acids arginine or lysine. The resulting mixture of peptides is separated by chromatography and electrophoresis. The *equivalent weight* of the polypeptide chain can be found from the arginine and lysine content of the protein and from the number of different peptides obtained. In theory the total number of peptides should be the sum of arginine and lysine residues plus one, unless, of course, the C-terminal residue is arginine or lysine. In practice there is often some deviation due to the slow cleavage of some arginine and lysine bonds, and thus both the daughter peptides as well as the slow-splitting parent peptides may be detached. If the tryptic digestion proceeds for many hours, traces of spurious peptides due to contaminating chymotrypsin may appear. Another difficulty is that peptide mapping has not been found to be an absolutely reproducible procedure. For this reason it is very important that comparative maps be made simultaneously.

The essential point, however, is that a protein molecule composed of identical subunits will yield a peptide map with much fewer fragments than predicted in theory on the basis of molecular weight alone. An isozyme composed of different polypeptide chains may yield a peptide map with more peptides than expected for identical polypeptide chains and fewer peptides on the basis of the molecular weight of the whole protein molecule because part of the different polypeptide chains is of identical primary structure. The enzymatic degradation may be replaced by nonenzymatic cleavage of peptide bonds, *e.g.*, by reaction of the protein with BrCN. In this case, cleavage of the polypeptide chains occur specifically at methionine residues.[78]

10.2.5 Hybridization

If, as is the case with bacteria, it is possible to obtain an isotopically labeled protein molecule, the hybridization technique may be used. If denaturation results in reversible dissociation of the protein molecule into its subunits, hybrids of "heavy" molecules and "light" molecules can be prepared and these hybrids can then be separated by centrifugation in a density gradient. The number of subunits formed on denaturation can be found directly from the differences in the densities of the heavy, the hybrid, and the light molecules.

Urea or guanidine hydrochloride cause reversible dissociation of β-galactosidase into four components which appear from diffusion and sedimentation measurements to be equal in size (molecular weight about 130,000, molecular weight of the native enzyme molecule 518,000).[79] It was possible to obtain hybrids of "light" enzyme molecules (^{12}C, ^{14}N) and "heavy" enzyme molecules (^{13}C, ^{15}N), and to compare the densities by density gradient centrifugation.[80] The results show again that the dissociation product obtained in the presence of urea is one quarter of the size of the whole molecule. With this technique it was proved that the DNA replication follows a semiconservative mechanism.[81]

10.2.6 Active Sites

A great number of enzyme or respiratory protein molecules contain several active sites[4, 45] (see also Table I). So far it has not been observed that an individual polypeptide chain (both in single-chain and multi-chain proteins) contains more than one active side. In some cases an active site is located in each polypeptide chain of the multi-chain enzyme (*e.g.*, alcohol dehydrogenase from horse liver,[39] lactate dehydrogenase from dogfish,[44] hemoglobin,[16, 35] catalase,[54, 55] and probably glyceraldehyde-3-phosphate dehydrogenase,[43, 48–50] alcohol dehydrogenase from yeast,[45, 83] and L-6-hydroxynicotine oxidase[40]). In other cases only part of the polypeptide chains contains active centers. Pyruvate carboxylase, for example, has only four biotine molecules and dissociates upon treatment at low temperatures into four subunits but each subunit is composed of three polypeptide chains.[70] In aspartate transcarbamylase ("composition" R_6C_6, *cf.* Table I) six polypeptide chains contain an active site whereas the other six function as regulatory polypeptide chains.[57]

The group of enzymes most intensively investigated to determine the relationship between quaternary structure and active site are the pyridine nucleotide-dependent dehydrogenases.[45, 82] The molecular weights of these dehydrogenases investigated so far vary from 19,000 to 310,000; most of the molecular weights are in the range of 50,000 to 200,000 (cf. Reference 45 and Table I). Dihydrofolate reductases[45, 84] and mannitol-1-phosphate dehydrogenase[45] possess the lowest and the glutamate dehydrogenases[9, 10] the highest molecular weights.

The majority of the dehydrogenases are composed of more than one polypeptide chain (two, four, six or eight, always an even number). In most cases the molecular weight of the polypeptide chains does not exceed 50,000. The molecular weights of the polypeptide chains generally approximate the equivalent weights of the coenzyme binding sites and, in addition, one active site is assigned to each coenzyme binding site. Therefore, with regard to the pyridine nucleotide-dependent dehydrogenases the estimation of the number of coenzyme binding sites (and also substrate binding sites) is important in elucidating the quaternary structure of these enzymes.

A few dehydrogenases deviate from the general properties mentioned above. (a) Malate dehydrogenase from *Neurospora crassa* has a molecular weight of 54,000 and is composed of four polypeptide chains ("composition" $\alpha_3\beta$; three of them identical), but contains only one coenzyme binding site.[85] At pH 2.8 one observes with inactivation a dissociation into the α_3 subunit and the β-polypeptide chain, which at pH 6.8 reassociates to the enzymic active tetramer $\alpha_3\beta$. (b) Isocitrate dehydrogenase from *Azotobacter vinelandii* has a molecular weight of 80,000 and does not dissociate in the presence of 6M guanidine hydrochloride.[86] (c) If dehydrogenases (*e.g.*, glutamate dehydrogenase from beef liver) contain in addition to the active site a regulatory site for purine nucleotide, the number of coenzyme binding sites may be higher than the number of polypeptide chains.

The following direct methods are used for estimating the number of pyridine nucleotide-binding sites.

10.2.6.1   Absorption spectrophotometry[87]

This method is based on the dehydrogenase–coenzyme difference spectra, which can be used for spectrophotometric titration.

NADH possesses a characteristic absorption band, the "dihydro band," at 340 nm due to the dihydronicotinamide moiety.[88] The maximum of this band is shifted from 340 to 325 nm in the presence of alcohol dehydrogenase from horse liver and human liver or lactate dehydrogenase from beef heart and rat liver.[45] The spectrophotometric analysis of the enzyme–NADH binding is restricted to these enzymes because the other pyridine nucleotide-dependent dehydrogenases which have been investigated do not cause a similar shift of the dihydro band. However, the characteristic absorption maximum of the addition compounds—NAD+ and NAD+ analogs with cyanide, hydroxylamine, sulfite, and sulfhydryl compounds structurally related to the substrate— is shifted to shorter wavelengths on binding to a number of pyridine nucleotide-dependent dehydrogenases which do not cause a spectral shift of NADH. These addition compounds can be regarded as dihydropyridine compounds and therefore as analogous to the reduced coenzyme molecule. The binary enzyme–NAD+ complex of D-glyceraldehyde-3-phosphate dehydrogenase from yeast can be analyzed by using

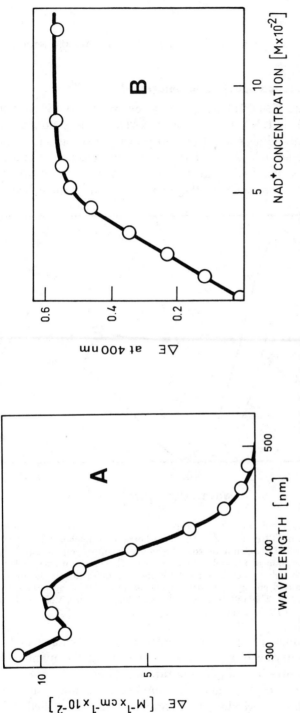

Figure 6. A: difference spectrum of the yeast glyceraldehyde-3-phosphate dehydrogenase (GAPDH)—NAD⁺ complex. Buffer, 50 mM sodium pyrophosphate, pH 8.5, 5 mM EDTA; enzyme, 0.143 mM; NAD⁺, 3 mM; temperature, 20°C. ε is calculated per mole of binding sites with four binding sites per enzyme molecule (molecular weight 140,000).

B: spectrophotometric titration of yeast glyceraldehyde-3-phosphate dehydrogenase with NAD⁺. Enzyme concentration and buffer as in Figure 6A. A micrometer syringe was used to add μl increments of a 0.1M solution of neutralized NAD⁺ to the enzyme in a cuvet (d = 2.0 cm). Readings were taken at 400 nm after completion of the slow relaxation (2–3 minutes) and where corrected for blank absorption and turbidity. [Reprinted, by permission, from K. Kirschner *et al.*, Proceedings of the National Academy of Sciences of the United States of America **56**: 1661 (1966). © ational Academy of Sciences.] In contrast to the yeast enzyme the binding of the fourth NAD⁺ molecule to enzymes from mammalian muscle (lobster, rabbit) causes no detectable change in absorbance.[90]

the specific difference spectrum of this complex (Figure 6, Reference 89).

10.2.6.2   Fluorescence spectrophotometry[45,87,88,91-93]

NADH (and also NADPH) fluoresces when irradiated by light in the 340-nm region, but NAD⁺ (and also NADP⁺) does not. Upon combination with pyridine nucleotide-dependent dehydrogenases, the fluorescence maximum of NADH shifts toward the ultraviolet and in most cases increases manyfold (Figure 7). On combination with NADH or with NAD⁺ the fluorescence of the enzyme protein is quenched.

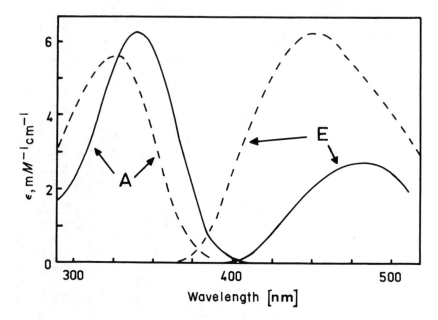

Figure 7.   Absorption, *i.e.*, excitation spectra (curves A) and emission spectra (curves E) of NADH, both free (—) and bound to horse liver alcohol dehydrogenase (---). The absorption spectrum of the enzyme-coenzyme complex has been corrected for the absorption of alcohol dehydrogenase. The emission spectra have been corrected for blank fluorescence but not for the characteristics of the instrument components. Emitted energy is in arbitrary units. (Reprinted, by permission, from A. Ehrenberg and H. Theorell, in *Comprehensive Biochemistry*, Vol. 3, ed. M. Florkin and E. Stotz, © 1962, Elsevier Publishing Company.)

The various changes of fluorescence intensity upon binding of the coenzyme to the enzyme can be used for estimating the number of coenzyme binding sites by fluorometric titration (Figure 8).

The determination of $K_{E,R}$ and the number of binding sites by graphic extrapolation (according to Figure 8) can be applied only if $K_{E,NADH}$ is 10^{-6}M or lower. If $K_{E,NADH}$ is higher, one has to take into

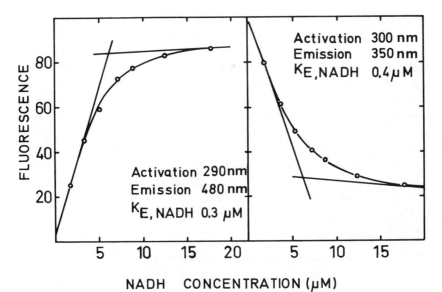

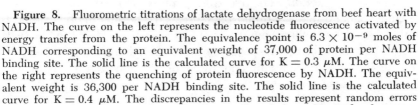

NADH CONCENTRATION (μM)

Figure 8. Fluorometric titrations of lactate dehydrogenase from beef heart with NADH. The curve on the left represents the nucleotide fluorescence activated by energy transfer from the protein. The equivalence point is 6.3×10^{-9} moles of NADH corresponding to an equivalent weight of 37,000 of protein per NADH binding site. The solid line is the calculated curve for K = 0.3 μM. The curve on the right represents the quenching of protein fluorescence by NADH. The equivalent weight is 36,300 per NADH binding site. The solid line is the calculated curve for K = 0.4 μM. The discrepancies in the results represent random errors and not systematic errors. The molecular weight of the polypeptide chain was found to be 35,000 (*cf.* Table I.) K is the dissociation constant of the enzyme-NADH complex ($K_{E,NADH} = \frac{[E] \, [NADH]}{[E-NADH]}$, [E] = concentration of coenzyme binding sites). Measurements on 0.1M tris-acetate buffer, pH 7.1, enzyme concentration 0.244 mg/ml. [Reprinted, by permission, from S. F. Velick, The Journal of Biological Chemistry **233:** 1455 (1958). © American Society of Biological Chemists, Inc.]

consideration that at low coenzyme concentrations also, only part of the coenzyme is bound to the enzyme and therefore the initial slope of the titration curve has to be corrected for the free coenzyme.

If the binding of NAD^+ is studied fluorometrically by competition with NADH according to Equation 2, it has to be considered that

$$E–NADH + NAD^+ \rightleftharpoons E–NAD^+ + NADH \qquad (2)$$

NAD^+ to some extent lowers the deflection ratio E–NADH/NADH.[95]

10.2.6.3 Optical rotatory dispersion and circular dichroism[96-98]

Upon binding of NADH the optical rotatory dispersion of alcohol dehydrogenase becomes anomalous due to appearance of the pronounced, single, negative cotton effect.[96] This *extrinsic* cotton effect is the result of the formation of an optically active complex between enzyme and coenzyme. The asymmetric binding of the chromophoric molecule to the native enzyme protein induces the optically active absorption band of the chromophore. The point of inflection of the cotton effect of horse liver alcohol dehydrogenase, at 327 nm, corresponds closely to the absorption maximum of the enzyme–NADH complex. The stoichiometry of the binding of NADH or coenzyme analogs with liver alcohol dehydrogenase can be estimated by means of rotatory dispersion titration using the change of the amplitude of the cotton effect after the successive addition of NADH.[99] The change of the circular-dichroism spectra was used for the estimation of the NAD^+ binding sites of glyceraldehyde-3-phosphate dehydrogenase from rabbit muscle.[50]

10.2.6.4 Electron paramagnetic resonance

Spin-labeled coenzyme analogs may be used for the study of enzyme–coenzyme interaction. The electron paramagnetic resonance of the analog adenosine 5'-diphosphate-4(2,2,6,6-tetramethylpiperidine-1-oxyl) broadened when bound to horse liver alcohol dehydrogenase.[100] The titration of the enzyme with the spin-labeled analog is followed by measuring the decrease in amplitude of the electron paramagnetic resonance spectrum.

10.2.6.5 Dialysis*

The standard procedure for the measurement of the binding of small molecules (ligands) by macromolecules is equilibrium dialysis.[87,101,102] During dialysis the concentration of the ligand being bound decreases after it has moved through the dialysis membrane into the macromolecule-containing compartment. From the decrease of the concentration of the ligand (*e.g.*, NAD) and the known concentration of the macromolecule (*e.g.*, enzyme), the number of binding sites as well as the dissociation constant of the enzyme–coenzyme complex can be calculated. The volumes of the "inner" and "outer" compartments must be determined with accuracy, and suitable control systems must be studied. This method was applied for the study of NAD^+ binding to glyceraldehyde-3-phosphate dehydrogenase from rabbit muscle.[90]

A disadvantage of the equilibrium dialysis method is that several hours are required for attainment of diffusion equilibrium across a membrane, although in most cases the actual chemical equilibrium for the binding reaction can be reached within a fraction of a second. This

---

Cf. also the next section on ultracentrifugal methods.

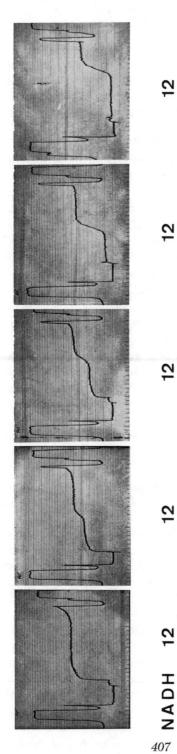

NADH	12	12	12	12	12
LDH	0	1	1.5	3	6

Figure 9. Representative sedimentation velocity patterns from various mixtures of lactate dehydrogenase (LDH) from chicken heart and NADH. For measurements in $0.1M$ phosphate buffer, pH 7, at 334 nm and 59780 rpm, the patterns were taken 40 minutes after reaching speed. NADH concentration in all experiments was 0.13 mM; the concentration of the enzyme was varied in the different experiments to give the molar ratios indicated below the pattern. The equivalent weight per coenzyme binding site was calculated to 35,000 (*cf.* also Table I). [Reprinted, by permission, from H. K. Schachman, Biochemistry **2**: 887 (1963). © American Chemical Society.]

407

disadvantage can be overcome by a rapid technique which is based on measuring the rate of dialysis of a radioactive ligand from an enzyme–ligand (NAD) equilibrium mixture.[103] Accurate results can be obtained quickly, thus permitting studies with labile substances.

10.2.6.6   Ultracentrifugal methods

With preparative and analytical ultracentrifuges enzyme–coenzyme binding can be analyzed quantitatively.[87, 104, 105, 107]

Using a preparative ultracentrifuge, various mixtures of coenzyme and enzyme along with a control containing only coenzyme are placed in the tubes of a preparative centrifuge rotor which is subjected to sustained operation at high speed for times sufficient to allow the enzyme to sediment to the bottom of the tubes. Then the supernatant layer is carefully removed and analyzed for the coenzyme concentration. The amount of coenzyme in the supernatant in each of the tubes which originally contained both components is then compared with the supernatant from the tubes which contain only the coenzyme. From these data the amount of bound coenzyme is calculated, and from the known initial concentrations of both enzyme and coenzyme the amount of moles of coenzyme bound per mole of enzyme is calculated for the various mixtures. From the data at different concentrations, the maximum number of binding sites and the dissociation constant of the enzyme–coenzyme complex can be determined according to Equation 3. r represents the average number of moles of bound ligand (L, e.g.,

$$r = n - K_{E,L} \cdot r/[L_{\text{free}}] \tag{3}$$

NAD^+) per mole enzyme E, n the maximum number of binding sites, $K_{E,L}$ the dissociation constant of the EL complex, and $[L_{\text{free}}]$ the concentration of free ligand.[87, 101] From the plot of $r/[L_{\text{free}}]$ versus r the maximum number of binding sites is obtained. The linearity of this plot provides evidence that the binding sites act independent from one another and therefore that all binding sites of one protein molecule have identical dissociation constants. The method was applied for the study of NAD-binding to yeast alcohol dehydrogenase.[106]

Interacting systems may also be investigated quantitatively by the analytical ultracentrifuge if the binding of the ligand can be followed by an absorption optical system.[105, 107] Figure 9 shows sedimentation velocity patterns from a series of experiments with different mixtures of NADH (DPNH) and lactate dehydrogenase (LDH). The pattern of NADH alone (on the left) shows that all of the light-absorbing material (at 340 nm) migrates slowly whereas upon addition of enzyme, part or all the coenzyme (depending on the enzyme concentration) sediments with a sedimentation coefficient the same as that of the native enzyme protein. From the absorption distribution in the ultracentrifuge cell during the sedimentatiton run the concentration of the bound and free coenzyme can be calculated; these data give the average number of coenzyme molecules bound to the enzyme molecule. Again the number of binding sites and the dissociation constants are calculated according to Equation 3.

10.2.7 Denaturation*

A protein molecule consisting of several polypeptide chains may dissociate into subunits or polypeptide chains on denaturation. The problem of finding the number of subunits or polypeptide chains then reduces to a molecular weight determination. Investigations of this nature are usually carried out with the aid of the ultracentrifuge,[104, 105, 110–115] by polyacrylamide gel electrophoresis,[60] or by light-scattering measurements.[116–118] If the polypeptide chains are not identical, they can be separated in the denatured state by electrophoresis,[75, 119] countercurrent distribution,[120] or chromatography.[75, 121–123]

In general, denaturation can be obtained in the presence of guanidine hydrochloride or urea both in high concentrations (see Section 10.2.7.5) or in the presence of sodium dodecylsulfate (see Section 10.2.7.4). To avoid formation of disulfide bonds from -SH groups during denaturation, which may cause a covalent linkage between polypeptide chains, mercaptoethanol is added to the solution. In some cases dissociation into polypeptide chains occurs after maleylation (*cf.* Section 10.2.7.2) or succinylation (*cf.* Section 10.2.7.3). If polypeptide chains are held together by disulfide bonds, reduction to -SH groups or oxidation to cysteic acid residues is necessary for dissociation into polypeptide chains.[75, 124]

In special cases protein molecules can be dissociated by interaction with heavy metals or with complex-forming agents. Yeast alcohol dehydrogenase, for instance, dissociates in the presence of silver ions or mercurials.[125, 126] The polypeptide chains of yeast alcohol dehydrogenase are also obtained in the presence of complex-forming agents like *o*-phenanthroline[127] because yeast alcohol dehydrogenase is a zinc enzyme. Complex-forming agents remove the zinc ions and cause inactivation in addition to dissociation. A strong correlation exists between enzymatic activity, zinc content, and undissociated enzyme (Figure 10). These results show that -SH groups and zinc ions are important factors in maintaining the native conformation.

In the case of yeast aldolase zinc ions are also absolutely necessary for enzymatic activity, but not for the stabilization of the dimeric

---

*For the general treatment of denaturation see References 18, 108, and 109.

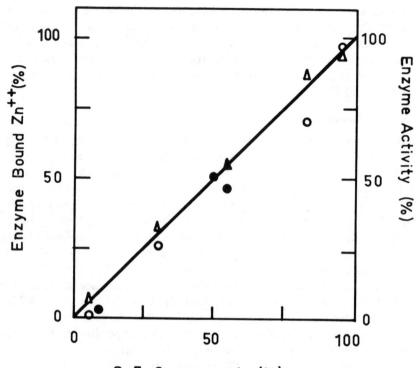

Figure 10. Correlation between the effects of *o*-phenanthroline and 8-hydroxy-quinoline-5-sulfonic acid on activity, enzyme-bound zinc, and structure of yeast alcohol dehydrogenase. All measurements are expressed in percentages of the control in $0.1M$ sodium phosphate buffer. The enzyme (0.05 mM), in $0.1M$ sodium phosphate buffer, pH 7.5, 0°C, was incubated with 7.5 mM *o*-phenanthroline or 20 mM 8-hydroxyquinoline-5-sulfonic acid. At various times aliquots were analyzed for activity and enzyme-bound zinc, and examined in the analytical ultracentrifuge. The percentage of the original enzyme remaining [S-7 protein, $s^0_{20,w}$ = 7.6S, molecular weight 150,000, with the polypeptide chain which is formed on treatment with the complex-forming agents having a sedimentation coefficient ($s^0_{20,w}$) of 2.8S, molecular weight 35,000[46]] is plotted on the abscissa, enzymatic activity after incubation with *o*-phenanthroline (o), or with 8-hydroxyquinoline-5-sulfonic acid (●) on one ordinate and enzyme-bound zinc after incubation with *o*-phenanthroline (△) or with 8-hydroxyquinoline-5-sulfonic acid (▲) on the other. [Reprinted, with permission, from J. H. R. Kägi and B. L. Vallee, The Journal of Biological Chemistry **235**:3188 (1960). © American Society of Biological Chemists, Inc.]

structure of this enzyme.[128] On the contrary, α-amylase from *Bacillus subtilis* needs zinc ions not for activity but for dimerization (Figure 11).[11, 129]

Under special conditions allosteric enzymes may be desensitized. Treatment of aspartate transcarbamylase with heat or with mercurials causes not only desensitization but also a dissociation into C and R polypeptide chains and therefore disappearance of the allosteric properties. However, the separated polypeptide chains retain their functional properties, namely, catalytic activity or affinity with the regulatory metabolite, cytidine triphosphate.[57, 130, 131]

Denaturation and dissociation may be performed also at extreme pH-values. At acid or alkaline pH-values aldolase,[52, 132] glutamate dehydrogenase,[133–135] and catalase[54] dissociate into subunits or polypeptide chains. However, these conditions may cause a hydrolytic cleavage of peptide bonds which yields an incorrect number of polypeptide chains.

After short-time incubation at pH 12.5–12.8, catalase dissociates into the polypeptide chains. Longer incubation causes splitting into several components.[54] Muscle aldolase, molecular weight 160,000, dissociates into the polypeptide chains upon exposure to cold alkaline borate buffer (pH 12.5)[132] or to pH 2.0.[52] If the native enzyme molecule is exposed to alkaline conditions at 20°C, the polypeptide chains undergo an initial rapid degradation which eventually yields alkali-resistant material. This resistance to further alkaline degradation suggests that hydrolysis is not completely random.[132] Analog results are obtained after long dialysis at pH 2.[52] On the other hand, glutamate dehydrogenase from beef liver does not undergo further degradation after long time dialysis at pH 12.[135]

10.2.7.1 Molecular weight determination

The molecular weight of the dissociation products can be determined by chemical (*cf.* especially Section 10.2.3 and 10.2.4) as well as by physical methods. Molecular weight determinations commonly used include ultracentrifugal methods (diffusion, sedimentation, sedimentation equilibrium,[104, 105, 110–115]) light-scattering measurements,[116–118] or osmotic pressure measurements.[136] The intrinsic viscosity of randomly coiled polymer chains corresponds exactly with molecular weight. Therefore, the molecular weight can be determined from viscosity measurements[137] (*cf.* Section 10.2.7.5). Densities and viscosities of various aqueous solutions are compiled in Table II. By comparison with known molecular weights the sizes of polypeptide chains under denaturing conditions may be estimated by polyacrylamide gel electrophoresis (*cf.* Section 10.2.7.4) or by gel filtration (*cf.* Section 10.2.7.5).

In the case of heterogeneity different average molecular weights are obtained. Osmotic pressure measurements yield the number-average molecular weight, $\overline{M}_n$ (Equation 4), and light-scattering or sedimentation equilibrium measurements yield the weight-average molecular weight $\overline{M}_w$ (Equation 5) or both. In these equations n_i is the number

$$\overline{M}_n = \frac{\Sigma n_i M_i}{\Sigma n_i} \tag{4}$$

$$\overline{M}_w = \frac{\Sigma g_i M_i}{\Sigma g_i} = \frac{\Sigma n_i M_i^2}{\Sigma n_i M_i} \tag{5}$$

and g_i the grams of molecules of kind i present in the mixture; M_i is their molecular weight. As a consequence of these equations, contaminating small molecules lead to erroneously low values in osmotic pressure measurements, while big-sized impurities give values too high if the weight-average molecular weight results.

The combination of sedimentation coefficient and diffusion coefficient for the calculation of molecular weight according to the Svedberg equation does not ordinarily yield the $\overline{M}_w$-values but a mixed average of the weight-average molecular weight ("$\overline{M}_{w,w}$"). Weight-average and number-average molecular weights are most frequently used. Higher averages are of interest only occassionally; they have been given the names z-average ($\overline{M}_z$), and ($z + 1$)-average ($\overline{M}_{z+1}$), etc. (Equation 6). If a sample is completely homogeneous, $\overline{M}_n = \overline{M}_w = \overline{M}_z$; if it is

Table II

Densities and Viscosities of Various Aqueous Solutions

No.	Composition	T °C	Density (g/ml)	Viscosity (cP)	Reference
1	H_2O	20	0.9982	1.002	
		25	0.9971	0.894	170
2	M/15 Potassium Sodium Phosphate Buffer, pH 7.6	20	1.0066°	1.034°	
3	0.05M Tris–HCl Buffer pH 7.6	20	0.9997°	1.037°	
4	6M Guanidine Hydrochloride - 0.1M β-Mercaptoethanol	20 25	1.1443° 1.1465	1.610° 1.476°	169
5	6M urea in H_2O	25	1.0886	1.254	169
6	6M Urea in M/15 Potassium Sodium Phosphate Buffer pH 7.6	20	1.0994°	1.388°	
7	5 plus 0.1M β-Mercaptoethanol	25	1.0934	1.270°	169
8	5 plus 2.0M NaCl	25	1.1631	1.749°	169
9	5 plus 0.02M KH_2PO_4	25	1.0973	1.260°	169

° Own measurements.

$$\overline{M}_z = \frac{\Sigma n_i M_i^3}{\Sigma n_i M_i^2} \quad \text{and} \quad \overline{M}_{z+1} = \frac{\Sigma n_i M_i^4}{\Sigma n_i M_i^3} \tag{6}$$

not homogeneous, $\overline{M}_n < \overline{M}_w < \overline{M}_z$.

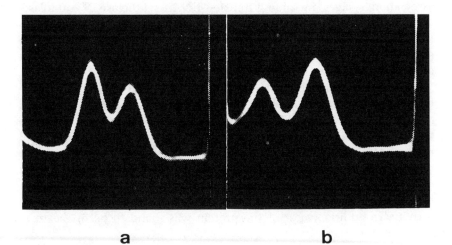

a   b

Figure 11. Sedimentation pattern of α-amylase from *Bacillus subtilis* with different zinc content: (a) 55% associated molecules ($s_{20} = 6.0S$, molecular weight 97,000), zinc content 1 gram-atom zinc per mole of enzyme, and 45% dissociated molecules ($s_{20} = 4,2S$, molecular weight 48,000). Zinc content about 0.6 gram-atom zinc per mole of enzyme; (b) 65% dissociated molecules, zinc content about 0.35 gram-atom zinc per mole of enzyme. Measurements in 0.02M sodium glycerophosphate buffer (+0.1M NaCl), pH 7.0, at 59,780 rpm, angle of the schlieren diaphragm 40° (a) or 35° (b). Pictures taken 120 minutes (a) or 150 minutes (b) after the rotor had reached full speed, with direction of sedimentation from right to left. [Reprinted, by permission, from E. A. Stein and E. H. Fischer, Biochimica et Biophysica Acta **39**: 287 (1960). © Elsevier Publishing Company.]

10.2.7.2 Maleylation

Maleic anhydride has many advantages as a reversible blocking reagent for amino groups of proteins or peptides. The introduction of negative charges (an increment of −2 per reacted amino group) according to Equation 7 is a mild procedure which favors the solubility of polypeptides at neutral pH. This reaction is reported to be specific for amino groups and reversible under mild conditions.

Proteins consisting of several polypeptide chains can be smoothly and completely dissociated at neutral pH or at pH not far from neutral pH by maleylation into soluble polypeptide chains because the predominant charges maximize electrostatic repulsion and minimize association (e.g., methionyl-transfer–RNA synthetase,[138] aldolase, transaldolase, fructose diphosphatase,[139] and glucose-6-phosphate dehydrogenase[140]; cf. also References 83 and 141). Denatured proteins are soluble at pH 8 in the absence of guanidine hydrochloride or urea after maleylation.

The maleyl amino bond is very stable above pH 6 but is readily hydrolyzed below pH 5 (the half-life of ϵ-maleyl lysine is 11 hours at 37°C and pH 3.5[141]). These mild conditions for hydrolysis minimize any other peptide bond cleavage. Therefore, maleylation is a useful method for "reversible" blocking of amino groups; treatment of the maleyl derivative of rabbit muscle aldolase at pH 4.5 and 25°C for 40 minutes in the presence of 10^{-3}M dithiothreitol resulted in the loss of half of the maleyl residues and reappearance of 46% of the enzymatic activity. In addition most of the protein was found to be in the reassociated (tetrameric) state.

At least in some cases the dissociation into polypeptide chains upon maleylation is based on the incorporation of relatively few maleyl groups. For instance, only 16 lysine residues of the aldolase molecule— 15% of the total amount of lysyl residues—react with maleic anhydride.[139] This small amount does not change the molecular weight to a great extent and therefore need not be taken into consideration in molecular weight calculation.

When maleylation is used to dissociate multichain protein molecules, sedimentation measurements are usually made to test for dissociation. If dissociation occurs, the sedimentation coefficient is markedly reduced, e.g., from 7.68 S to 1.71 S for rabbit muscle aldolase, from 6.2 S to 1.7 S for spinach aldolase, from 7.04 S to 2.04 S for rabbit liver fructose–diphosphatase, from 4.18 S to 1.34 S for transaldolase,[139] and from 9.3 S to 3.4 S for glucose-6-phosphate dehydrogenase.[140] A decrease of the sedimentation coefficient is not necessarily a proof of dissociation because highly charged proteins exhibit marked electrostatic effects on sedimentation. Therefore experiments must be carried out at increasing ionic strengths (up to one). In addition, a denaturation may cause unfolding and therefore a decrease of sedimentation coefficient. If the resulting solution is not homogeneous, a second maleylation can be performed or the dissociation product can be separated by gel filtration on Sephadex G-100.[140]

The determination of the molecular weight has to be carried out within a few days after the reaction with maleic anhydride because the maleyl derivatives subsequently show a tendency to (nonspecific?) reassociation and, therefore, to production of higher molecular weight components.

Procedure (for methionyl-transfer–RNA synthetase[138]): Redistilled maleic anhydride (2.5 mg) (thirtyfold excess over all free amino groups) is added to 1.5 mg of protein in 200 μl of 0.5M sodium borate buffer, pH 9.5, and incubated at room temperature for 1 hour. It is then dialyzed against 0.1M phosphate buffer, pH 7.0, containing 1.0 to 0.4M sodium chloride. During the reaction the pH of the solution is maintained between 8.5 and 9.5 by the addition of 0.1N sodium hydroxide or ammonia. Maleic anhydride dissolved in benzene can also be added to the protein.

In order to avoid the inaccuracies inherent in the determination of the molecular weights of proteins in either 8M urea or 6M guanidine hydrochloride, -SH groups containing proteins (*e.g.*, alcohol dehydrogenase from liver or yeast and glyceraldehyde-3-phosphate dehydrogenase[83]) can be carboxymethylated and then maleylated by reaction with maleic anhydride.[83] The solution of carboxymethylated protein in urea or guanidine hydrochloride is adjusted to pH 9 by the addition of 0.1N sodium hydroxide and is then reacted with maleic anhydride in a twentyfold excess over the total amino groups, with the pH being maintained between 8.5 and 9.0 by the addition of 0.1N sodium hydroxide. When the reaction is complete, the protein is dialyzed exhaustively against 0.5% w/v ammonium hydrogen carbonate and then finally against this buffer containing in addition 0.4M sodium chloride.

10.2.7.3 Succinylation

Succinylation with succinic anhydride[142] (or its derivatives, *e.g.*, tetrafluorosuccinic anhydride[143]) (Equation 8) is an analogous acyla-

$$-NH_3^{\oplus} + \underset{O}{\overset{O}{\underset{\diagdown}{\text{C}}}}\overset{\text{CH}_2}{\underset{\text{CH}_2}{\big|\big|}} \longrightarrow -NH-CO-CH_2-CH_2-COO^{\ominus} + 2\,H^{\oplus}$$

tion reaction to maleylation. Depending on the ratio of moles of the added reagent to the moles of free amino groups in the protein, the number of groups introduced can vary from few to the maximum number of the amino groups. The reagent shows a strong preference for amino groups but will also react with hydroxyl and sulfhydryl groups under suitable circumstances.

Procedure (for catalase[54]): To 15–18 mg catalase dissolved in 1 ml 67 mM phosphate buffer, pH 7.6, 10 mg of solid succinic anhydride is added; after 10 and 30 minutes, 10 mg succinic anhydride is added again. During the succinylation the solution is stirred and the pH is maintained between 7.6 and 8.0 by addition of 0.5N sodium hydroxide. After 2 hours incubation at room temperature the solution is dialyzed against phosphate buffer and examined in the analytical ultracentrifuge. The gradient of the native protein decreases and the gradient of the polypeptide chain appears; with sufficient succinic anhydride the former disappears completely (Figure 12).

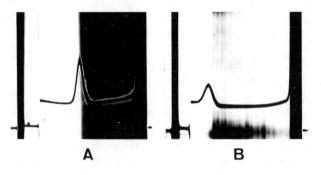

A   **B**

Figure 12. Sedimentation pattern of native catalase (A, 4.8 mg/ml) and succinylated catalase (B, 2.7 mg/ml). Measurements in 67 mM potassium sodium phosphate buffer, pH 7.6, at 59,780 rpm and 20.0°C, angle of the schlieren diaphragm 60°. Picture taken 32 (A) and 48 (B) minutes after the rotor had reached full speed. [Reprinted, by permission, from H. Sund, K. Weber, and E. Mölbert, European Journal of Biochemistry 1: 400 (1967). © Springer-Verlag.]

After succinylation, beef liver catalase ($s^0_{20,w} = 11.29$ S, molecular weight 243,000) is dissociated into four polypeptide chains ($s^0_{20,w} = 3.22$ S),[54] hemerythrin from *Golfingia gouldii* ($s^0_{20,w} = 6.75$ S, molecular weight 107,000) is dissociated into eight polypeptide chains ($s^0_{20,w} = 1.95$ S).[144]

10.2.7.4   Molecular weight determination of polypeptide chains by dodecylsulfate–polyacrylamide gel eletrophoresis[60]

Electrophoresis in polyacrylamide gel in the presence of sodium dodecylsulfate (SDS) has proved to be a useful tool for the separation and characterization of polypeptide chains.[60, 145] This technique can be used for rapid and simple estimation of molecular weights of polypeptide chains. The method is widely applicable and indicates that the molecular weight of a particular protein can be obtained within a day.

The method is based on the premise that separation of proteins on polyacrylamide gel is dependent not only on the charge but very strongly on the size of the molecule. There exists a relationship between electrophoretic mobilities and molecular weights of various proteins. Comparison of the mobility of an unknown polypeptide chain with the mobilities of marker proteins yields the unknown molecular weight. In addition, the electrophoretic mobility depends on the amount of methylenebisacrylamide used for the preparation of the gel. For example, the mobility of fumarase (molecular weight 50,000) changes from 0.28 to 0.63 if the amount of the cross-linker is reduced from the normal amount to half of this value. This mobility change suggests that gels with half the normal amounts of cross-linker are useful in studying proteins over a wide range of molecular weights. Therefore, the procedure may be adapted to study different molecular weight ranges.

Usually 10 μg of protein is applied per gel. The amount can be lowered if the gel is stained for a long period. Two procedures can be used to examine the relationship between electrophoretic mobilities and molecular weights of various proteins. Either several SDS-denatured proteins are mixed and run together on one gel or the proteins are run in a parallel manner on different gels. Identical results are obtained by both methods.

The results obtained with 37 different polypeptide chains in the molecular weight range 10,000–70,000 are shown in Figure 13. When the electrophoretic mobilities are plotted against the logarithm of the known polypeptide chain molecular weights, a straight line is obtained. The maximum deviation from the predicted values in this range is less than 10% and for most proteins the agreement is better. A hyperbolic curve instead of a straight line is obtained if polypeptide chains in the higher molecular-weight range 50,000–200,00 are studied using gels with half the normal amount of cross-linkers. With these proteins an accuracy of ±10% also still seems possible.

In connection with these methods some theoretical problems arise. The binding of SDS to proteins has been shown for several proteins and is assumed to be the basis of separation of the denatured proteins upon performance of SDS electrophoresis on polyacrylamide. If so, one must assume that the individual charge pattern of each protein molecule is totally changed by the binding of SDS anions, rendering all molecules negatively charged. At present it is difficult to see why proteins that differ widely in amino acid composition and isoelectric points should all follow the general pattern. It is possible that the sieving, which is an exponential function, overcomes the charge effect; this factor may be of minor importance. In their investigation Weber and Osborn[60] have shown that close to 40 different proteins have electrophoretic mobilities which are independent of the isoelectric point and the amino acid composition (and perhaps of the amount of

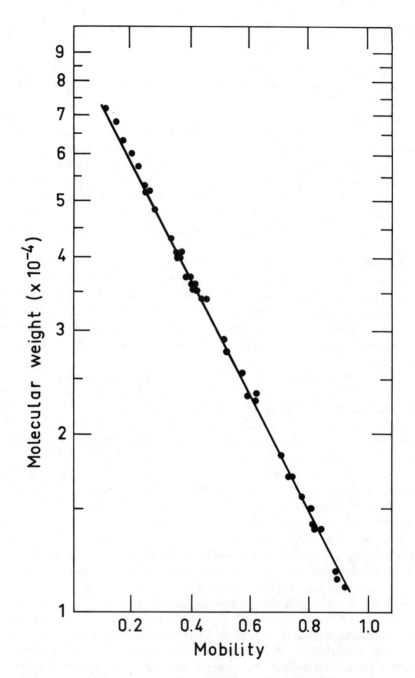

Figure 13. Comparison of the molecular weights of 37 different polypeptide chains in the molecular weight range 11,000–70,000 with their electrophoretic mobilities with the normal amount of cross-linker. [Reprinted, by permission, from K. Weber and M. Osborn, The Journal of Biological Chemistry **244**: 4406 (1969). © American Society of Biological Chemists, Inc.]

bound SDS) and which seem governed solely by the molecular weights of their polypeptide chains.

This method is a strong competitor with others commonly employed, due to the excellent resolving power of the gels over a wide range of molecular weights, the fact that a molecular weight can be obtained within a day by means of a simple method, and the small amount of protein needed. The theoretically fully developed ultracentrifuge method, for instance, needs special and expensive equipment. The limitation of this method is more or less the accuracy of the determination of the partial specific volume. An uncertainty of 0.02 in this value (that is, 2–3%) introduces a deviation up to 10%, even if all the ultracentrifuge measurements are performed exactly. In the molecular weight range between 15,000 and 100,000 the accuracy for the molecular weight determination by SDS–polyacrylamide gel electrophoresis is better than ±10%. This range can be covered with numerous commercially available proteins as standards. The difficulty with higher molecular weights is probably due only to the fact that fewer markers are available for the range 90,000–200,00. Molecular weights of polypeptide chains are generally less than 100,000.

Procedure[60]

Preparation of protein solutions. The proteins are incubated at 37°C for 2 hours in 0.01M sodium phosphate buffer, pH 7.0, 1% in SDS, and 1% in β-mercaptoethanol. (In the cases of tropomyosin, paramyosin, and myosin, proteins were dissolved in this buffer in the presence of 8M urea.) The protein concentration was normally between 0.2 and 0.6 mg per ml. After incubation the protein solutions are dialyzed for several hours at room temperature against 500 ml of 0.01M sodium phosphate buffer, pH 7.0, containing 0.1% SDS and 0.1% β-mercaptoethanol. In most cases the dialysis step may be omitted and the protein dissolved directly in the dialysis buffer.

Preparation of gels. Gel buffer contains 7.8 g $NaH_2PO_4 \cdot H_2O$, 38.6 g $Na_2HPO_4 \cdot 7H_2O$, 2 g SDS per liter. For the 10% acrylamide solution, 22.2 g of acrylamide and 0.6 g of methylenebisacrylamide are dissolved in water to give 100 ml of solution. Insoluble material is removed by filtration through Whatman No. 1 filter paper. The solution is kept at 4°C in a dark bottle. Gels with increased and decreased cross-linker contain twice and half the concentration of cross-linker, respectively.

The glass gel tubes are 10 cm long with an inner diameter of 6 mm. Before use they are soaked in cleaning solution, rinsed, and ovendried. For a typical run of 12 gels, 15 ml of gel buffer is deaerated and mixed with 13.5 ml of acrylamide solution. After further deaeration, 1.5 ml of freshly made ammonium persulfate solution (15 mg per ml) and 0.045 ml of *N,N,N',N'*-tetramethylethylenediamine are added. After mixing, each tube is filled with 2 ml of the solution. Before the gel hardens a few drops of water are layered on top of the gel solution. After 10–20 minutes an interface can be seen indicating that the gel has solidified. Gels with the normal amount of cross-linker remain clear, those with doubled cross-linker turn opaque. Just before use the water layer is sucked off and the tubes are placed in the electrophoresis apparatus.

Preparation of samples.   For each gel, 3 μl of tracking dye (0.05% Bromophenol Blue in water), 1 drop of glycerol, 5 μl of mercapto-ethanol, and 50 μl of dialysis buffer are mixed in a small test tube. Then 10–50 μl of the protein solution is added. After mixing, the solutions are applied on the gels. Gel buffer, diluted 1:1 with water, is carefully layered on top of each sample to fill the tubes. The two compartments of the electrophoresis apparatus are filled with gel buffer diluted 1:1 with water. Electrophoresis is performed at a constant current of 8 mA per gel with the positive electrode in the lower chamber. Under these conditions the marker dye moved three-quarters through the gel in approximately 4 hours. The time taken to run the gel may be decreased by decreasing the molarity of the gel buffer.

After electrophoresis, the gels are removed from the tubes by squirting water from a syringe between gel and glass wall and by using a pipet bulb to exert pressure. The length of the gel and the distance moved by the dye are measured.

Staining and destaining.   The gels are placed in small tubes filled with staining solution prepared by dissolving 1.25 g of Coomassie Brilliant Blue in a mixture of 454 ml 50% methanol and 46 ml glacial acetic acid, and removing insoluble material by filtration through Whatman No. 1 filter paper. Staining is performed at room temperature. The time varies from 2–10 hours. The gels are removed from the staining solution, rinsed with distilled water, and placed in destaining solution (75 ml acetic acid, 50 ml methanol, and 875 ml water) for a minimum of 30 minutes. The gels are then further destained electrophoretically for 2 hours in a gel electrophoresis apparatus using destaining solution. The length of the gels after destaining and the positions of the blue protein zones are recorded. The gels are stored in 7.5% acetic acid solution.

In the acidic solution used for staining and destaining the gels swell some 5%. Gels with lower amount of cross-linker show more swelling. Therefore, the calculation of the mobility has to include the length of the gel before and after staining as well as the mobility of the protein and of the marker dye. Assuming even swelling of the gels, the mobility is calculated as

$$\text{Mobility} = \frac{\text{distance of protein migration}}{\text{length after destaining}} \times \frac{\text{length before staining}}{\text{distance of dye migration}}$$

The mobilities are plotted against the known molecular weights expressed on a semilogarithmic scale.

10.2.7.5 Denaturation by guanidine hydrochloride

A variety of denaturing agents are known. Generally guanidine hydrochloride or urea, both in high concentrations, are used for denaturation of proteins. The result from the experimental evidence is that guanidine hydrochloride in connection with a reducing agent (to break disulfide bonds and avoid formation of disulfide bonds from -SH groups during the denaturation) is the preferred reagent whenever complete dissociation of a protein molecule into its constituent polypeptide chains is desired.

All proteins which have been studied in high concentrations of the denaturing agent guanidine hydrochloride have been found to contain no residual noncovalent structure. If all existing disulfide bonds have been broken, each polypeptide chain is completely unfolded and behaves as a linear random coil retaining no elements of the original native conformation. The molecular radius of a randomly coiled polypeptide chain is a function only of the number of amino acid residues in the chain, and hence, a function of the molecular weight of the chain regardless of its composition. Therefore, molecular weight and hydrodynamic properties like sedimentation coefficient and intrinsic viscosity can be correlated direcly.[137, 146, 147]

After denaturation in the presence of guanidine hydrochloride the molecular weight of the dissociation product may be estimated by different methods: (a) sedimentation equilibrium measurements, (b) combination of sedimentation velocity and diffusion measurements, (c) viscosity measurements, and (d) gel filtration.

Sedimentation and diffusion experiments are performed with the analytical ultracentrifuge. In the presence of high guanidine hydrochloride concentration the sedimentation coefficient ($s_{20,w}^0$) is generally 1–2 S.[51, 54, 84, 146] In the case of catalase the $s_{20,w}^0$-value of the native enzyme molecule (11.29 S, molecular weight 243,000) decreases to 2.1 S (molecular weight 59,000) in the presence of 5M guanidine hydrochloride and 0.1M β-mercaptoethanol (Figure 14). Catalase is dissociated into four polypeptide chains.[54] In the case of aldolase (molecular weight 158,000), the sedimentation coefficient decreases from 8.0 S

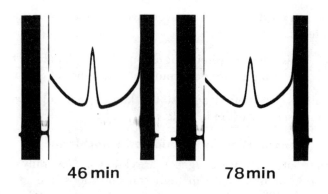

46 min   **78 min**

Figure 14. Sedimentation pattern of beef liver catalase (4.63 mg/ml) in 5M guanidine hydrochloride–0.1M β-mercaptoethanol. Measurement in 67 mM potassium sodium phosphate buffer, pH 7.6, at 59,780 rpm and 20°C in the synthetic boundary cell (valve type), diaphragm angle 60°. Picture was taken 46 and 78 minutes after formation of the boundary, respectively. [Reprinted, by permission, from H. Sund, K. Weber, and E. Mölbert, European Journal of Biochemistry 1: 400 (1967). © Springer-Verlag.]

to 1.85 S (molecular weight 40,000) in the presence of guanidine hydrochloride.[51] In any case, however, the sedimentation coefficient will decrease upon treatment with guanidine hydrochloride—in the absence of dissociation, also, because this reagent causes unfolding.

If β-mercaptoethanol is absent during the denaturation process, it is possible that -SH groups are oxidized to disulfide groups which link polypeptide chains together covalently. This factor may explain why, after treatment of catalase with guanidine hydrochloride in the absence of β-mercaptoethanol, the molecular weight is found to be about 144,000 instead of 59,000 which is the molecular weight of the polypeptide chains[54] and is obtained in the presence of β-mercaptoethanol.

At guanidine concentrations lower than 5M the dissociation into polypeptide chains may not be complete. For instance, in the presence of increasing concentrations of guanidine hydrochloride (in the presence of 0.1M β-mercaptoethanol at pH 6.3 and 20°C) between 0.25M and 2.0M, a progressive decrease of the $s_{20,w}^0$-values of ATP-creatine transphosphorylase from 5.1–1.9 S occurred. At 0.25M or below ($s_{20,w} = $ 5.1 S, which is characteristic for the associated molecule, the native enzyme molecule has a $s_{20,w}$ value of 5.3 S) and above 2.0M guanidine hydrochloride ($s_{20,w} = 1.4$ S), only a single sedimenting species was evident; between these concentrations the system shows a tendency toward aggregate formation with species sedimenting in the neighborhood of 20 S.[148]

The calculation of molecular weight requires the knowledge of the partial specific volume (v) of the protein. The accuracy of the estimation of this value determines largely the accuracy of the molecular weight determination. Results in the literature conflict regarding the effect of concentrated guanidine hydrochloride on v. The apparent partial specific volume of a number of proteins is unaffected or only slightly diminished in the presence of guanidine hydrochloride; there are reasons for believing that v for proteins in 6M guanidine hydrochloride will not in general differ significantly from its value in dilute aqueous salt solutions (*cf.* Table III and References 149–152). However, the assumption of preferential hydration is still often used to account for deviations and therefore, if possible, it has to be proved if v is changed in guanidine hydrochloride solutions. Based on the high

Table III

Partial Specific Volume (v) of Proteins in the Native State
and in Guanidine Hydrochloride Solution at 25°C

Protein	v in the Native State	v in 6M Guanidine Hydrochloride Solution*	Reference
Beef Liver Glutamate Dehydrogenase	0.751	0.726	153
Bovine Serum Albumin	0.734	0.728	152
β-Lactoglobulin	0.752	0.756	154
Rabbit γ-Globulin	0.74	0.72	155
Rabbit Muscle Aldolase	0.739	0.733	152
Ribonuclease	0.709	0.709	154

*In the case of beef liver glutamate dehydrogenase the guanidine concentration was 5.7M.

density of the guanidine solution compared to buffer solutions, inaccurate values of v cause relatively large errors in the term $(1 - v\rho)$ in the Svedberg equation and, therefore, on the molecular weight. For example, a deviation in v of 0.01 ml/g causes an error of 7% in the molecular weight if determined in 5M guanidine hydrochloride but only 3% for the conditions in buffer solutions.

Measurements of viscosities of polypeptide chains in concentrated aqueous solutions of guanidine hydrochloride and in the presence of β-mercaptoethanol show that the intrinsic viscosities ($[\eta]$) depend on molecular weight exactly as predicted for randomly coiled polymer chains.[137, 146] The molecular weight can be estimated according to Equation 9: n is the number of monomer units (here amino acid resi-

$$[\eta] = K'n^a = K''(M_o n)^a \tag{9}$$

dues) per chain; M_o is the average molecular weight per monomer unit; and K', K'', and a are constants. The constant a usually lies in the range 0.5–0.8 for random coils. For polypeptide chains in the molecular weight range 3,000–197,000 the viscosities follow Equation 9 with a value of $a = 0.67$ at 25°C.

Gel filtration of reduced polypeptide chains in 6M guanidine hydrochloride permits molecular weight estimations between the limits of 80,000 and 1,000.[147] The accuracy is 7% in the molecular weight range 40,000–10,000, and 10% at the extreme calibration limits. A column of Bio-Gel A-5M with a nominal agarose content of 6% was used as the gel filtration medium. The molecular weight is obtained by comparison with marker proteins of known molecular weights by plotting the logarithm of the molecular weight against the distribution coefficients or against V_e/V_o, where V_e is the elution volume and V_o the void volume (see Chapter 5, Sections 5.3 and 5.5).

The gel filtration method may be useful as a substitute for viscosity measurements. As with the hydrodynamic properties mentioned above, the gel filtration elution position of a linear random coil is a function of its chain length and therefore represents a measure of molecular weight. If limited amounts of material prevent viscosity measurements, gel filtration may provide an adequate substitute. A molecular weight estimated by gel filtration which is significantly lower than that obtained by sedimentation equilibrium measurement would suggest that cross-links (*e.g.*, disulfide bonds) still remain.

Procedure*

Preparation of samples (for catalase[54]). One ml of about 10M guanidine hydrochloride solution (dissolved in 67mM potassium sodium phosphate buffer, pH 7.6, containing 0.2M mercaptoethanol) is layered with 1 ml catalase solution (10–20 mg/ml dissolved in 67mM phosphate buffer, pH 7.6) followed by rapid mixing (to bring the solution rapidly from zero concentration to 5M guanidine hydrochloride) and dialysis against 100 ml 5M guanidine hydrochloride (dissolved in 67 mM potassium sodium phosphate buffer, pH 7.6, containing 0.1M mercaptoethanol) for 48 hours.

---

*Densities and viscosities of various aqueous solutions of guanidine hydrochloride and urea are compiled in Table II, Section 10.2.7.1.

During dialysis it is necessary to watch that water does not evaporate from any part of the system. After dialysis, all operations have to be performed rapidly and evaporation of water must be carefully avoided. Otherwise, one can observe in the analytical ultracentrifuge a so-called salt gradient in addition to the protein gradient because the guanidine concentration may be different in the protein solution and in the solvent used for formation of the synthetic boundary.

Purification of guanidine hydrochloride[156]

The salt should be recrystallized by the following procedure: 250 g of guanidine hydrochloride is dissolved in 1 liter of hot absolute ethanol and the solution is destained with charcoal and filtered through a heated large funnel if necessary. Benzene (500 ml) is added to the still hot ethanol solution, and the mixture is kept in the cold for several hours before the crystal needles are collected. Sometimes a yellow, sticky impurity can be removed from the original sample by washing with acetone.

The recrystallized sample should be further recrystallized either from methanol by dissolving in nearly boiling methanol and chilling in Dry Ice–acetone mixture, or from water by vacuum evaporation of an aqueous solution nearly saturated at 40°C. It is finally dried in a vacuum desiccator under P_2O_5.

An alternative procedure is to neutralize a slurry of guanidinium carbonate, which has been recrystallized from 50% ethanol, with distilled 20% hydrochloric acid to pH 4. The solution is evaporated to obtain crystals of guanidine hydrochloride. The product is fairly pure but still requires recrystallization as described above.

Criteria of purity specify (1) that the absorption spectrum of a 6M solution should show gradually increasing absorbance and no peak from 350 to near 230 nm, where a sharp rise is usually observed to approximately 0.15 absorbance unit at 225 nm; and (2) that a minute amount of standard acid or base should cause a large pH shift.

10.2.8 Association-Dissociation Equilibrium

In general, protein molecules behave under physiological conditions as monodisperse substances with a defined molecular weight, though as shown in the preceding sections, a great number of proteins do not consist of a single polypeptide chain. However, in some cases an association-dissociation equilibrium exists between monomers and polymers. With this the monomer can represent the polypeptide chain (chymotrypsin,[157, 158] hemerythrin[159]) or subunits consisting of more polypeptide chains (glutamate dehydrogenase[9, 10]).

Several mechanisms for association-dissociation equilibria are possible.[2, 3, 5, 9] Two limiting cases exist: (a) the closed (one step) association-dissociation equilibrium with no intermediates (Equation 10, where M_1 denotes the monomer); and (b) the open association-

$$nM_1 \xrightleftharpoons{K_{1,n}} M_n; \quad K_{1,n} = \frac{[M_n]}{[M_1]^n} \tag{10}$$

dissociation equilibrium with the consecutive association of monomers according to Equation 11 without limit and with identical equilibrium constants for all steps. In other cases the equilibrium constants may be

$$M_1 + M_1 \xrightleftharpoons{K_{1,2}} M_2; \quad K_{1,2} = \frac{[M_2]}{[M_1]^2}$$

$$M_2 + M_1 \xrightleftharpoons{K_{2,3}} M_3; \quad K_{2,3} = \frac{[M_3]}{[M_2][M_1]}$$

$$M_3 + M_1 \xrightleftharpoons{K_{3,4}} M_4; \quad K_{3,4} = \frac{[M_4]}{[M_3][M_1]} \tag{11}$$

$$M_i + M_1 \xrightleftharpoons{K_{i,i+1}} M_{i+1}; \quad K_{i,i+1} = \frac{[M_{i+1}]}{[M_i][M_1]}$$

different; i.e., one constant for the dimerization and a second constant for all other steps. Between these limiting cases a great number of special cases are possible, for instance, after the formation of dimers from monomers further association only of dimers.

While the investigation of an association-dissociation equilibrium may be carried out mainly by sedimentation equilibrium or light scattering measurements, it also may proceed under osmotic pressure measurements or molecular sieve chromatography. Methods for the interpretation of experimental data are described by several authors (References 2, 3, 5, 6, and 160–164). The best way to elucidate the mechanism of association seems to be comparison of the experimental data with calculated curves for different mechanisms.

An example of a closed association-dissociation equilibrium is hemerythrin (azid complex) from *Golfingia gouldii*. Based on sedimentation equilibrium measurements, the equilibrium constant for the association-dissociation equilibrium between monomers and octamers (Equation 12) was found at 5°C to be 3.4×10^{36} M^{-7}.[159]

$$8M_1 \rightleftharpoons M_8, \quad K_{1,8} = \frac{[M_8]}{[M_1]^8} \tag{12}$$

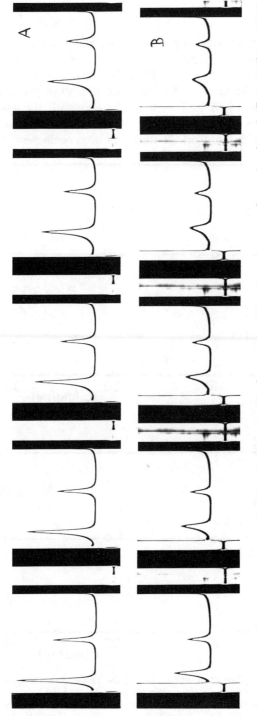

Figure 15. Sedimentation pattern of beef liver glutamate dehydrogenase in 67 mM potassium sodium phosphate buffer, pH 7.6 at 20°C. Measurements at 50,740 rpm in the synthetic boundary cell (value type), angle of the schlieren diaphragm 65°. Time difference between the exposures 2 minutes. Protein concentrations 8.39 mg/ml (A, layering with c = 5.59 mg/ml) and 5.60 mg/ml (B layering with c = 3.73 mg/ml). (Reprinted by permission, from H. Sund, in *Mechanismen enzymatischer Reaktionen*, 14. Colloquium der Gesellschaft für Physiologische Chemie, 1963, © 1964, Springer-Verlag.)

The association-dissociation equilibrium of glutamate dehydrogenase from beef liver can be described as an open association-dissociation equilibrium according to Equation 11. This enzyme exhibits an anomalous sedimentation behavior. The concentration gradient is not symmetrical (in Figure 15, A and B, the solutions with the low protein concentration); it shows a trailing edge on the solvent side, and at concentrations lower than 5 mg/ml the sedimentation coefficient drops with

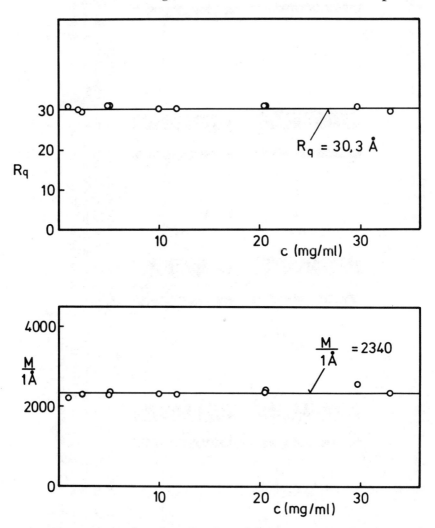

Figure 16. Radius of gyration (R_q) of the cross-section and mass per 1 Å length of the glutamate dehydrogenase particles (M/1Å) as a function of protein concentration (c). Measurements in 67 mM potassium sodium phosphate buffer pH 7.6. [Reprinted, by permission, from H. Sund, I. Pilz, and M. Herbst, European Journal of Biochemistry 7: 517 (1969). © Springer-Verlag.]

decreasing protein concentrations.[166, 167] These anomalies are due to the dissociation of the enzyme molecule into subunits. The trailing edge, which is observed during the sedimentation, is caused by the decrease of the protein concentration across the boundary between the solution and the solvent corresponding to the concentration distribution in the ultracentrifuge cell. This favors further dissociation during the run. Figure 15 shows that the dissociation during the sedimentation can be prevented if the experiment is performed in a synthetic boundary cell not layered in the usual way with buffer but with glutamate dehydrogenase solution of lower concentration. In this case the boundary is formed between glutamate dehydrogenase solutions of different concentrations, and the concentration gradient is symmetrical. In the top solution, as with normal layering, a boundary is formed between the solution and the solvent, and therefore the concentration gradient in this part of the cell is not symmetrical.

From diffusion, sedimentation, and viscosity data it was concluded that the glutamate dehydrogenase molecule in the associated state has a prolate shape and that the dissociation involves a transverse cleavage.[165, 167] This conclusion was confirmed by X-ray small-angle measurements[8] (*cf.* also Reference 10). The radius of gyration of the cross section and the mass per unit length were independent of protein concentration and therefore independent of molecular weight (Figure 16). In agreement with this, a linear relationship is obtained if the length of the glutamate dehydrogenase particles is plotted versus their molecular weights (Figure 17).

The dependence of the molecular weight as a function of protein concentration was investigated by light-scattering measurements.[9, 10] Figure 18 shows that the molecular weights reach a maximum at about 9 mg/ml. At higher protein concentrations the molecular weights decrease. The increase of molecular weight at protein concentrations below the maximum can be attributed to self-association; however, the decrease at higher concentrations cannot be ascribed to equilibrium constants, which are always positive. Thus, the decrease of the molecular weights is due to effects of nonideality. With consideration of a second virial coefficient $[A_2 = 8 \times 10^{-9}(\text{Mole} \cdot l/g^2)]$, the experimental data can be fitted assuming an open association-dissociation equilibrium according to Equation 11 with identical equilibrium constants for all steps $(K_{i, i+1} = 1.1 \times 10^6 \ M^{-1})$ and without limit. A theoretical curve calculated with the molecular weight of the monomer (M_1) of 310,000* and with consideration of the second virial coefficient $(\overline{M}_{w(\text{app})}$ in Figure 18) corresponds well with the experimental data. A second curve, shown in Figure 18, is calculated without consideration

---

*The monomeric subunit, having a molecular weight of 310,000, as well as the associated particles are enzymatically active.

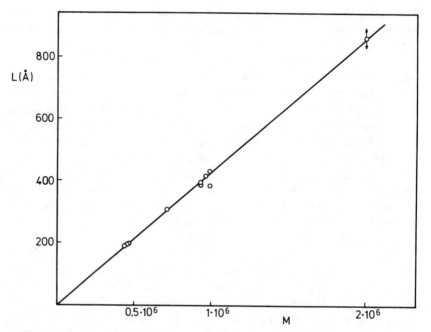

Figure 17. Length L of glutamate dehydrogenase as a function of molecular weight M. Measurements in 67 mM potassium sodium phosphate buffer pH 7.6. [Reprinted, by permission, from H. Sund, I. Pilz, and M. Herbst, European Journal of Biochemistry 7: 517 (1969). © Springer-Verlag.]

of the second virial coefficient, according to Equation 13, where c is the protein concentration in mg/ml. The second curve represents the

$$\frac{1}{M_{app}} = \frac{1}{M_{tr}} + 2A_2 c \tag{13}$$

true weight average molecular weight ($\overline{M}_{w(tr)}$). This mechanism seems to be the only reasonable one because the associated molecule is formed by an interaction of end groups of the subunits which gives rodlike particles. A closed association-dissociation equilibrium according to Equation 10 can be excluded.

It is interesting to recognize that the properties of glutamate dehydrogenase from pig and beef liver are very similar. These two enzymes from different sources can form hybrid-associated particles in a statistical manner.[168] The contact areas of these enzymes are therefore the same or similar.

10.3 Concluding Remarks

The elucidation of the quaternary structure of a certain protein is not as simple as it looks. The methods which are usually available do not always give unequivocal results. Difficulties arise because (a) no quantitative yield is obtained if chemical methods are used; (b) dissociation may not be complete; (c) denatured proteins are often investigated in solutions which are far from ideal (the effects of nonideality in any real macromolecule is often significant and sometimes overlooked); and (d) under the conditions of the applied procedures pep-

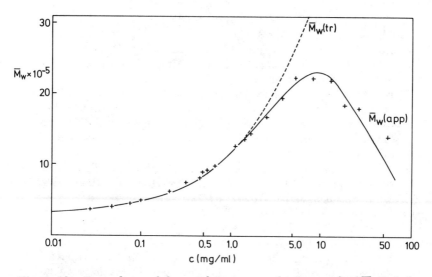

Figure 18. Dependence of the weight average molecular weight ($\overline{M}_w$) of glutamate dehydrogenase as a function of protein concentration (c). Measurements in 67 mM potassium sodium phosphate buffer, pH 7.6, at 20°C. Experimental data from light scattering measurements at 436 nm (x), for c > 8 mg/ml extrapolated for $\lambda \rightarrow \infty$. Curves are calculated for an open association-dissociation equilibrium according to Equation 11 with $K_{i,\,i+1} = 1.1 \times 10^6\ M^{-1}$, $M_1 = 310{,}000$, and $A_2 = 8 \times 10^{-9}$ [Mole l/g^2] ($\overline{M}_{w\,(app)}$) or without consideration of the second virial coefficient ($\overline{M}_{w\,(tr)}$). (Reprinted, by permission, from J. Krause *et al.*, in *Pyridine Nucleotide-Dependent Dehydrogenases*, ed. H. Sund, p 279, © 1970, Springer-Verlag.)

tide bonds may be hydrolyzed. In the latter case the number of polypeptide chains found is too high, whereas in cases (a) and (b) the opposite is true. Indeed, the limiting factor in the determination of molecular weights is usually the degree of homogeneity of the material under investigation.

The greater the difference between the molecular weight of the native protein molecule and the polypeptide chains, the greater are the difficulties in determining the exact number of polypeptide chains which constitute the native protein molecule. The best method for getting an exact and unequivocal answer is the use of a number of different methods.

REFERENCES

1. Reithel, F. J.   Advan. Protein Chem. 18, 123 (1963).
2. Nichol, L. W., J. L. Bethune, G. Kegeles, and E. L. Hess, in The Proteins, Vol. 2, 2nd ed., Ed. by H. Neurath (New York and London: Academic Press, 1964), p 305.
3. Casassa, E. F., and H. Eisenberg.   Advan. Protein Chem. 19, 287 (1964).
4. Sund, H., and K. Weber.   Angew. Chem. 78, 217 (1966); Angew. Chem. Intern. Ed. 5, 231 (1966).
5. Elias, H. G., and R. Bareiss.   Chimia 21, 53 (1967).
6. Adams, E. T.   Fractions, No. 3 (Palo Alto, California: Spinco Division of Beckman Instruments Inc., 1967).
7. Engström, A., and B. Strandberg (eds.).   Symmetry and Function of Biological Systems at the Macromolecular Level (Stockholm: Almqvist & Wiksell; and New York, London: John Wiley, 1969).
8. Sund, H., I. Pilz, and M. Herbst.   European J. Biochem. 7, 517 (1969).
9. Krause, J., K. Markau, M. Minssen, and H. Sund, in Pyridine Nucleotide-Dependent Dehydrogenases, Ed. by H. Sund (Heidelberg, New York: Springer-Verlag, 1970), p 279.
10. Eisenberg, H., in Pyridine Nucleotide-Dependent Dehydrogenases, Ed. by H. Sund (Heidelberg, New York: Springer-Verlag, 1970), p 293.
11. Kakiuchi, K.   J. Phys. Chem. 69, 1829 (1965).
12. Linderstrøm Lang, K. U., and J. A. Schellman, in The Enzymes, Vol. 1, 2nd ed., Ed. by P. D. Boyer, H. Lardy, and K. Myrbäck (New York: Academic Press, 1959), p 443.
13. Bernal, J. D.   Discussions Faraday Soc. 25, 7 (1958).
14. Pauling, L., and R. B. Corey.   Fortschr. Chem. Org. Naturstoffe 11, 180 (1954); Pauling, L.   The Nature of the Chemical Bond, 3rd ed. (Ithaca, New York: Cornell University Press, 1960), p 498.
15. Kendrew, J. C.   Angew. Chem. 75, 595 (1963).
16. Perutz, M. F.   Angew. Chem. 75, 589 (1963); Sci. Am. 211, No. 5, 64 (1964).
17. Monod, J., J. Wyman, and J. P. Changeux.   J. Mol. Biol. 12, 88 (1965).
18. Kauzmann, W.   Advan. Protein Chem. 14, 1 (1959).
19. Scheraga, H. A., in The Proteins, Vol. 1, 2nd ed., Ed. by H. Neurath (New York, London: Academic Press, 1963), 477.
20. Némethy, G.   Angew. Chem. 79, 260 (1967); Angew. Chem. Intern. Ed. 6, 195 (1967).
21. Timasheff, S. N., and G. D. Fasman (eds.).   Structure and Stability of Biological Macromolecules (New York: Marcel Dekker, 1969).
22. Lumry, R., and H. Eyring.   J. Phys. Chem. 58, 110 (1954).
23. Pauling, L., in Symposium on Protein Structure, Ed. by A. Neuberger (London: Methuen, 1958), p 17.

24. Crick, F. H. C., in *The Biological Replication of Macromolecules,* Symposia of the Society for Experimental Biology, No. 12 (Cambridge: The University Press 1958), p 138.
25. Canfield, R. E., and C. B. Anfinsen, in *The Proteins,* Vol. 1, 2nd ed., Ed. by H. Neurath (New York, London: Academic Press, 1963), p 311.
26. Phillips, D. C. Sci. Am. **215,** No. 5, 78 (1966).
27. Edelman, G. M., and W. E. Gall. Ann. Rev. Biochem. **38,** 415 (1969).
28. Neurath, H. Sci. Am. **211,** No. 6, p. 68 (1964).
29. Stryer, L. Ann. Rev. Biochem, **37,** 25 (1968).
30. Steiner, D. F., and J. L. Clark. Proc. Nat. Acad. Sci. U.S. **60,** 622 (1968).
31. Schmidt, D. D., and A. Arens. Z. Physiol. Chem. **349,** 1157 (1968).
32. Bornstein, P., and K. A. Piez. Biochemistry **5,** 3460 (1966).
33. Piez, K. A. Ann. Rev. Biochem. **37,** 547 (1968).
34. Blumenfeld, O. O., and P. M. Gallop. Proc. Nat. Acad. Sci. U.S. **56,** 1260 (1966).
35. Bolton, W., J. M. Cox, and M. F. Perutz. J. Mol. Biol. **33,** 283 (1968).
36. Harris, C. E., R. D. Kobes, D. C. Teller, and W. J. Rutter. Biochemistry **8,** 2442 (1969).
37. Yue, R. H., R. H. Palmieri, O. E. Olson, and S. A. Kuby. Biochemistry **6,** 3204 (1967).
38. Theorell, H., in *Pyridine Nucleotide-Dependent Dehydrogenases,* Ed. by H. Sund (Heidelberg, New York: Springer-Verlag, 1970), p 121.
39. Brändén, C.-I., E. Zeppezauer, T. Boiwe, G. Söderlund, B.-O. Söderberg, and B. Nordström, in *Pyridine Nucleotide-Dependent Dehydrogenases,* Ed. by H. Sund (Heidelberg, New York: Springer-Verlag, 1970), p 129.
40. Dai, V. D., K. Decker, and H. Sund. European J. Biochem. **4,** 95 (1968).
41. Stegink, L. D., and C. S. Vestling. J. Biol. Chem. **241,** 4923 (1966).
42. Millar, D. B., V. Frattali, and G. E. Willick. Biochemistry **8,** 2416 (1969).
43. Jaenicke, R., in *Pyridine Nucleotide-Dependent Dehydrogenases,* Ed. by H. Sund (Heidelberg, New York: Springer-Verlag, 1970), p 71.
44. Adams, M. J., A. McPherson, M. G. Rossmann, R. W. Schevitz, I. E. Smiley, and A. J. Wonacott, in *Pyridine Nucleotide-Dependent Dehydrogenases,* Ed. by H. Sund (Heidelberg, New York: Springer-Verlag, 1970), p 157.
45. Sund, H., in *Biological Oxidations,* Ed. by T. P. Singer (New York, London: Interscience Publishers, 1968), p 641.
46. Bühner, M., and H. Sund. European J. Biochem. **11,** 73 (1969).
47. Harrington, W. F., and G. M. Karr. J. Mol. Biol. **13,** 885 (1965).
48. Harris, J. I., in *Pyridine Nucleotide-Dependent Dehydrogenases,* Ed. by H. Sund (Heidelberg, New York: Springer-Verlag, 1970), p 57.
49. Kirschner, K., and I. Schuster, in *Pyridine Nucleotide-Dependent De-*

hydrogenases. Ed. by H. Sund (Heidelberg, New York: Springer-Verlag, 1970), p 217.

50. De Vijlder, J. J. M., W. Boers, A. G. Hilvers, B. J. M. Harmsen, and E. C. Slater, in *Pyridine Nucleotide-Dependent Dehydrogenases*, Ed. by H. Sund (Heidelberg, New York: Springer-Verlag, 1970), p. 233.

51. Kawahara, K., and C. Tanford.   Biochemistry **5**, 1578 (1966).

52. Sia, C. L., and B. L. Horecker.   Arch. Biochem. Biophys. **123**, 186 (1968).

53. Rechler, M. M.   J. Biol. Chem. **244**, 551 (1969).

54. Sund, H., K. Weber, and E. Mölbert.   Eur. J. Biochem. **1**, 400 (1967).

55. Kiselev, N. A., D. J. De Rosier, and A. Klug.   J. Mol. Biol. **35**, 561 (1968).

56. Leuzinger, W., M. Goldberg, and E. Cauvin.   J. Mol. Biol. **40**, 217 (1969).

57. Weber, K.   Nature **218**, 1116 (1968).

58. Wiley, D. C., and W. N. Lipscomb.   Nature **218**, 1119 (1968).

59. Jovin, T. M., P. T. Englund, and LeRoy L. Bertsch.   J. Biol. Chem. **244**, 2996 (1969).

60. Weber, K., and M. Osborn.   J. Biol. Chem. **244**, 4406 (1969).

61. Crick, F. H. C., and L. E. Orgel.   J. Mol. Biol. **8**, 161 (1964).

62. Catcheside, D. D.   Brookhaven Symp. Biol. **17**, 1 (1964).

63. Wickerson, R. E., in *The Proteins*, Vol. 2, 2nd ed., Ed. by H. Neurath (New York, London: Academic Press, 1964), 603.

64. Caspar, D. L. D.   Advan. Protein Chem. **18**, 37 (1963).

65. Birbeck, M. S. C., in *A Laboratory Manual of Analytical Methods of Protein Chemistry*, Vol. 3, Ed. by P. Alexander and R. J. Block (Oxford: Pergamon Press, 1961), p 1.

66. Kay, D. (ed.).   *Techniques for Electron Microscopy*, 2nd ed. (Oxford: Blackwell Scientific Publications, 1965). (Note especially Chapter 11 by R. W. Horne on p 328).

67. Reimer, L.   *Elektronenmikroskopische Untersuchungs- und Präparationsmethoden*, 2nd ed. (Berlin, Heidelberg, New York: Springer-Verlag, 1967).

68. Slayter, E. M., in *Physical Principles and Techniques of Protein Chemistry*, Ed. by S. J. Leach (New York, London: Academic Press, 1969), Part A, p 1.

69. Levin, Ö.   J. Mol. Biol. **6**, 95 (1963); Arkiv Kemi. **21**, 1 (1963).

70. Valentine, R. C., N. G. Wrigley, M. C. Scrutton, J. J. Irias, and M. F. Utter.   Biochemistry **5**, 3111 (1966).

71. Josephs, R., in *Pyridine Nucleotide-Dependent Dehydrogenases*, Ed. by H. Sund (Heidelberg, New York: Springer-Verlag, 1970), p 301.

72. Finch, J. T., and A. Klug.   J. Mol. Biol. **15**, 344 (1966).

73. Harris, J. I., and V. M. Ingram, in *A Laboratory Manual of Analytical Methods of Protein Chemistry*, Vol 2, Ed. by P. Alexander and R. J. Block (Oxford: Pergamon Press, 1960), p 421.

74. *Methods in Enzymology*, Ed. by S. P. Colowick and N. O. Kaplan, Vol. 11, Ed. by C. H. W. Hirs (New York, London: Academic Press, 1967), Section II, 125–166.

75. Leggett Bailey, J.   *Techniques in Protein Chemistry*, 2nd ed. (Amsterdam: Elsevier Publishing Company, 1967).
76. Ingram, V. M., in *Methods in Enzymology*, Vol. 6, Ed. by S. P. Colowick and N. O. Kaplan (New York, London: Academic Press, 1963), p 831.
77. Bennett, J. C., in *Methods in Enzymology*, Ed. by S. P. Colowick and N. O. Kaplan, Vol. 11, Ed. by C. H. W. Hirs (New York, London: Academic Press, 1967), p 330.
78. Gross, E., and B. Witkop.   J. Am. Chem. Soc. **83**, 1510 (1961).
79. Wallenfels, K., H. Sund, and K. Weber.   Biochem. Z. **338**, 714 (1963).
80. Zipser, D.   J. Mol. Biol. **7**, 113 (1963).
81. Meselson, M., and F. W. Stahl.   Proc. Nat. Acad. Sci. U.S. **44**, 671 (1958).
82. Sund, H. (ed.).   *Pyridine Nucleotide-Dependent Dehydrogenases* (Heidelberg, New York: Springer-Verlag, 1970).
83. Butler, P. J. G., H. Jörnvall, and J. I. Harris.   FEBS Letters **2**, 239 (1969).
84. Albrecht, A. M., F. K. Pearce, W. J. Suling, and D. J. Hutchison.   Biochemistry **8**, 960 (1969).
85. Munkres, K. D.   Biochemistry **4**, 2180, 2186 (1965).
86. Chung, A. E., and J. S. Franzen.   Biochemistry **8**, 3175 (1969).
87. Vestling, C. S.   Methods Biochem. Analy. **10**, 137 (1962).
88. Sund, H., in *Biological Oxidations*, Ed. by T. P. Singer (New York, London: Interscience Publishers, 1968), p 603.
89. Kirschner, K., M. Eigen, R. Bittman, and B. Voigt.   Proc. Nat. Acad. Sci. U.S. **56**, 1661 (1966).
90. Conway, A., and D. E. Koshland.   Biochemistry **7**, 4011 (1968).
91. Ehrenberg, A., and H. Theorell, in *Comprehensive Biochemistry*, Vol. 3, Ed. by M. Florkin and E. Stotz (Amsterdam, New York: Elsevier Publishing Company, 1962), p 169.
92. Udenfriend, S.   *Fluorescence Assay in Biology and Medicine* (New York, London: Academic Press, 1962).
93. Sund, H., H. Diekmann, and K. Wallenfels.   Advan. Enzymol. **26**, 115 (1964).
94. Velick, S. F.   J. Biol. Chem. **233**, 1455 (1958).
95. Winer, A. D., and H. Theorell.   Acta Chem. Scand. **14**, 1729 (1960).
96. Ulmer, D. D., and B. L. Vallee.   Advan. Enzymol. **27**, 37 (1965).
97. Velluz, L., M. Legrand, and M. Grosjean.   *Optical Circular Dichroism* (Weinheim: Verlag Chemie; and New York, London: Academic Press, 1965).
98. Jirgensons, B.   *Optical Rotatory Dispersion of Proteins and other Macromolecules* (Berlin, Heidelberg, New York: Springer-Verlag, 1969).
99. Li, T. K., D. D. Ulmer, and B. L. Vallee.   Biochemistry **1**, 114 (1962).
100. Weiner, H.   Biochemistry **8**, 526 (1969).
101. Rosenberg, R. M., and I. M. Klotz, in *A Laboratory Manual of Analyt-*

ical Methods of Protein Chemistry, Vol. 2, Ed. by P. Alexander and R. J. Block (Oxford: Pergamon Press, 1960), p 131.

102. Craig, L. C., and T. P. King. Methods Biochem. Analy. **10**, 175 (1962).

103. Colowick, S. P., and F. C. Womack. J. Biol. Chem. **244**, 774 (1969).

104. Schachman, H. K. *Ultracentrifugation in Biochemistry* (New York, London: Academic Press, 1959).

105. Schachman, H. K. Biochemistry **2**, 887 (1963).

106. Hayes, J. E., and S. F. Velick. J. Biol. Chem. **207**, 225 (1954).

107. Schachman, H. K. Cold Spring Harbor Symp. Quant. Biol. **28**, 409 (1963).

108. Joly, M. *A Physicochemical Approach to the Denaturation of Proteins* (London, New York: Academic Press, 1965).

109. Tanford, C. Advan. Protein Chem. **23**, 121 (1968).

110. Elias, H. G. *Ultrazentrifugen-Methoden* (München: Beckman Instruments GmbH., 1961).

111. Claesson, S., and I. Moring-Claesson, in *A Laboratory Manual of Analytical Methods of Protein Chemistry*, Vol. 3 Ed. by P. Alexander and R. J. Block, (Oxford: Pergamon Press, 1961), p 119.

112. Sund, H., and K. Weber. *Beckman Report* (München, Beckman Instruments GmbH., 1965), 3/4, p 7.

113. Steinberg, I. Z., and H. K. Schachman. Biochemistry **5**, 3728 (1966).

114. van Holde, K. E. *Fractions*, No. 1 (Palo Alto, California: Spinco Division of Beckman Instruments, Inc., 1967).

115. Chervenka, C. H. *A Manual of Methods for the Analytical Ultracentrifuge*, (Palo Alto, California: Spinco Division of Beckman Instruments Inc., 1969).

116. Stacey, K. A. *Light Scattering in Physical Chemistry* (London: Butterworths, 1956).

117. Stacey, K. A., in *A Laboratory Manual of Analytical Methods of Protein Chemistry*, Vol. 3, Ed. by P. Alexander and R. J. Block (Oxford: Pergamon Press, 1961), p 245.

118. Burchard, W., and J. M. Cowie, in *Light Scattering from Polymer Solutions*, Ed. by M. B. Huglin (New York, London: Academic Press, 1971), in press.

119. Cann, J. R., in *Physical Principles and Techniques of Protein Chemistry*, Ed. by S. J. Leach (New York, London: Academic Press, 1969), Part A, p 369.

120. Craig, L. C., in *A Laboratory Manual of Analytical Methods of Protein Chemistry*, Vol. 1, Ed. by P. Alexander and R. J. Block (Oxford: Pergamon Press, 1960), p 121.

121. Keller, S., R. J. Block, E. A. Peterson, and H. A. Sober, in *A Laboratory Manual of Analytical Methods of Protein Chemistry*, Vol. 1, Ed. by P. Alexander and R. J. Block (Oxford: Pergamon Press, 1960), p 65.

122. Turba, F. Advan. Enzymol. **22**, 417 (1960).

123. Determann, H. *Gelchromatographie*, (Heidelberg, New York: Springer-Verlag, 1967).

124. *Methods in Enzymology*, Ed. by S. P. Colowick and N. O. Kaplan, Vol. 11, Ed. by C. H. W. Hirs (New York, London: Academic Press, 1967), Section IV, p 197.
125. Snodgrass, P. J., B. L. Vallee, and F. L. Hoch. J. Biol. Chem. **235**, 504 (1960).
126. Sund, H. Biochem. Z. **333**, 205 (1960).
127. Kägi, J. H. R., and B. L. Vallee. J. Biol. Chem. **235**, 3188 (1960).
128. Harris, C. E., R. D. Kobes, D. C. Teller, and W. J. Rutter. Biochemistry **8**, 2442 (1969).
129. Stein, E. A., and E. H. Fischer. Biochim. Biophys. Acta **39**, 287 (1960).
130. Gerhart, J. C. Brookhaven Symp. Biol. **17**, 222 (1964).
131. Gerhart, J. C., and H. Holoubek. J. Biol. Chem. **242**, 2886 (1967).
132. Sine, H. E., and L. F. Hass. J. Biol. Chem. **244**, 430 (1969).
133. Fisher, H. F., L. L. McGregor, and D. G. Cross. Biochim. Biophys. Acta **65**, 175 (1962).
134. Fisher, H. F., L. L. McGregor, and U. Power. Biochem. Biophys. Res. Commun. **8**, 402 (1962).
135. Minssen, M., and H. Sund. Abstracts, 6th Meeting of the Federation European Biochemical Societies, Madrid 1969, p 330.
136. Adair, G. S., in *A Laboratory Manual of Analytical Methods of Protein Chemistry*, Vol. 3 Ed. by P. Alexander and R. J. Block (Oxford: Pergamon Press, 1961), p 24.
137. Tanford, C., K. Kawahara, and S. Lapanje. J. Biol. Chem. **241**, 1921 (1966).
138. Bruton, C. J., and B. S. Hartley. Biochem. J. **108**, 281 (1968).
139. Sia, C. L., and B. L. Horecker. Biochem. Biophys. Res. Commun. **31**, 731 (1968).
140. Cohen, P., and M. A. Rosemeyer. Eur. J. Biochem. **8**, 8 (1969).
141. Butler, P. J. G., J. I. Harris, B. S. Hartley, and R. Leberman. Biochem. J. **103**, 78P (1967).
142. Klotz, I. M., in *Methods in Enzymology*, Ed. by S. P. Colowick and N. O. Kaplan, Vol. 11, Ed. by C. H. W. Hirs (New York, London: Academic Press, 1967), p 576.
143. Braunitzer, G., K. Beyreuther, H. Fujiki, and B. Schrank. Z. Physiol. Chem. **349**, 265 (1968).
144. Klotz, I. M., and S. Keresztes-Nagy. Biochemistry **2**, 445 (1963).
145. Shapiro, A. L., E. Viñuela, and J. V. Maizel. Biochem. Biophys. Res. Commun. **28**, 815 (1967).
146. Tanford, C., K. Kawahara, and S. Lapanje. J. Am. Chem. Soc. **89**, 729 (1967).
147. Fish, W. W., K. G. Mann, and C. Tanford. J. Biol. Chem. **244**, 4989 (1969).
148. Yule, R. H., R. H. Palmieri, O. E. Olson, and S. A. Kuby. Biochemistry **6**, 3204 (1967).
149. Kirby Hade, E. P., and C. Tanford. J. Am. Chem. Soc. **89**, 5034 (1967).

150. Katz, S.   Biochim. Biophys. Acta **154**, 468 (1968).
151. Ullmann, A., M. E. Goldberg, D. Perrin, and J. Monod.   Biochemistry **7**, 261 (1968).
152. Reisler, E., and H. Eisenberg.   Biochemistry **8**, 4572 (1969).
153. Eisenberg, H., and G. M. Tomkins.   J. Mol. Biol. **31**, 37 (1968).
154. Reithel, F. J., and J. D. Sakura.   J. Phys. Chem. **67**, 2497 (1963).
155. Marler, E., C. A. Nelson, and C. Tanford.   Biochemistry **3**, 279 (1964).
156. Nozaki, Y., and C. Tanford, in *Methods in Enzymology*, Ed. by S. P. Colowick and N. O. Kaplan, Vol. 11, Ed. by C. H. W. Hirs (New York, London: Academic Press, 1967), p 715.
157. Schwert, G. W.   J. Biol. Chem. **179**, 655 (1949).
158. Rao, M. S. N., and G. Kegeles.   J. Am. Chem. Soc. **80**, 5724 (1958).
159. Langerman, N. R., and I. M. Klotz.   Biochemistry **8**, 4746 (1969).
160. Ackers, G. K.   J. Biol. Chem. **243**, 2056 (1968).
161. Chun, P. W., S. J. Kim, C. A. Stanley, and G. K. Ackers.   Biochemistry **8**, 1625 (1969).
162. Adams, E. T., and M. S. Lewis.   Biochemistry **7**, 1044 (1968).
163. Steiner, R. F.   Arch. Biochem. Biophys. **39**, 333 (1952).
164. Steiner, R. F.   Biochemistry **7**, 2201 (1968).
165. Sund, H., in *Mechanismen enzymatischer Reaktionen*, 14th Colloq. Ges. Physiol. Chem., 1963 (Berlin, Heidelberg: Springer-Verlag, 1964), p 318.
166. Olson, J. A., and C. B. Anfinsen.   J. Biol. Chem. **197**, 67 (1952).
167. Sund, H.   Acta Chem. Scand. **17**, S 102 (1963).
168. Dessen, P., and D. Pantaloni.   Eur. J. Biochem. **8**, 292 (1969).
169. Kawahara, K., and C. Tanford.   J. Biol. Chem. **241**, 3228 (1966).
170. Umstätter, H.   Einführung in die Viskosimetrie und Rheometrie (Berlin-Göttingen-Heidelberg: Springer-Verlag, 1952).

Index

This book was typeset at SSPA Typesetting, Inc., Carmel, Indiana in 10-point Caledonia, printed and bound by LithoCrafters, Inc., Ann Arbor, Michigan.